NATO ASI Series

Advanced Science Institutes Series

A series presenting the results of activities sponsored by the NATO Science Committee, which aims at the dissemination of advanced scientific and technological knowledge, with a view to strengthening links between scientific communities.

The Series is published by an international board of publishers in conjunction with the NATO Scientific Affairs Division

A Life Sciences B Physics	Plenum Publishing Corporation London and New York
C Mathematical and Physical Sciences D Behavioural and Social Sciences E Applied Sciences	Kluwer Academic Publishers Dordrecht, Boston and London
F Computer and Systems Sciences G Ecological Sciences H Cell Biology	Springer-Verlag Berlin Heidelberg New York London Paris Tokyo Hong Kong

Series H: Cell Biology Vol. 35

The ASI Series Books Published as a Result of Activities of the Special Programme on CELL TO CELL SIGNALS IN PLANTS AND ANIMALS

This book contains the proceedings of a NATO Advanched Research Workshop held within the activities of the NATO Special Programme on Cell to Cell Signals in Plants and Animals, running from 1984 to 1989 under the auspices of the NATO Science Committee.

The books published as a result of the activities of the Special Programme are:

Vol. 1: **Biology and Molecular Biology of Plant-Pathogen Interactions.** Edited by J.A. Baily. 1986.

Vol. 2: **Glial-Neuronal Communication in Development and Regeneration.** Edited by H.H. Althaus and W. Selfert. 1987.

Vol. 3: **Nicotinic Acetylcholine Receptor: Structure and Function.** Edited by A. Maelicke. 1986.

Vol. 4: **Recognition in Microbe-Plant Symbiotic and Pathogenic Interactions.** Edited by B. Lugtenberg. 1986.

Vol. 5: **Mesenchymal-Epithelial Interactions in Neural Development.** Edited by J.R. Wolff, J. Sievers, and M. Berry. 1987.

Vol. 6: **Molecular Mechanisms of Desensitization to Signal Molecules.** Edited by T.M. Konjin, P.J.M. Van Haastert, H. Van der Starre, H. Van der Wel, and M.D. Houslay. 1987.

Vol. 7: **Gangliosides and Modulation of Neuronal Functions.** Edited by H. Rahmann. 1987.

Vol. 9: **Modification of Cell to Cell Signals During Normal and Pathological Aging.** Edited By S. Govoni and F. Battaini. 1987.

Vol. 10: **Plant Hormone Receptors.** Edited by D. Klämbt. 1987.

Vol. 11: **Host-Parasite Cellular and Molecular Interactions in Protozoal Infections.** Edited by K.-P. Chang and D. Snary. 1987.

Vol. 12: **The Cell Surface in Signal Transduction.** Edited by E. Wagner, H. Greppin, and B. Millet. 1987.

Vol. 19: **Modulation of Synaptic Transmission and Plasticity in Nervous Systems.** Edited by G. Hertting and H.-C. Spatz. 1988.

Vol. 20: **Amino Acid Availability and Brain Function in Health and Disease.** Edited by G. Huether. 1988.

Vol. 21: **Cellular and Molecular Basis of Synaptic Transmission.** Edited by H. Zimmermann. 1988.

Vol. 23: **The Semiotics of Cellular Communication in the Immune System.** Edited by E.E. Sercarz, F. Celada, N.A. Mitchison, and T. Tada. 1988.

Vol. 24: **Bacteria, Complement and the Phagocytic Cell.** Edited by F.C. Cabello and C. Pruzzo. 1988.

Vol. 25: **Nicotinic Acetylcholine Receptors in the Nervous System.** Edited by F. Celementi, C. Gotti, and E. Sher. 1988.

Vol. 26: **Cell to Cell Signals in Mammalian Development.** Edited by S.W. de Laat, J.G. Bluemink, and C.L. Mummery. 1989.

Vol. 27: **Phytotoxins and Plant Pathogenesis.** Edited by A. Graniti, R.D. Durbin, and A. Ballio. 1989.

Vol. 31: **Neurobiology of the Inner Retina.** Edited by R. Weiler and N.N. Osborne. 1989.

Vol. 32: **Molecular Biology of Neuroreceptors and Ion Channels.** Edited by A. Maelicke. 1989.

Vol. 33: **Regulatory Mechanisms of Neuron to Vessel Communication in the Brain.** Edited by F. Battaini, S. Govoni, M.S. Magnoni, and M. Trabucchi. 1989.

Vol. 35: **Cell Separation in Plants: Physiology, Biochemistry and Molecular Biology.** Edited by D.J. Osborne and M.B. Jackson. 1989.

Cell Separation in Plants

Physiology, Biochemistry and Molecular Biology

Edited by

Daphne J. Osborne

Department of Plant Sciences
University of Oxford, Oxford, U.K.

Michael B. Jackson

Department of Agricultural Sciences
Long Ashton Research Station
University of Bristol, Bristol, U.K.

Springer-Verlag Berlin Heidelberg New York
London Paris Tokyo Hong Kong
Published in cooperation with NATO Scientific Affairs Division

Proceedings of the NATO Advanced Research Workshop on Signals for Cell Separation in Plants held at Turin, Italy, September 25–30, 1988

ISBN-13: 978-3-642-74163-0 e-ISBN-13: 978-3-642-74161-6
DOI: 10.1007/978-3-642-74161-6

Library of Congress Cataloging-in-Publication Data. NATO Advanced Research Workshop on Signals for Cell Separation in Plants (1988 : Turin, Italy). Cell separation in plants : physiology, biochemistry, and molecular biology / edited by Daphne J. Osborne, Michael B. Jackson. p. cm.—(NATO ASI series. Series H, Cell biology ; vol. 35) Proceedings of the NATO Advanced Research Workshop on Signals for Cell Separation in Plants held at Turin, Italy, September 25–30, 1988."—CIP t.p. verso.

1. Plant cell differentiation—Congresses. 2. Plant cellular control mechanisms—Congresses. I. Osborne, Daphne J., 1930– . II. Jackson, Michael B. III. Title. IV. Series. QK725.N375 1988 581.87'612—dc20
89-21648

Softcover reprint of the hardcover 1st edition 1989

Printing: Druckhaus Beltz, Hemsbach; Binding: J. Schäffer GmbH & Co. KG, Grünstadt
2131/3140-543210 – Printed on acid-free-paper

FOREWORD

This NATO Advanced Research Workshop held 25-30 September, 1988 at the Villa Gualino, Turin, Italy, was the first international meeting of its kind to be devoted solely to cell separation in plants. The partial or complete dissociation of one cell from another is an integral process of differentiation. Partial cell separations are basic physiological components of the overall programme of plant development. Complete cell separations are major events in the ripening of fruits, and the shedding of plant parts. Unscheduled cell separations commonly occur when tissues are subjected to pathogenic invasion. Environmental stresses too, evoke their own separation responses.

Over the past five years much new knowledge has been acquired on the regulation of gene expression in specific stages of cell differentiation. Specific molecular markers have been identified that designate the competence of cells for achieving separation. Certain of the chemical signals (hormones, elicitors) that must be emitted or perceived by cells to initiate and sustain separation, are now known to us, and the resulting cell wall changes have come under close chemical scrutiny.

The Turin meeting was a focus for those currently involved in such investigations. It assessed factors controlling cell separation in a wide spectrum of different cell types under a variety of conditions. The cell biologists, biochemists, physiologists and microbiologists who took part in the deliberations, each probed and dissected the status of the current knowledge on cell separation from the view point of their own different areas of expertise. Each has committed his or her words of wisdom to these published Proceedings.

Understanding the signals and the controls that determine the loss of cell adhesion with a knowledge of the means by which the cellular processes could be manipulated, offers exciting potential for future research and for new efforts in practical crop production. It is our belief that the information and ideas contained herein will prove a valuable contribution towards these goals.

We are indebted to NATO as the major sponsor and to the counsel and assistance of the Secretariat of the NATO Scientific Affairs Division. We thank the University of Turin and the Italian Society for Horticultural Science for the very considerable support that they gave the meeting. The generosity and interest of the Italian organizations who are listed hereafter, is greatly appreciated. To Professor Roberto Jona, as Local Secretary, and to his able assistants, we owe a special debt of gratitude for their tireless smoothing of the path for all the local arrangements. The Organizing Committee thanks everyone who contributed so splendidly. It is now the pleasure of the Editors to commit this volume to the reader on behalf of all those who took part in the Turin Workshop.

Daphne J. Osborne
Michael B. Jackson
10 April, 1989

SUPPORTING SPONSORS

Assessorato alla Cultura Regione Piemonte
Assessorato alla Tutela dell;Ambiente Regione Piemonte
Assessorato all'Agrocoltura Regione Piemonte
Assessorato all'Agricoltura Provincia di Torino
Presidenza della Provincia di Torino
Assessorato all'Agricoltura Provincia di Asti
Assessorato all'Agricoltura Provincia di Cuneo
Assessorato all'Ecologia Provincia di Cuneo
Camera di Commercio, Industria, Agricoltura ed Artigianato - Torino
Ente per la Valorizzazione dei Vini Astigiani - Asti
Carl Zeiss s.r.l. - Milano
Martini e Rossi - Torino
Tenimenti di Barolo e Fontanafredda - Serralugna d'Alba
Federagrario - Torino
Cassa di Risparmio - Torino

PARTICIPANTS AND THEIR AFFILIATIONS

I.T. Agar,
Versuchsstation Bavendorf, der Universität Hohenheim, D-7980 Ravensburg 1, West Germany

C. Balagué,
Laboratoire de Biotechnologie Végétale, Ecole Nationale Supérieure Agronomique de Toulouse, 145 Avenue de Muret, 31076 Toulouse Cédex, France

R. Ben-Arie,
Agricultural Research Organization, The Volcani Center, P.O.B. 6, Bet Dagan 50250, Israel

A.B. Bennett,
Department of Vegetable Crops, University of California, Davis, California 95616, U.S.A.

G. Bounous,
Istituto di Coltivaziono Arboree, Dell'Universita di Torino, Via Pietro Giuria, 10126 Torino TO, Italy

J. Bruinsma,
Department of Plant Physiology of the Agricultural University, Arboretumlaan 4, 6703 BD Wageningen, The Netherlands

J. Burdon,
Department of Biological Sciences, University of Stirling, Stirling, FK9 4LA, U.K.

R.E. Christoffersen,
Department of Biological Sciences, University of California, Santa Barbara, California 93106, U.S.A.

E.C. Cocking,
Department of Plant Sciences, University of Nottingham, University Park, Nottingham, NG7 2RD, U.K.

C.P. Connern,
Department of Botany and Microbiology, School of Biological Sciences, University College of Wales, Aberystwyth SY23 3DA, U.K.

R. Cooper,
Department of Biological Sciences, University of Bath, Claverton Down, Bath BA2 7AY, U.K.

F. Ergenoglu,
Department of Horticulture, Faculty of Agriculture, University of Cukurova, Adana 01330, Turkey

M.T. Esquerré-Tugayé,
Université Paul Sabatier, Centre Physiologie Végétale, 118 Route de Narbonne, 31062 Toulouse Cédex, France

J. Friend,
Department of Plant Biology & Genetics, The University of Hull, Hull, HU6 7RX, U.K.

R. Goren,
Faculty of Agriculture, Hebrew University of Jerusalem, P.O.B. 12, Rehovot 76-100, Israel

I. Gribaudo,
Centro Miglioramento Genetico della Vite CNR, Via Pietro Giuria 15, 10126 Torino TO, Italy

D. Grierson,
Department of Physiology and Environmental Sciences, University of Nottingham, Sutton Bonington, Loughborough, Leics, LE12 5RD, U.K.

A.M. Haigh,
Department of Plant Physiology of the Agricultural University, Arboretumlaan 4, 6703 BD Wageningen, The Netherlands

A.H. Halevy,
Department of Horticulture and Ornamental Crops, Faculty of Agriculture, Hebrew University of Jerusalem, P.O.B. 12, Rehovot 76-100, Israel

H. Imaseki,
Research Institute for Biochemical Regulation, Faculty of Agriculture, Nagoya University, Chikusa, Nagoya 464, Japan

M.B. Jackson,
Long Ashton Research Station, University of Bristol, Long Ashton, Bristol BS18 9AF, U.K.

C. Jeffree,
Department of Botany, University of Edinburgh, The King's Buildings, Mayfield Road, Edinburgh EH9 3JH, Scotland, U.K.

R. Jona,
Istituto di Coltivazioni Arboree, Dell'Universita di Torino, Via Pietro Giuria 15, 10126 Torino TO, Italy

A. Kargiolaki,
Department of Plant Sciences, University of Oxford, South Parks Road, Oxford OX1 3RA, U.K. Permanent address: Department of Forestry and Natural Environment, Aristotelian University of Thessaloniki, Thessaloniki 54006, Greece

N. Kaska,
Department of Horticulture, Faculty of Agriculture, University of Cukurova, Adana 01330, Turkey

C.J. Keijzer,
Department of Cytology and Morphology, Agricultural University, Arboretumlaan 4, 6703 BD Wageningen, The Netherlands

N. Kieliszewski,
Laboratory 206, Plant Biology Building, Wilson Road, Michigan State University, East Lansing, Michigan, U.S.A.

M. Knee,
Department of Horticulture, Ohio State University, Colombus, Ohio 43210-1293, U.S.A.

D. Labavitch,
Department of Pomology, University of California, Davis, California 95616, U.S.A.

D.T.A. Lamport,
MSU-DOE Plant Research Laboratory, Michigan State University, East Lansing, Michigan 48824-1312, U.S.A.

G.A. Lang,
Department of Horticulture, Louisiana State University, Baton Rouge, Louisiana 70803, U.S.A.

E. Marré,
Dipartimento di Biologia, Universita Státale, Via Celoria 26, 20133 Milano MI, Italy

G.C Martin,
Department of Pomology, 1045 Wickson Hall, University of California, Davis, California 95616, U.S.A.

A.K. Mattoo,
Plant Molecular Biology Laboratory, USDA Agricultural Research Service, BARC-West, Beltsville, Maryland 20705, U.S.A.

M.E. McCully,
Biology Department, Carleton University, Ottawa, Ontario K1S 5B6, Canada

M.T. McManus,
Department of Biochemistry, Royal Holloway and Bedford New College, Egham Hill, Egham, Surrey TW20 0EX, U.K.

R. Mertens,
Schering AG, Agrochemical Research, Postfach 650311, D-1000 Berlin 65, West Germany

D.J. Morré,
Department of Medicinal Chemistry and Pharmacognosy, Life Sciences Research Building, Purdue University, West Lafayette, Indiana 47907, U.S.A.

G. Nicolás,
Departamento de Biologia Vegetal, Fisiologia Vegetal, Facultad de Biologia, Universidad de Salamanca, 37008 Salamanca, Spain

D.J. Osborne,
University of Oxford, Department of Plant Sciences, South Parks Road, Oxford, OX1 3RA, U.K.

R. Paglietta,
Instituto di Coltivazioni Arboree, Dell'Universita di Torino, Via Pietro Giuria 15, 10126 Torino TO, Italy

A.F. Pow,
Bristol Polytechnic, Coldharbour Lane, Frenchay, Bristol BS16 1QY, U.K.

A. Ramina,
Instituto di Coltivazioni Arboree Dell'Università, Via Gradenigo 6, 35121 Padova, Italy

N. Rascio,
Dipartimento di Biologia Orto Botanico, Via del Santo, 35100 Padova, Italy

J. Riov,
Faculty of Agriculture, Hewbrew University of Jerusalem, P.O.B. 12, Rehovot 76-100, Israel

J.A. Roberts,
Department of Physiology and Environmental Sciences, Univerrsity of Nottingham, Sutton Bonington, Loughborough, Leics, LE12 5RD, U.K.

R. Sexton,
Department of Biological Science, University of Stirling, Stirling, FK9 4LA, Scotland, U.K.

G. Sirju-Charran,
Department of Plant Biochemistry and Physiology, University of the West Indies, St. Augustine, Trinidad

G. Syzmkowiak,
Department of Biology, Yale University, P.O.B. 6666, New Haven, Connecticut 06511-7444, U.S.A.

D. Watanabe,
Research Institute for Biochemical Regulation, Faculty of Agriculture, Nagoya University, Chikusa, Nagoya 464, Japan

L. Wyckham,
Department of Plant Biochemistry and Physiology, University of the West Indies, St. Augustine, Trinidad

T. Yamaki,
3-4-14, Yagumo, Meguro-ku, Tokyo 152, Japan

SYMPOSIUM ORGANIZATION

ORGANIZING COMMITTEE

D.J. Osborne, Workshop Director (U.K.)

J. Bruinsma (The Netherlands)

R. Goren (Israel)

H. Imaseki (Japan)

M.B. Jackson, Secretary & Treasurer (U.K.)

R. Jona, Local Secretary (Italy)

LOCAL ORGANIZING COMMITTEE, TURIN, ITALY

R. Jona, Secretary

G. Bounous

R. Paglietta

CONTENTS

SECTION VII. ADDITIONAL CONTRIBUTIONS

SECTION I

Gene Regulation and Gene Expression

SIGNALS FOR GENE EXPRESSION IN RIPENING TOMATO FRUIT

D. Grierson, C.J.S. Smith, P.C. Morris, C. Watson, C.R. Bird[1], J. Ray[1], W. Schuch[1], J.E. Knapp, K. Davies, S.J. Picton
Department of Physiology and Environmental Science
University of Nottingham School of Agriculture
Sutton Bonington
Loughborough LE12 5 RD
U.K.

INTRODUCTION

cDNA clones for a number of mRNAs expressed in ripening tomato fruit have been characterised (Grierson *et al.*, 1986a; Maunders *et al.*, 1987; Holdsworth *et al.*, 1987; Ray *et al.*, 1987, 1988). One of these (pTOM 6) has been shown by DNA and protein sequence analysis to encode the enzyme polygalacturonase (PG) (Grierson *et al.*, 1986b), which is synthesised *de novo* during ripening (Tucker and Grierson, 1982) and is involved in the solubilisation of the pectin fraction of fruit cell walls. cDNA clones for PG have also been isolated by a number of other groups (DellaPenna *et al.*, 1986; Sheehy *et al.*, 1987; Lincoln *et al.*, 1987). Ethylene stimulates the accumulation of ripening-related mRNAs (Maunders *et al.*, 1987) and PG mRNA is inhibited by silver ions (Davies *et al.*, 1988), which are thought to interfere with ethylene perception or action. Production of PG and, to a lesser extent, other mRNAs is substantially reduced in the tomato ripening mutants *rin* and *Nr* (DellaPenna *et al.*, 1987; Knapp *et al.*, 1989), and is also inhibited in normal fruit at 35°C (Picton and Grierson, 1988).

The sequence of events leading to expression of the PG gene probably involves: 1) activation of the ethylene-synthesis pathway 2) transition of the PG gene to a state where it is *potentially* active 3) perception of ethylene by a receptor 4) consequent activation of the signal transduction pathway, which is unknown 5) production or modification of *trans*-acting protein factors that bind to a regulatory region(s) of the PG gene 6) activation of transcription of the PG gene, followed by mRNA processing and transport 7) mRNA translation 8) the production of a PG pre-protein that is subsequently cleaved and glycosylated as it is transported through the cell endomembrane system and secreted into the wall 9) solubilisation of cell wall pectins by PG. Results of experiments relating to steps 6) 7) and 8)

[1] ICI Seeds, Plant Biotechnology Section, Jealott's Hill Research Station, Bracknell, Berkshire, RG12 6EY, UK

NATO ASI Series, Vol. H 35
Cell Separation in Plants
Edited by D. J. Osborne and M. B. Jackson

are discussed by Bennett *et al.* (this volume). To test other aspects of this scheme further, we have isolated and sequenced a complete PG gene from tomato (Bird *et al.*, 1988). Genetic engineering experiments have enabled us to show that *cis*-acting control sequences determining ripening-specific expression are located within a 1450 base pair sequence to the 5′ side of the PG gene (Bird *et al.*, 1988). In a novel approach to plant genetic engineering, we have also expressed a stably-inherited PG antisense gene in tomato that reduces expression of the endogenous gene by 90% (Smith *et al.*, 1988).

RESULTS AND DISCUSSION

We have used the cDNA clone pTOM 6 as a hybridisation probe to show that PG mRNA is not present in significant quantities in stems, leaves, roots, or unripe fruit but accumulates at the onset of ripening. The synthesis of PG enzyme follows the production of mRNA. In the fruit of the *rin* mutant however, only about 0.5% of the normal amount of PG mRNA is produced (Fig. 1), despite the fact that the PG gene is present in the mutant (DellaPenna *et al.*, 1987; Knapp *et al.*, 1989). The *rin* fruit are deficient in several aspects of ripening, and the *rin* gene has been mapped to chromosome 5, whereas the PG gene is on chromosome 10 (Mutschler *et al.*, 1988). These observations are consistent with the idea that a gene product affected by the *rin* mutation is required for expression of the PG gene.

Exogenous ethylene stimulates the accumulation of PG mRNA in mature green fruit, but the response to ethylene takes much longer in immature fruit and does not occur in other organs. The role of ethylene in stimulating expression of the PG gene in fruit was studied using silver ions, which are believed to interfere with ethylene perception or part of the response mechanism. If applied to mature green fruit *before* ripening, the production of PG mRNA was inhibited (Fig. 1) and ripening did not occur. There was no *general* inhibitory effect of silver on protein synthesis or mRNA production, but rather a specific inhibition of production of mRNAs associated with ripening. Application of silver to fruit after ripening had begun led to a rapid loss of PG mRNA, suggesting that the continued perception of ethylene during ripening is required in order for the gene to be expressed at a high level (Davies *et al.*, 1988). It is not clear whether the decline in PG mRNA production at later stages of ripening in normal fruit not treated with ethylene (Fig. 1) is due to lack of ethylene perception or a change in the signal transduction mechanism.

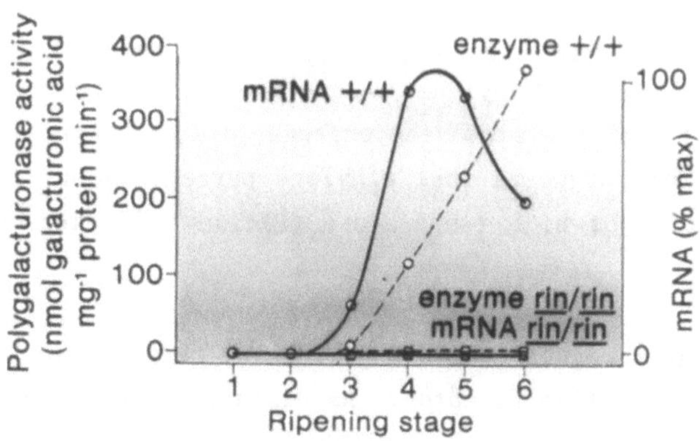

In wild type fruit, the increase in PG enzyme activity associated with softening follows an increase in the corresponding mRNA. In the rin mutant, which does not ripen normally, both mRNA and enzyme levels are substantially reduced. The PG gene is, however, present in rin fruit.

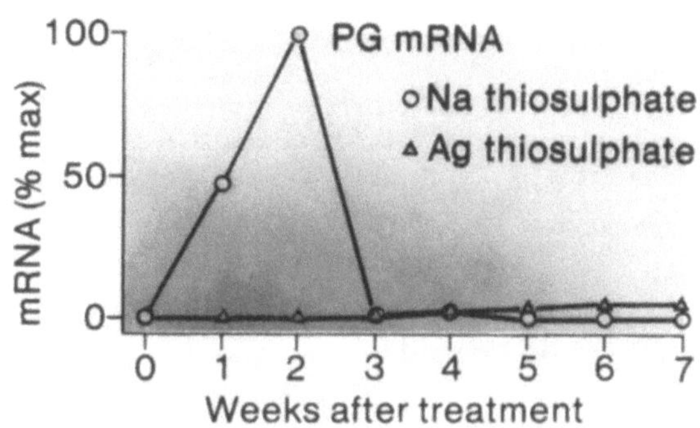

PG gene expression is strongly inhibited by treatment with silver ions. It is thought that silver interferes with ethylene reception. Other aspects of ripening are also inhibited.

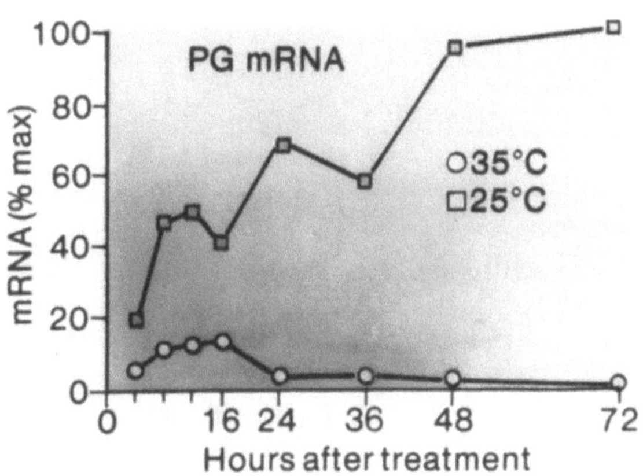

When ripening fruit are transferred from 25°C to 35°C, PG mRNA production stops. If green fruit are placed at 35°C they do not subsequently accumulate PG mRNA

FIGURE 1. Environmental and genetic factors affecting polygalacturonase mRNA levels in tomato fruit

It is known that ripening can be retarded at elevated temperatures and this has been show recently to be associated with the reduced accumulation of some ripening-related mRNAs (Picton and Grierson, 1988). The mRNA for PG is particularly sensitive, and at 35°C there is a sharp decline in mRNA content, even after fruit have begun to ripen (Fig. 1). This is not explained simply by the decline in ethylene production at 35°C, since supplying the gas to fruit at elevated temperature did not restore expression of the PG gene. This suggests that ethylene perception, or the response mechanism, is disrupted at high temperature (Picton and Grierson, 1988).

Our working model for PG gene regulation during ripening begins with the interaction of ethylene with a putative receptor. As discussed above, there must be factors regulating the production of ethylene, and perhaps also the

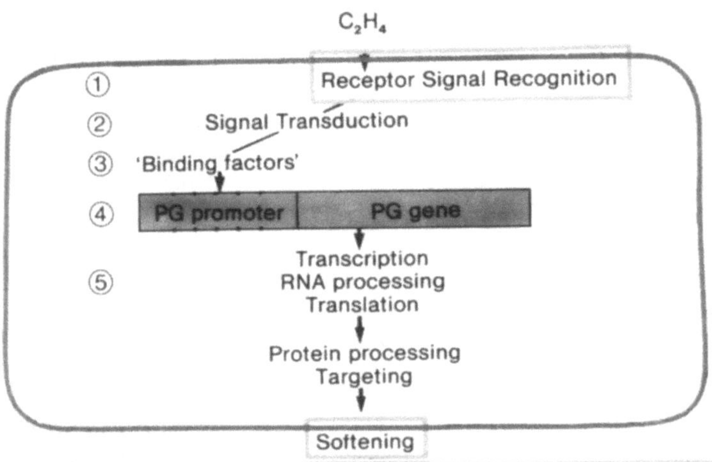

● Expression of the PG gene can be influenced at various points along this pathway.

1 Silver ions are thought to interfere with ethylene - receptor interactions

2 Heat stress switches off the PG gene. The mechanism remains unknown. It may be due to lack of ethylene perception or failure of the signal transduction mechanism.

3 A genetic lesion in the rin mutant inhibits PG gene expression. The mutation is possibly a gene coding for a regulatory factor.

4 The 5′ DNA sequences of the PG gene (called the PG promoter) direct ripening specific expression of foreign reporter genes in fruit of transgenic plants

5 PG gene expression can be down-regulated in genetically engineered plants expressing antisense RNA. It is not yet clear whether inhibition of gene expression occurs at the level of mRNA production, processing and transport, or translation.

FIGURE 2. A working model for polygalacturonase gene expression in tomato fruit

potential for the PG gene to be expressed, but we know nothing about these parts of the process. Binding of ethylene to the receptor is assumed to lead to the propagation of a signal (the transduction mechanism) that leads to synthesis and/or changes in the binding properties of factors (probably proteins) that interact with the PG gene to regulate transcription (Fig. 2). The first step in testing this model was to isolate the PG gene (Bird *et al.*, 1988). Genomic cloning, sequencing, and Southern hybridisation indicate that there may be only one PG gene. It covers approximately 8,000 base pairs and is interrupted by eight introns, ranging in size from 99 to 954 base pairs (see Fig. 3). The normal concensus sequences associated with plant gene transcription, processing, and translation are present, and there is a 28 base pair direct repeat sequence between -475 and -535 base pairs upstream from the transcription start point.

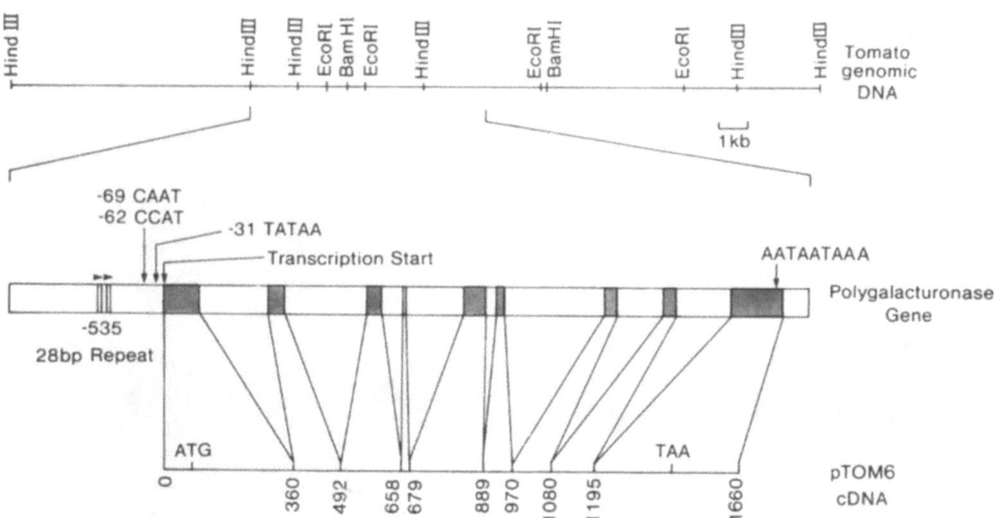

FIGURE 3. Structure of the polygalacturonase gene

DNA sequences regulating the expression of specific plant genes have, in a number of cases, been located to the 5′ side of genes. However, situations are known where control regions occur within an intron, and also to the 3′ side of genes. We tested the hypothesis that *cis*-acting DNA sequences determining ripening-specific expression were located to the 5′ side of the PG gene. A reporter gene construct was made using a 1450 base pair genomic sequence 5′ to the PG structural gene, fused to a chloramphenicol acetyl transferase (CAT) coding region, and terminated by the 3′ end of the nopaline synthase (*nos*) gene. This chimaeric gene was used to transform tomato plants using the *Agrobacterium* Ti binary plasmid vector system Bin 19, containing a gene for kanamycin-resistance. Transformed cells were selected on kanamycin and whole plants regenerated. A number were allowed to flower and set fruit and expression of the CAT reporter gene was analysed in different organs. CAT activity, manifested by the presence of an enzyme that acetylated chloramphenicol, was detected in ripening fruit of genetically engineered plants, but not unripe fruit, leaves, stems, or roots (Fig. 4). No CAT activity was detected in any organ of control plants. Thus, the 1450 base pair sequence to the 5′ side of the PG gene confers on the bacterial CAT gene the property of being regulated like a fruit ripening gene when transferred to tomato plants.

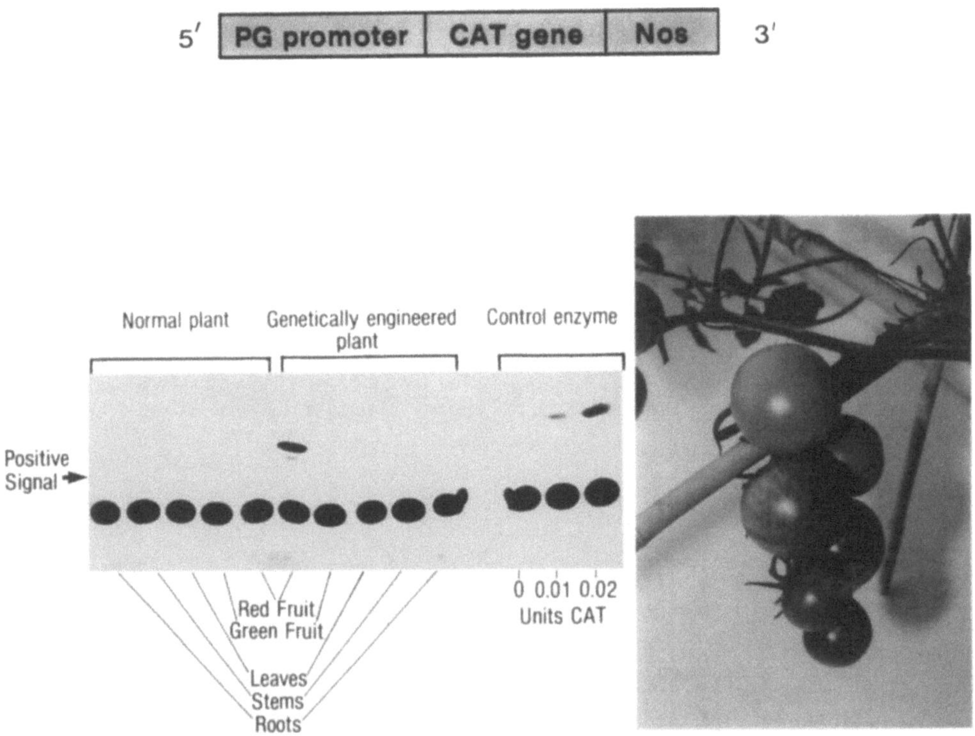

FIGURE 4. The 5′ region of the polygalacturonase gene controls ripening specific expression. A diagram of the chimaeric CAT reporter gene is shown, together with results demonstrating the specific expression of CAT activity in ripening fruit

Having demonstrated successful transfer and expression of chimaeric genes in tomato, we investigated the possibility of being able to down-regulate expression of the endogenous PG gene, using antisense RNA technology (Smith *et al.*, 1988). A chimaeric gene construct was made from the cauliflower mosaic virus (CaMV) 35S transcription promoter, fused to an inverted 730 base pair fragment of the PG cDNA pTOM 6, and terminated with the 3′ end of *nos*. This construct was transferred to tomato plants using Bin 19 and PG protein and enzyme activity was measured in the fruit of transgenic tomato plants during ripening. The levels of PG proteins were substantially reduced in the transformed fruit and PG enzyme activity was approximately 10% of normal in plant GR16. However, antisense fruit with reduced levels of PG developed full colour (Fig. 5). Northern blotting experiments, in which mRNA from control and antisense plants was hybridised

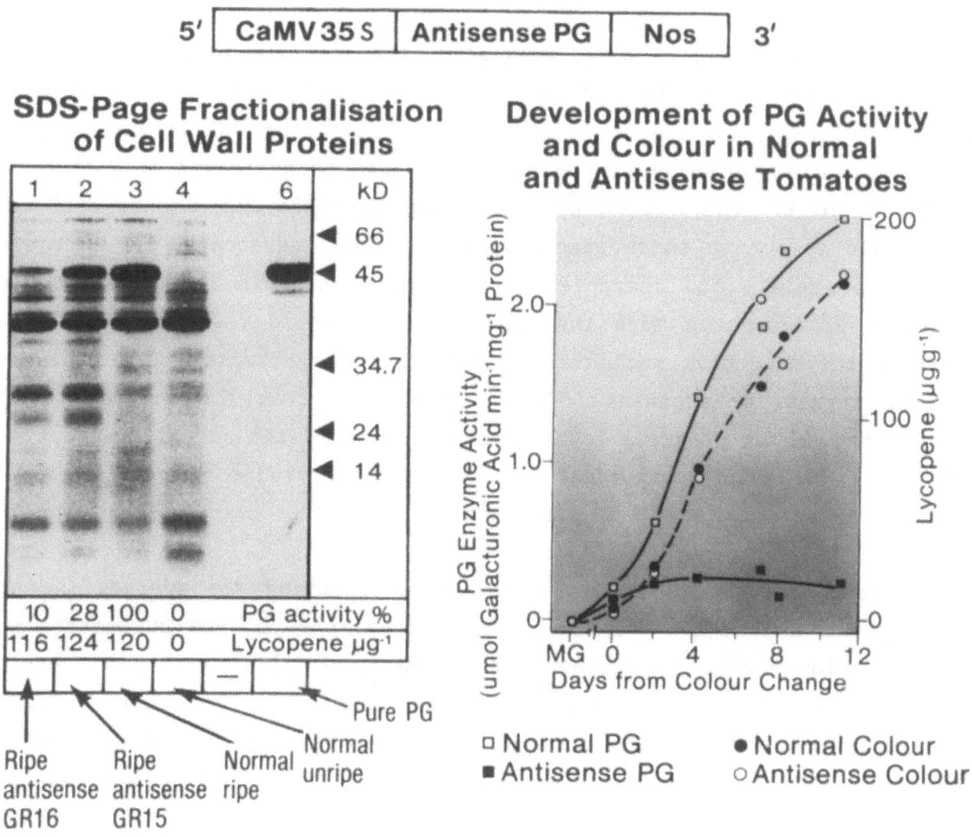

FIGURE 5. Reduction of polygalacturonase enzyme activity in ripe tomato fruit induced by PG antisense RNA expression. A diagram of the chimaeric antisense gene is shown, together with measurements of PG protein and enzyme activity during ripening. Note that control and antisense fruit develop normal colour (lycopene), although the latter have only 10% of normal PG levels

to strand-specific probes, indicated that the antisense RNA led to a reduction in the amount of PG mRNA. The precise mechanism by which antisense RNA causes this reduction is not yet clear (Smith *et al.*, 1988). Fruit containing reduced levels of PG induced by antisense RNA still underwent softening as judged by compressibility tests. However, it is possible that they contained sufficient PG activity (10% of normal) to allow enzymic reactions important for softening to continue. Alternatively, it is possible that PG does not play as important a role in softening as was once thought. Further analysis of antisense plants should resolve this question.

CONCLUSIONS

These results show that problems related to the regulation of expression and function of PG in ripening fruits can be addressed by techniques of plant molecular biology and genetic engineering. *Cis*-acting DNA control regions governing ripening-specific expression have been demonstrated to lie to the 5′ side of the PG gene. Promoter-deletion/mutagenesis experiments will make it possible to locate these precisely on the DNA sequence. The next step will be to isolate and characterise *trans*-acting protein factors that bind to the PG promoter and turn the gene on. It should then be possible to determine the molecular mechanisms governing the synthesis and/or activity of these regulatory proteins. Thus, by working from the enzyme, to the gene, we are making progress towards the question of the signal transduction mechanism. It would be of great benefit to these studies if the nature of the ethylene receptor could be elucidated.

The demonstration that a synthetic gene directing the synthesis of PG antisense RNA can reduce the normal expression of PG by 90% suggests a way to test various ideas about the role of PG in ethylene synthesis, pectin solubilisation, fruit softening, and other aspects of ripening. Recent experiments in our laboratory indicate that expression of the PG gene can be reduced to 1% of its normal levels, suggesting this is now a feasible proposition.

These approaches, adding genes so that they are expressed in a specific location or developmental stage, and reducing or abolishing the expression of specific endogenous genes using antisense RNA, are generally applicable to studies on many aspects of plant growth and development. They also have important implications for plant biotechnology.

REFERENCES

Bird CR, Smith CJS, Ray JA, Moureau P, Bevan MW, Bird AS, Hughes S, Morris PC, Grierson D, Schuch W (1988) The tomato polygalacturonase gene and ripening-specific expression in transgenic tomato plants. Plant Molec Biol 11: 651-662

Davies KM, Hobson GE, Grierson D (1988) Silver ions inhibit the ethylene-stimulated production of ripening-related mRNAs in tomato. Plant, Cell & Environment 11: 729-738

DellaPenna D, Alexander DC, Bennett AB (1986) Molecular cloning of tomato fruit polygalacturonase: analysis of polygalatcuronase mRNA levels during ripening. Proc Natl Acad Sci 83: 6420-6424

DellaPenna D, Kates DS, Bennett AB (1987) Polygalacturonase gene expression in Rutgers, *rin*, *nor*, and *Nr* tomato fruits. Plant Physiol 85: 502-507

Grierson D, Maunders MJ, Slater A, Ray J, Bird CR, Schuch W, Holdsworth MJ, Tucker GA, Knapp JE (1986a) Gene expression during tomato ripening. Phil Trans R Soc Lond B314: 399-410

Grierson D, Tucker GA, Keen J, Ray J, Bird CR, Schuch W (1986b) Sequencing and identification of a cDNA clone for tomato polygalacturonase. Nucl Acids Res 14: 8595-8603

Holdsworth MJ, Bird CR, Ray J, Schuch W, Grierson D (1987) Structure and expression of an ethylene-related mRNA from tomato. Nucl Acids Res 15: 731-739.

Knapp J, Moureau P, Schuch W, Grierson D (1989) Organisation and expression of polygalacturonase and other ripening-related genes in Ailsa Craig, 'Never ripe' and 'ripening inhibitor' mutant tomatoes. Plant Molec Biol (in press)

Lincoln JE, Cordes S, Read E, Fischer RL (1987) Regulation of gene expression by ethylene during *Lycopersicon esculentum* (tomato) fruit development. Proc Natl Acad Sci USA 84: 2793-2797

Maunders MJ, Holdsworth MJ, Slater A, Knapp JE, Bird CR, Schuch W, Grierson D (1987) Ethylene stimulates the accumulation of ripening-related mRNAs in tomatoes. Plant, Cell & Environment 10: 177-184

Mutschler M, Guttieri M, Kinzer S, Grierson D, Tucker G (1988) Changes in ripening-related processes in tomato conditioned by the *alc* mutant. Theor Appl Genet 76: 285-292

Picton S, Grierson D (1988) Inhibition of expression of tomato-ripening genes at high temperature . Plant, Cell & Environment 11: 265-272

Ray J, Bird C, Maunders M, Grierson D, Schuch W (1987) Sequence of pTOM5, a ripening-related cDNA from tomato. Nucl Acids Res 15: 10587

Ray J, Knapp J, Grierson D, Bird C, Schuch W (1988) identification and sequence determination of a cDNA clone for tomato pectin esterase. Eur J Biochem 174: 119-124

Sheehy RE, Pearson J, Brady CJ, Hiatt WR (1987) Molecular characterization of tomato fruit polygalacturonase. Molec Gen Genet 208: 30-36

Smith CJS, Watson CF, Ray J, Bird CR, Morris PC, Schuch W, Grierson D (1988) Antisense RNA inhibition of polygalacturonase gene expression in transgenic tomatoes. Nature 334: 724-726

Tucker GA, Grierson D (1982) Synthesis of polygalacturonase during tomato fruit ripening. Planta 155: 64-67

REGULATION, MATURATION AND FUNCTION OF TOMATO FRUIT POLYGALACTURONASE

Alan B. Bennett, Dean DellaPenna, Robert L. Fischer[1], James Giovannoni[1], James Lincoln[1]

Mann Laboratory
Department of Vegetable Crops
University of California
Davis, CA 95616 USA

INTRODUCTION

Ripening represents a major transition in tomato fruit development. This developmental transition is manifested by alterations in metabolism that contribute to final fruit colour, flavour and texture. It has been proposed, and a great deal of supporting evidence advanced, that these alterations in metabolism result from specific alterations in gene expression (Briggs *et al.*, 1986; DellaPenna *et al.*, 1986; Lincoln *et al.*, 1987; Mansson *et al.*, 1985; Maunders *et al.*, 1987; Rattanapanone *et al.*, 1978; Speirs *et al.*, 1984).

Two subcellular compartments, the plastids and cell wall, are sites of major metabolic transitions that occur during ripening. The most apparent aspect of tomato fruit ripening is the red colouration which results from a combination of chlorophyll degradation, lycopene biosynthesis and dramatic structural reorganization of the chloro/chromoplast. Equally apparent, although perhaps less well appreciated, are metabolic transitions within the cell wall which contribute to textural changes and softening of the fruit.

Numerous subtle changes in cell wall structure occur during fruit ripening. However the major structural change is the degradation of polyuronides (Wallner and Bloom, 1977). While polyuronide degradation is best examined biochemically, the degradation is so extensive that it can be observed as a loss of the darkly staining middle lamella in electron micrographs (Crookes and Grierson, 1983). A single enzyme, polygalacturonase (PG), has been implicated as the sole agent of polyuronide degradation in ripening tomato fruit (Huber, 1981; Themmen *et al.*, 1982; Wallner and Bloom, 1977). In addition, PG-dependent polyuronide

---

[1]Division of Molecular Plant Biology, University of California, Berkeley, CA 94720 USA

NATO ASI Series, Vol. H 35
Cell Separation in Plants
Edited by D. J. Osborne and M. B. Jackson

degradation has been suggested to be the major determinant of fruit softening (Huber, 1983; Hobson, 1964; Brady *et al.*, 1982), and it has also been speculated that PG-dependent polyuronide degradation may release pectic fragments that regulate other components of the ripening process (Baldwin and Pressey, 1988; Bennett and DellaPenna, 1987b; Brady, 1987). If these proposed roles of PG are true, then the importance of PG in contributing to a major component of fruit quality (i.e. softening) would certainly be surpassed by its role as an endogenous regulator of the ripening process. The goal of our research has been to understand the regulation of PG gene expression at several levels in order to design and implement molecular genetic strategies to critically assess the physiological function of PG in the ripening process. In this paper, we will review our progress in this endeavour.

REGULATION OF PG GENE EXPRESSION

PG was first characterized in tomato fruit in 1964 (Hobson) and its activity was shown to increase with ripening of the fruit. Subsequent studies indicated that the increase in PG activity resulted from de novo synthesis of the protein (Brady *et al.*, 1982; Tucker and Grierson, 1982) and increased levels of PG mRNA (DellaPenna *et al.*, 1986; Sato *et al.*, 1984; Maunders *et al.*, 1987). We have previously shown that PG mRNA is quite abundant, accumulating to 1.2% of the poly $(A)^+$ RNA mass in ripe fruit (cv. Castlemart) (Bennett and DellaPenna, 1987a). In subsequent experiments, we have observed levels of PG mRNA exceeding 4.6% in ripe fruit of a different genotype (cv. Rutgers).

Transcriptional activity of the PG gene was assayed by nuclear run-on transcription to determine whether the accumulation of PG mRNA in ripening tomato fruit resulted from transcriptional activation of the PG gene. Our results demonstrated that transcription of the PG gene was undetectable in mature green fruit, and became detectable only at the mature green-3 stage of ripening. The gene then remained transcriptionally active in ripe fruit (DellaPenna *et al.*, 1989). In addition, transcriptional activity of the PG gene was depressed in the ripening-impaired mutants, rin, nor and Nr. Together these results demonstrate that transcription of the PG gene is an important control point regulating PG expression in ripening of wild type tomato fruit and that PG expression in the ripening-impaired mutants is blocked at the transcriptional level.

In the in vitro transcription assays, the transcriptional activity of other genes was assessed for comparison to PG gene transcription. Interestingly, in ripe fruit the rate of PG gene transcription was quite low in relation to its mRNA abundance as compared to the other genes examined. This observation suggests that, while the developmental timing of PG expression is transcriptionally regulated, post transcriptional processes, such as high mRNA stability, may contribute to the high levels of accumulation of PG mRNA.

MATURATION OF THE PG PROTEIN

In addition to processes regulating PG mRNA levels in ripening fruit, we have begun to examine potential sites of post-translational regulation of PG. This possible level of regulation is suggested: 1) by of the existence of three isozymes of PG that may differ in their function in cell wall degradation and 2) by the fact that PG is a secreted protein, therefore undergoing post-translational processing associated with translocation through the endomembrane system.

Because antibodies to one PG isozyme (2A) cross-react with the other two isozymes (1 and 2B) (Ali and Brady, 1982) and because a single mRNA species appears to give rise to the three PG isozymes, it has been proposed that all three isozymes may be derived from a single gene (DellaPenna *et al.*, 1986). We provided further support for this by showing that PG 2A and 2B differ from one another by the degree of glycosylation, which apparently results from heterogeneous glycosylation of the primary translation product upon entry of the nascent polypeptide into the endoplasmic reticulum (DellaPenna and Bennett, 1988). Conclusive evidence that PG I, 2A & 2B are derived from one gene has now been obtained by forcing expression of a single PG gene in transgenic rin tomato fruit and observing the production of all three PG isozymes (Giovannoni, DellaPenna, Bennett and Fischer, unpublished).

Early studies of in vitro translation of PG mRNA indicated that the primary PG translation product was 54 kD as compared to the 45/46 kD of the mature PG2 protein (Sato *et al.*, 1984; DellaPenna *et al.*, 1986). Isolation and sequencing of full length cDNA clones indicated that the size discrepancy was due to a 71 amino acid N-terminal extension (Bennett and DellaPenna, 1987a; DellaPenna and Bennett, 1988; Grierson *et al.*, 1986; Sheehy *et al.*, 1987) and a 13 amino acid C-terminal extension (Sheehy *et al.*, 1987) both of which are encoded by the cDNA but not found on the mature protein.

We carried out an analysis of in vitro processing of the primary PG in vitro translation product to determine if the 71 amino acid N-terminal extension represented one or multiple proteolytic processing domains (DellaPenna and Bennett, 1988). The results of this analysis indicated that at least two processing domains could be identified. The first was a typical hydrophobic signal sequence cleaved co-translationally between Ser(24) and Asn(25) (Fig. 1). By analogy with many other proteins, this domain is responsible for directing the nascent polypeptide into the endoplasmic reticulum. A second domain of 47 amino acids is apparently cleaved after entry into the endoplasmic reticulum but prior to maturation of the protein in the cell wall. This second processing domain is cleaved between Asn(71) and Gly(72) (Fig. 1), a cleavage site similar to that observed in post-translational proteolytic cleavage of many vacuolar proteins. The function of this second processing domain, referred to as the pro-sequence, is unknown but may be involved in maintaining the protein in an inactive state or in facilitating translocation of the protein to the cell wall. In either case, processing of this domain may contribute to a further post-translational level of regulation of PG expression.

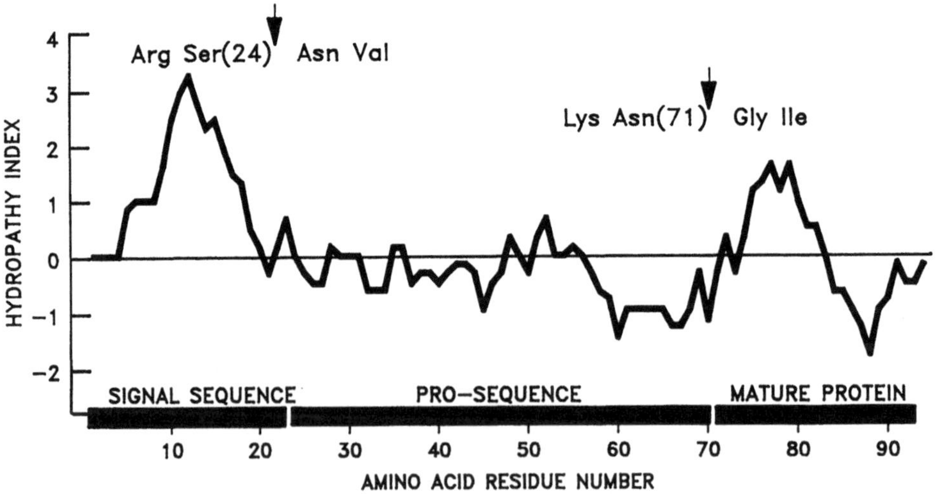

FIGURE 1. Hydropathy plot of the first 93 amino acids of PG showing the relative size, cleavage sites, and hydropathy of the signal sequence and pro-peptide.

FUNCTION OF PG

As discussed above, PG has been proposed to play an important role in pectin degradation and fruit softening during tomato ripening and it has been suggested that PG activity may release wall fragments that contribute to activation of some components of the ripening process. To critically assess these proposed functions of PG it is necessary to design molecular genetic strategies which specifically modify PG expression and then carefully monitor the phenotypic consequences of this modified PG expression. There are basically two approaches to carry out such an experiment; the first being the use of antisense RNA to depress PG expression in wild type tomato fruit and observe the consequence of reduced PG expression, and the second being to force to PG expression in a genetic background that is null for PG expression and observe the phenotypic consequences resulting from this expression. The two approaches generate complimentary information that, together, should provide a comprehensive assessment of the function of PG. The use of antisense RNA to depress PG expression has been reported by Grierson and colleagues (Smith *et al.*, 1988) and by Sheehy *et al.*, (1988). We have instead adopted the second strategy, that of forcing PG expression in a PG null genetic background (Giovannoni et al., 1989).

Our experiments have employed the ripening-impaired mutant <u>rin</u> as the PG null genotype and a ripening-associated gene referred to as E8 (Lincoln *et al.*, 1987). This gene (E8) is coordinately regulated with PG in normally ripening fruit, although we have recently observed that is rate of

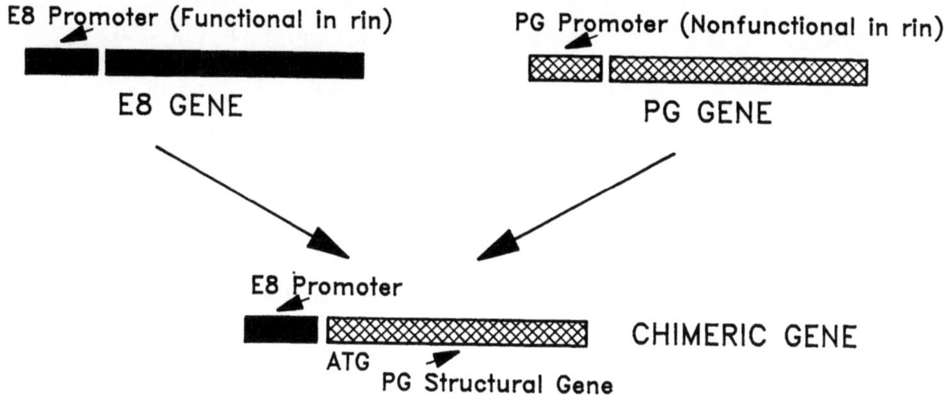

FIGURE 2. Diagram illustrating the construction of the chimeric gene comprised of a 5′ upstream region of E8 and the coding sequence of the PG gene. This chimeric gene was transfected into <u>rin</u> tomato plants by Agrobacterium-mediated transformation.

transcription is approximately 7 times greater than that of the PG gene (DellaPenna et al., 1989). Fortuitously, the E8 gene is transcriptionally active in rin fruit whereas the PG gene is transcriptionally inactive in rin. This observation suggested a strategy of constructing a chimeric gene by fusing the E8 5'promoter region to the structural PG gene (Fig. 2). Our previous data suggested that such a chimeric gene should be functional in rin, resulting in the expression of PG at the appropriate developmental time.

The E8/PG chimeric gene was constructed and transformed into rin (genetic background cv. Ailsa Craig) tomato plants and the fruit were analyzed by harvesting after 35 days and holding in air or 500 ppm propylene. Protein extracted from the fruit cell wall was analyzed by SDS-PAGE and immunoblotting and it was found that after 11 days in propylene, fruit from each of the three transgenic plants were expressing PG (Fig. 3). The PG detected on Western blots was isolated from a cell wall fraction and appeared to be processed to its normal size, suggesting the chimeric gene was competent in producing PG protein that was appropriately processed and targeted to the cell wall in rin fruit.

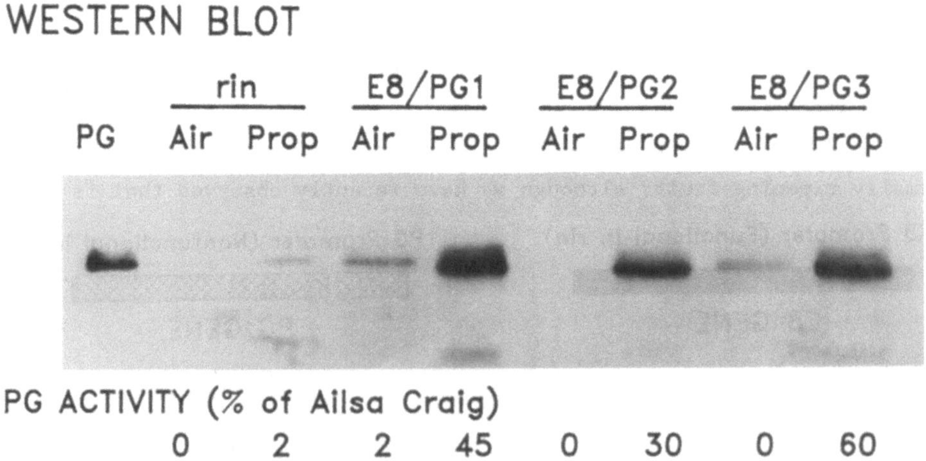

FIGURE 3. PG expression in rin fruit was analyzed by western blot analysis using antibodies raised against pure PG. Pure PG protein was run in lane 1, lanes 2-9 are cell wall protein extracts from rin fruit or from fruit of three different transgenic plants that contain the E8/PG chimeric gene. Fruit were harvested and held in air or 500 ppm propylene for 11 days prior to extraction. PG protein was detected in transgenic fruit (labeled E8/PG1,2,&3) if the fruit were held in 500 ppm propylene but not when held in air. The immunologically-detectable protein was enzymically active when assayed in vitro as indicated in the lower panel.

Our analysis of the transgenic rin fruit indicates that:

1) Expression of PG in rin fruit does result in the degradation of polyuronides (as indicated by the production of EDTA-soluble polyuronides) at levels comparable to wild-type fruit, suggesting that the enzyme is active in vivo.

2) Expression of PG in the rin genotype does not result in ethylene production or the enhancement of ripening, at least as indicated by lycopene production.

3) Softening of the transgenic rin fruit is significantly less than in wild-type fruit, suggesting that PG is not the sole determinant of fruit softening.

Overall these results indicate that PG is likely to be the sole enzyme responsible for polyuronide degradation in ripening tomato fruit. However, our results do not support a primary role for PG in fruit softening or in activating other components of the ripening process.

CONCLUSIONS

PG was first characterized in ripening tomato fruit nearly 25 years ago. During the last quarter century the effects of this enzyme on tomato fruit physiology have been extensively studied and in the last 5 years the molecular biology of this enzyme and its gene has also been characterized. We are now initiating a period of research using the molecular characterization of PG to implement strategies that modify PG expression. These experiments are in turn providing novel plant material with which we can critically assess the physiological function of PG. In general, this process is a demonstration of the power and elegance of using molecular genetic tools and strategies to address important and persistent physiological problems. In more specific terms, this process should lead to a clear view of the functional importance of PG in tomato fruit ripening. The results of this analysis should dictate whether PG remains the focus of attention for another quarter century or fades into the background as we focus on other "key" enzymes in the ripening process.

ACKNOWLEDGEMENTS

This research was supported by United States Department of Agriculture-Competitive Research Grants Office grant 87-CRCR-1-2525 and by gifts from Chesebrough-Ponds and Beatrice/Hunt Wesson.

REFERENCES

Ali ZM, Brady CJ (1982) Purification and characterization of the polygalacturonase of tomato fruits. Aust J. Plant Physiol 9:155-169

Baldwin EA, Pressey R (1988) Tomato polygalacturonase elicits ethylene production in tomato fruit. J Amer Soc Hort Sci 113:92-95

Bennett, AB, DellaPenna D (1987a) Polygalacturonase gene expression in ripening tomato fruit. In: D Nevins, R Jones (ed) Tomato biotechnology, pp 299-308

Bennett AB, DellaPenna D (1987b) Polygalacturonase: its importance and regulation in ripening. In: WW Thompson, E Nothnagle, R Huffaker (eds) Plant senescence: Its biochemistry and physiology. Amer Soc of Plant Physiol, Rockville, pp 98-107

Biggs MS, Harriman RW, Handa AK (1986) Changes in gene expression during tomato fruit ripening. Plant Physiol 81:395-403

Brady CJ (1987) Fruit ripening. Annu Rev Plant Physiol 38:155-178.

Brady CJ, MacAlpine G, McGlasson WB, Ueda Y (1982) Polygalacturonase in tomato fruits and the induction of ripening. Aust J Plant Physiol 9:171-178

Crookes PR, Grierson D (1983) Ultrastructure of tomato fruit ripening and the role of polygalacturonase isoenzymes in cell wall degradation. Plant Physiol 72:1088-1093

DellaPenna D, Alexander DC, Bennett AB (1986) Molecular cloning of tomato fruit polygalacturonase: Analysis of polygalacturonase mRNA levels during ripening. Proc Natl Acad Sci USA 83:6420-6424

DellaPenna D, Lincoln J, Fischer R, Bennett AB (1989) Transcriptional analysis of polygalacturonase and other ripening-associated genes in normal and mutant tomato fruit. Plant Physiol (in press)

DellaPenna D, Bennett AB (1988) In vitro synthesis and processing of tomato fruit polygalacturonase. Plant Physiol 86:1057-1063

Giovannoni J, DellaPenna D, Bennett AB, Fischer R (1989) Expression of a chimeric polygalacturonase gene in transgenic rin fruit results in pectin degradation but not fruit softening. Plant Cell 1:53-63

Grierson D, Tucker GA , Keen J, Ray J, Bird CR, Schuch W (1986) Sequencing and identification of a cDNA clone for tomato polygalac turonase. Nucl Acids Res 14:8595-8603

Hobson GE (1964) Polygalacturonase in normal and abnormal tomato fruit. Biochem J 92:324-332

Hobson DJ (1965) The firmness of tomato fruit in relation of polygalacturonase activity. J Hort Sci 40:66-72

Huber DJ (1981) Polyuronide degradation and hemicellulose modifications in ripening tomato fruit. J Amer Soc Hort Sci 108:405-409

Huber DJ (1983) The role of cell wall hydrolases in fruit softening. Hort Rev 5:169-219

Lincoln JE, Cordes S, Read E, Fischer RL (1987) Regulation of gene expression by ethylene during tomato fruit development. Proc Natl Acad Sci USA 84:2793-2797

Mansson PE, Hsu D, Stalker D (1985) Characterization of fruit specific cDNAs from tomato. Mol Gen Genet 200:356-361

Maunder MJ, Holdsworth MJ, Slater A, Knapp JE, Bird CR, Schuch W, Grierson D (1987) Ethylene stimulates the accumulation of ripening-related mRNAs in tomatoes. Plant Cell and Environ 10:177-184

Rattanapanone N, Spiers J, Grierson D (1978) Evidence for changes in messenger RNA content related to tomato fruit ripening. Phytochem 17:1485-1486

Sato T, Kusaba S, Nakagawa H, Ogura N (9184) Cell free synthesis of a putative precursor of polygalacturonase in tomato fruits. Plant and Cell Physiol 25:1069-1071

Sheehy RE, Pearson J, Brady CJ, Hiatt WR (1987) Molecular characterization of tomato fruit polygalacturonase. Mol Gen Genet 208:30-36

Sheehy RE, Kramer M, Hiatt WR (1988) Reduction of polygalacturonase activity in tomato fruit by antisense RNA. Proc Natl Acad Sci USA 85:8805-8809

Smith CJS, Watson CF, Ray J, Bird CR, Morris PC, Schuch W, Grierson D (1988) Antisense RNA inhibition of polygalacturonase gene expression in transgenic tomatoes. Nature 334:724-726

Speirs J, Brady CJ, Grierson D, Lee E (1984) Changes in ribosome organization and mRNA abundance in ripening tomato fruit. Aust J Plant Physiol 11:225-233

Themmen APN, Tucker GA, Grierson D (1982) Degradation of isolated tomato cell walls by purified polygalacturonase _in vitro_. Plant Physiol 69:122-124

Tucker D (1982) Synthesis of polygalacturonase during tomato fruit ripening. Planta 155:64- 67

Wallner SJ, Bloom HL (1977) Characteristics of tomato cell wall degradation _in vitro_. Implication for the study of fruit-softening enzymes. Plant Physiol 60:207-210 [illegible] (1977) [illegible] by [illegible] during [illegible] development [illegible] 1975-1977

[illegible] (19[illegible]) [illegible] Mol. Gen. Genet. [illegible]

[illegible] (19[illegible]) [illegible] RNA [illegible]

[illegible] (1978) [illegible] 1966

[illegible] (1981) [illegible] Genetics [illegible] 1967

[illegible]

CHARACTERIZATION AND EXPRESSION OF A RIPENING-INDUCED CELLULASE GENE FROM AVOCADO

Rolf E. Christoffersen, Laura G. Cass, Douglas J. McGarvey
Frank W. Percival[1], and Kristin R. Bozak
Department of Biological Sciences
University of California
Santa Barbara, CA 93106
U.S.A

INTRODUCTION

Softening of avocado fruit during ripening is mediated primarily by autodigestion of the cell wall. Two enzymes, polygalacturonase [poly(1,4-α-D-galacturonide) glycanohydrolase. E.C. 3.2.1.15] and cellulase [endo-1,4-β-D-glucan 4-glucanohydrolase, E.C. 3.2.1.4], are thought to be the principal enzymes involved in cell wall breakdown. In fact, increases in these enzyme activities are closely correlated with the softening of avocado fruit (Pesis *et al.*, 1979; Awad and Young, 1979).

Cellulase enzyme activity during avocado ripening is regulated by the accumulation of cellulase mRNA and *de novo* synthesis of cellulase protein throughout the climacteric rise in respiration and ethylene production (Christoffersen *et al.*, 1984; Tucker and Laties, 1984; Tucker *et al.*, 1985; Christoffersen, 1987). Unripe fruit have a very low level of cellulase message which increases during ripening to become one of the major species in the mRNA population at the climacteric peak of ethylene production. Previous studies have shown that avocado fruit accumulate an *in vitro* translation product of 52,000 molecular weight within 24 hours after ethylene treatment of preclimacteric fruit and it was inferred that this represented the cellulase mRNA (Tucker and Laties, 1984; Tucker *et al.*, 1985).

In the present study, we have shown that mature avocado fruit are responsive to exogenous propylene when applied continuously from one day

[1]Permanent adress: Department of Biology, Westmont College
Montecito, CA 93108

NATO ASI Series, Vol. H 35
Cell Separation in Plants
Edited by D. J. Osborne and M. B. Jackson

after harvest. The responsiveness of fruit was monitored by initiation of autocatalytic ethylene production and accumulation of cellulase mRNA and protein during the first 24 hours of propylene treatment. In addition, we have characterized two distinct regions of the avocado genome which hybridize to a cellulase cDNA clone. One of these genomic regions was shown to be a ripening-related cellulase gene.

MATERIALS AND METHODS

Plant material. Avocado fruit (*Persea americana* Mill. cv Hass) were locally grown in Santa Barbara, CA. Fruit were harvested, washed, and immediately placed in jars with a constant stream of ethylene-free humidified air. For propylene treatment, a stream of propylene (500 $\mu l/l$) mixed in air was introduced to the jars 24 hours after harvest. Ethylene synthesis was monitored by gas chromatography at 12 hour intervals . Each day during the treatment period mesocarp tissue of three representative fruit was frozen in liquid nitrogen and stored at -70°C.

Light microscopy. Small cubes (5 mm) of unfrozen tissue from either unripe or fully ripe (3 days postclimacteric) avocado mesocarp were fixed in 4% glutaraldehyde in 100 mM Na cacodylate, pH 6.8, for 3 h followed by overnight incubation in 4% OsO_4 in 100 mM Na cacodylate, pH 6.8. The tissues were then dehydrated and embedded in epoxy resin. Sections were cut with a glass knife on a Porter-Blum MTZ-B microtome and then stained with 1% toluidine blue followed by counterstaining with saturated paraphenylene diamine. Sections were photographed on a Zeiss Ultraphot microscope.

Western immunoblot electrophoresis. The relative amount of cellulase antigen was determined in each fruit sample by immunoblot SDS-PAGE.. Frozen tissue was extracted in SDS sample buffer, boiled for 3 min and centrifuged at 15,000 x g for 30 min. Equivalent amounts were separated by electrophoresis, transferred to nitrocellulose, and then challenged with antiserum to avocado cellulase. Detection of antigen-antibody complexes was accomplished as described previously (Della-Penna *et al.*, 1986).

RNA and DNA hybridization. RNA extraction from fruit at various stages of ripening, DNA isolation from avocado fruit, and agarose gel hybridization methods have been described previously (Christoffersen, 1987). The nearly full length cDNA probe, pAVe6, used to detect cellulase mRNA and DNA has been described previously (Christoffersen, 1987).

RESULTS AND DISCUSSION

Structural changes associated with ripening. The basic structure of the avocado fruit has been investigated previously at the level of both the light and electron microscope (Scott *et al.*, 1963). Studies of the developmental changes specifically associated with ripening have primarily focused on the cell wall (Platt-Aloia *et al.*, 1980) and were limited to electron microscopy investigations. In the present study, we investigated the possibility of using light microscopy to study the developmental changes associated with ripening. The mature mesocarp from unripe fruit consists primarily of parenchyma cells which are filled with multiple lipid bodies (Fig. 1). These cells are surrounded by a thin primary wall with no secondary wall deposition. Comparison of cells from

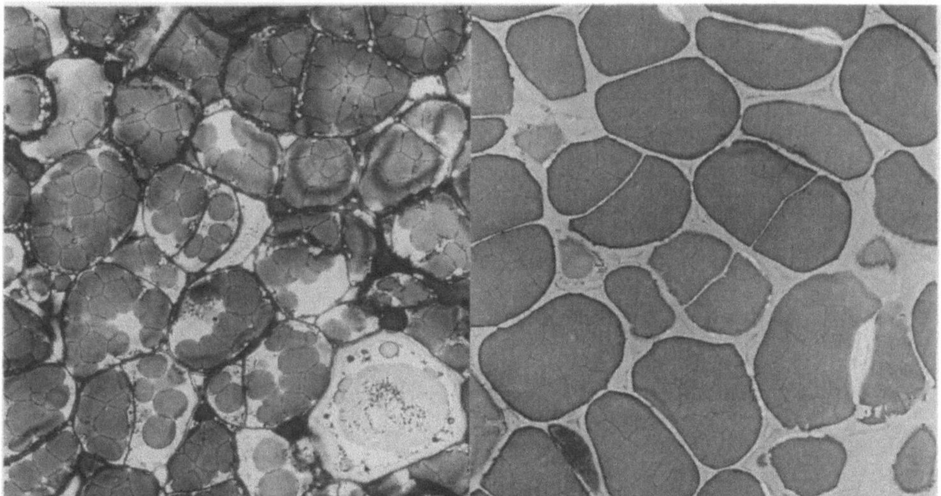

FIGURE 1. *Light micrographs of avocado mesocarp cells from a unripe fruit (left) or a fully ripe fruit (right). The atypical cell in the lower right corner of the unripe fruit is designated an idioblast. Oil bodies, nuclei, and cell walls are the prominent feature of the unripe mesocarp parenchyma cells.*

unripe and fully ripe (three days postclimacteric) fruit showed a dramatic difference in the structure and staining properties of the cell wall material. Whereas toluidine blue darkly stained the unripe wall, the fully ripe tissue showed a much lower tendency to take up the stain. In addition, the walls from ripe cells were diffuse, with the appearance of being swollen relative to the unripe cells, and having greatly reduced intracellular gas spaces in the ripe tissue. Also observed was the loss of cytoplasmic contents in cells of fully ripe tissue with the notable exception of lipid bodies, which seem to have filled the entire cell space (Fig. 1, right panel). Cellulase is thought to play a role in ripening-associated disruption of the cell wall matrix, possibly by cleavage of uncharacterized cross-links rather than degradation of cellulose fibrils (Hatfield and Nevins, 1986).

Ethylene synthesis during propylene-induced ripening. When freshly harvested avocado fruit are treated with either ethylene or propylene, the onset of the climacteric ripening is initiated within 1-2 days whereas untreated fruit begin ripening 6-10 days after harvest (Adato and Gazit, 1974; Eaks, 1980; Sitrit *et al.*, 1986). In order to relate the temporal expression of cellulase mRNA and antigen with initiation of ethylene biosynthesis, fruit were treated continuously with propylene to initiate ripening. Ethylene synthesis of the treated fruit was monitored at 12 hour

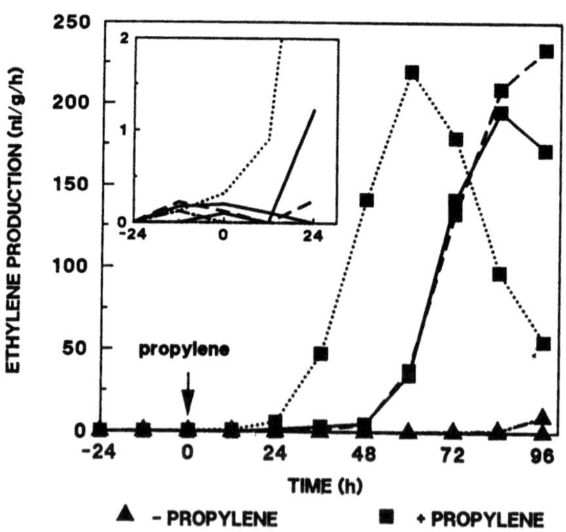

FIGURE 2. *Time course of ethylene production from propylene-treated and control avocado fruit. Fruit were harvested 3 hours before the first ethylene measurements at -24 hours. Each curve is from a single representative fruit.*

intervals (Fig. 2). The majority of fruit (16 of 18 total) showed a measurable increase in ethylene synthesis within 24 hours of initial propylene treatment. In contrast, there was considerable variability in the time that individual fruit took to reach the climacteric peak of ethylene synthesis. A few fruit were extremely resistant to propylene treatment and did not reach a climacteric peak of ethylene synthesis during the course of our experiments (data not shown).

In the avocado, ripening is normally not initiated until the vascular system is cut off by abscission or detachment of fruit from the tree (Biale and Young, 1971; Brady, 1987). The suggestion has been made that a ripening inhibitor coming from the stem may be involved in suppressing initiation of ripening (Tingwa and Young, 1975) and that this factor continues to function for a limited time after harvest (Adato and Gazit, 1974). Consequently, ripening can proceed only after degradation of the inhibiting factor has rendered the fruit susceptible to ethylene-induced ripening. Differing levels of ripening inhibitor within the fruits at harvest could account for the varying responses to propylene observed here (Fig. 2). Alternatively, each fruit could have differing capacities for removal of the inhibitory factor. Variation was observed repeatedly even though fruit were obtained from the same tree within each experiment. Because of this variation, it is imperative that the ripening stage of individual avocado fruit be monitored with a physiological marker, such as system 2 ethylene synthesis (Yang, 1987), rather than simply by duration of ethylene treatment.

Propylene-induced accumulation of cellulase protein and mRNA. Fruit, freshly harvested from the tree, had no detectable cellulase antigen. After 24 hours of propylene treatment, a small amount of cellulase antigen was observed (Fig. 3). Upon further propylene treatment, the cellulase protein continues to accumulate throughout the climacteric, eventually becoming one of the major polypeptides of the ripe fruit (Christoffersen *et al.*, 1984, Bennett and Christoffersen, 1986).

Two molecular weight antigens are observed by cellulase immunoblot SDS-PAGE of total protein extracts (Christoffersen *et al.*, 1984; Fig. 3). The lower molecular weight antigen co-migrates with the form of cellulase purified by the method of Awad and Lewis (1980). The larger cellulase

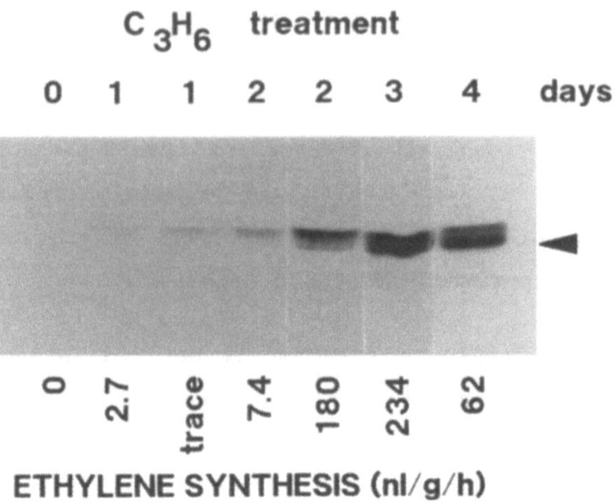

FIGURE 3. *Cellulase immunoblot of total protein extracts from propylene-treated avocado fruit fractionated by SDS-PAGE. The arrowhead indicates an* M_r *of 53,000, the molecular weight of purified cellulase. Antigen-antibody complexes were detected with goat anti-rabbit immunoglobulin conjugated to horse radish peroxidase.*

antigen may correspond to a previously described membrane-associated form of cellulase, and it was suggested that the higher molecular weight species of cellulase could be a precursor of the lower molecular weight species (Bennett and Christoffersen, 1986). In support of this hypothesis, we observed a relatively high abundance of the larger antigen in the fruit after one day of propylene treatment, at which time synthesis of cellulase protein is in its early stages (Fig. 3, day 1). As ripening progresses, the lower molecular weight cellulase antigen became the predominant form (Fig. 3, days 3 and 4), as would be expected if it were the final product of a post-translational processing pathway.

The temporal expression of cellulase mRNA during propylene-induced ripening was investigated by northern analysis using two different probes. One probe, a full-length cellulase cDNA, was used to obtain the autoradiogram shown in Fig. 4. The alternate probe was a *Cel-1* (see below) gene-specific fragment of the cDNA which gave a pattern of expression identical to that with the full length probe (data not shown). Because three characterized cellulase cDNA clones (Tucker *et al.*, 1987; Christoffersen, 1987) are completely homologous to the *Cel-1* gene, we infer that the *Cel-1* gene is responsible for a major portion, if not all, of the cellulase mRNA transcripts in the ripe fruit.

The level of cellulase mRNA in fruit frozen immediately after harvest was compared to the level in fruit treated with air for either 1 or 6 days. All showed a low, basal level of cellulase mRNA expression which did not change throughout the preclimacteric period. The propylene-treated fruit, on the other hand, showed a rapid accumulation of cellulase mRNA during the first day of treatment (Fig. 4). Further accumulation of cellulase RNA transcripts continued throughout the climacteric rise in ethylene production.

Cellulase gene family. To study the mechanism of ethylene-induced expression of the cellulase gene in ripening fruit, we have cloned and characterized genomic DNA from avocado which hybridizes to a cellulase cDNA clone (Cass and Christoffersen, unpublished results). Two different hybridizing regions of the avocado genome have been identified by Southern analysis and restriction endonuclease mapping. One region, designated *Cel-1*, is completely homologous to the ripening-induced cDNA sequence (Fig. 5).

FIGURE 4. *Expression of cellulase mRNA in propylene-treated avocado fruit. Total RNA was separated by agarose gel electrophoresis and probed with pAVe6, a cellulase cDNA. Fruit were either frozen immediately after harvest (H), treated with propylene 24 hours after harvest for 0, 1, 2, and 3 days, or kept in ethylene-free air for 6 days. The arrowhead indicates the position of 2.1 kB RNA as determined by stained RNA standards.*

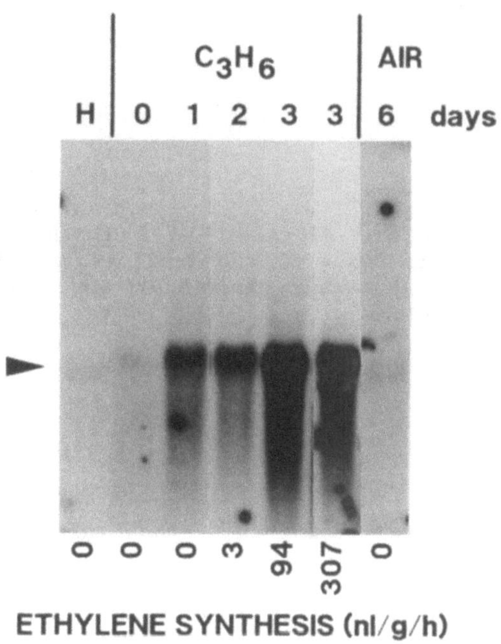

The other hybridizing region, designated *Cel-2*, is similar to *Cel-1* but lacks the first 550 bp of the cellulase cDNA, a region which includes 270

bp of coding sequence. Thus, *Cel-2* is a related member of a small cellulase gene family but is divergent from *Cel-1* at its 5′ end.

We have no evidence that the *Cel-2* gene is expressed in ripening fruit, or in any other tissue or developmental stage. It may represent a truncated pseudogene which has lost the N-terminal coding region of the cellulase protein through a genetic translocation event. The uncloned 4 kb EcoR1 fragment observed in the genomic Southern blots (Fig. 5) hybridizes to the 5′ half of the cellulase cDNA and therefore could be the missing 5′ exons of the *Cel-2* gene. We can not, as yet, exclude the possibility that the *Cel-2* gene is actively transcribed. A functional gene could have originated by recombination with divergent 5′ exons which do not hybridize to our cellulase cDNA clone. Resolution of whether *Cel-2* is a functional gene or a pseudogene awaits further experimental data.

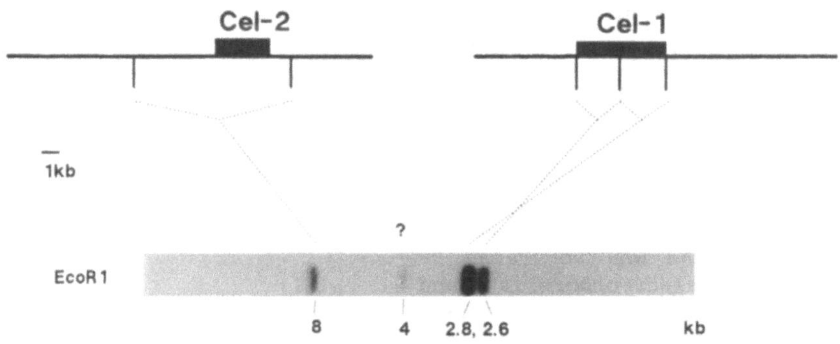

FIGURE 5. *Alignment of a Southern blot of EcoR1 endonuclease digested genomic DNA with the EcoR1 restriction map of Cel-1 and Cel-2 λ genomic clones. Avocado genomic DNA was probed with pAVe6, a full-length cellulase cDNA clone.*

CONCLUSIONS

We have shown that the initiation of ethylene synthesis and the accumulation of cellulase mRNA begins within 24 hr after treatment of unripe avocado fruit with the ethylene analog, propylene. In addition, we propose that RNA transcripts from the *Cel-1* gene are likely to represent most of the ripening-induced mRNAs which code for the cellulase protein.

ACKNOWLEDGEMENTS

The authors thank Deborah D. Fisher for excellent work in microscopy. This research was funded by USDA-CRGO #86-CRCR-1-1976 and a U.C. Santa Barbara faculty research grant.

REFERENCES

Adato I, Gazit S (1974) Postharvest response of avocado fruits of different maturity to delayed ethylene treatments. Plant Physiol 53:899-902

Awad M, Lewis LN (1980) Avocado cellulase: Extraction and purification. J Food Sci 45:1625-1628

Awad M, Young RE (1979) Postharvest variation in cellulase, polygalacturonase, and pectinmethylesterase in avocado (Persea americana Mill. cv Fuerte). Plant Physiol 64:306-308

Bennett AB, Christoffersen RE (1986) Synthesis and processing of cellulase from ripening avocado fruit. Plant Physiol 81:830-835

Biale JB, Young RE (1971) The avocado pear. *In* Hulme AC (ed) The biochemistry of fruits and their products, part I. Academic Press, New York, p 1-63

Brady CJ (1987) Fruit ripening. Ann Rev Plant Physiol 38:155-178

Christoffersen RE (1987) Cellulase gene expression during fruit ripening. *In* Thomson WW, Nothnagel EA, Huffaker RC (eds) Plant senescence: its biochemistry and physiology. Amer Soc Plant Physiol, Rockville, pp 89-97

Christoffersen RE, Tucker ML, Laties GG (1984) Cellulase gene expression in ripening avocado fruit: the accumulation of cellulase mRNA and protein as demonstrated by cDNA hybridization and immunodetection. Plant Molec Biol 3:385-391

Della-Penna D, Christoffersen RE, Bennett AB (1986) Biotinylated proteins as molecular weight standards on western blots. Anal Biochem 52:329-332

Eaks IL (1980) Respiratory rate, ethylene production, and ripening response of avocado fruit to ethylene or propylene following harvest at different maturities. J Amer Soc Hort Sci 105:744-747

Hatfield R, Nevins DJ (1986) Characterization of the hydrolytic activity of avocado cellulase Plant Cell Physiol 27:541-552

Pesis E, Fuchs G, Zauberman G (1978) Cellulase activity and fruit softening in avocado. Plant Physiol 61:416-419

Platt-Aloia KA, Thomson WW, Young RE (1980) Ultrastructural changes in the walls of ripening avocados: transmission, scanning, and freeze fracture microscopy. Bot Gaz 141:366-373

Scott FM, Bystrom BG, Bowler E (1963) Persea americana, mesocarp cell structure, light and electron microscope study. Bot Gaz 124:423-428

Sitrit Y, Riov J, Blumenfeld A (1986) Regulation of ethylene biosynthesis in avocado fruit during ripening. Plant Physiol 81:130-135

Tingwa PO, Young RE (1975) Studies on the inhibition of ripening in attached avocado (*Persea americana* Mill.) fruits. 'J Amer Soc Hort Sci 100:447-449

Tucker ML, Laties GG (1984) Interrelationship of gene expression, polysome prevalence, and respiration during ripening of ethylene and/or cyanide-treated avocado fruit (*Persea americana*). Plant Physiol 74:307-315

Tucker ML, Christoffersen RE, Woll L, Laties GG (1985) Induction of cellulase by ethylene in avocado fruit. In: Roberts JA, Tucker GA (eds) Ethylene and plant development. Butterworths, London, pp 163-171

Tucker ML, Durbin ML, CLegg MT, Lewis LN (1987) Avocado cellulase: nucleotide sequence of a putative full-length cDNA clone and evidence for a small gene family. Plant Molec Biol 9:197-203

Yang SF (1987) The role of ethylene and ethylene synthesis in fruit ripening. *In* Thomson WW, Nothnagel EA, Huffaker RC (eds) Plant senescence: its biochemistry and physiology. Amer Soc Plant Physiol, Rockville, pp 156-165

GENE EXPRESSION IN SENESCING LEAVES

Akira Watanabe, Naoto Kawakami[1] and Yoshitaka Azumi
Research Institute for Biochemical Regulation
Faculty of Agriculture
Nagoya University
Chikusa-ku, Nagoya 464-01
Japan

INTRODUCTION

Molecular events occurring in senescing leaves are a necessary pre-requisite for leaf abscission. Fruit ripening also involves similar molecular changes prior to the softening of the tissue. An understanding of cellular senescence is therefore critical to understand the initiation of these processes of cell separation.

Leaf senescence has been believed for a long time to involve activation of nuclear genes. Early supporting evidence was provided by observation of the delaying effects of enucleation by hypertonic treatment of *Elodea* leaf cells (Yoshida, 1961) and by applying inhibitors of protein synthesis to senescing oat leaves (Martin and Thimann, 1972). We have tried to confirm this notion more directly by analyzing in various ways the mRNA population in senescing radish cotyledons.

CHANGE IN THE POPULATION OF mRNA IN SENESCING RADISH COTYLEDONS

Radish cotyledons grown at 25°C under continuous light start to senesce when transferred to darkness; chlorophyll content does not change for the first 24 hours but decreases rapidly to 20% of the initial level during the subsequent 48 hours (Kawakami and Watanabe, 1988a). Changes in the population of mRNA in the senecsing cotyledons were analyzed by separating *in vitro* translation products from poly(A)$^+$RNA by two-dimensional polyacrylamide gel electrophoresis (2-D PAGE) by the method of O'Farrell *et al.* (1977). We were able to detect approximately 1,000 polypeptides on the fluorogram of a gel. Most did not change as judged by the extent of darkening on the fluorogram, which means that the content of translatable

---

[1] Present address: Kihara Institute for Biological Research, Yokohama City University, Mutsukawa, Minami-ku, Yokohama 232, Japan.

NATO ASI Series, Vol. H 35
Cell Separation in Plants
Edited by D. J. Osborne and M. B. Jackson

mRNA for those polypeptides stayed at a constant level during the dark-treatment. However, after close inspection of the fluorograms we detected a number of polypeptides which increased, decreased, or increased temporarily during 72 hours of dark-treatment. Typical polypeptides which changed in these ways are presented in Fig. 1. Fig. 2 summarizes the map of such polypeptides on the 2-D gel and the timing of the observed changes. Some of the changes in the content of specific mRNA were observed as early as 12 hours after exposure to darkness and before the cotyledons showed any visible symptoms (Kawakami and Watanabe, 1988a).

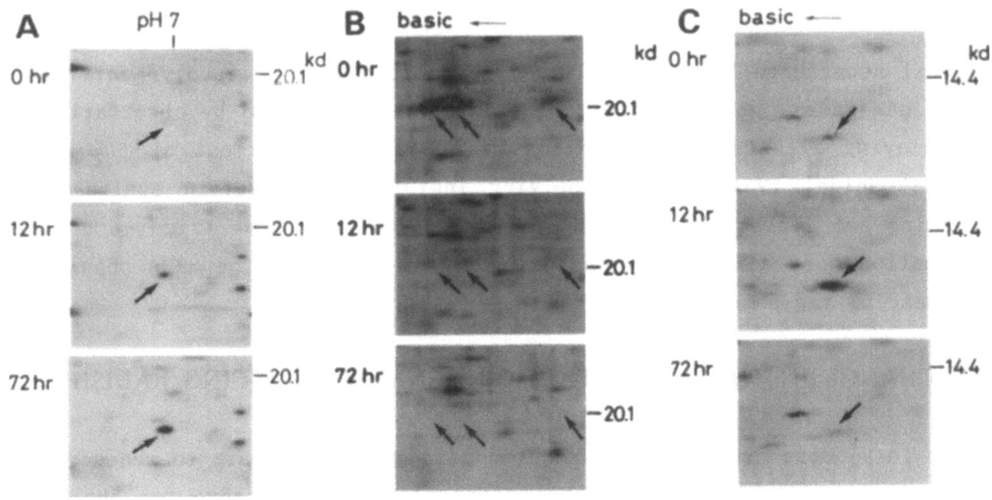

FIGURE 1. Typical translation products, the mRNA content of which increased (A), decreased (B), or increased temporarily (C) in radish cotyledons during dark-induced senescence. Poly(A)$^+$ RNA was isolated from cotyledons kept in the dark for designated period and translated in a wheat germ system in the presence of ^{35}S-methionine. The translation products were separated by 2-D PAGE with horizontal separation by isoelectric focusing in A, and by non-equilibrated pH gradient electrophoresis in B and C. Portions of fluorograms were compared in each panel.

The observations indicate that a drastic change occurs in the expression of nuclear genes when a green leaf is transferred to the dark to accelerate senescence. The level of some of the mRNA species was regulated simply by the light conditions themselves since changes induced by darkness were reversed by re-illumination of the seedlings. However, changes in others were irreversible, suggesting closer association with the progress of senescence (Kawakami and Watanabe, 1988b).

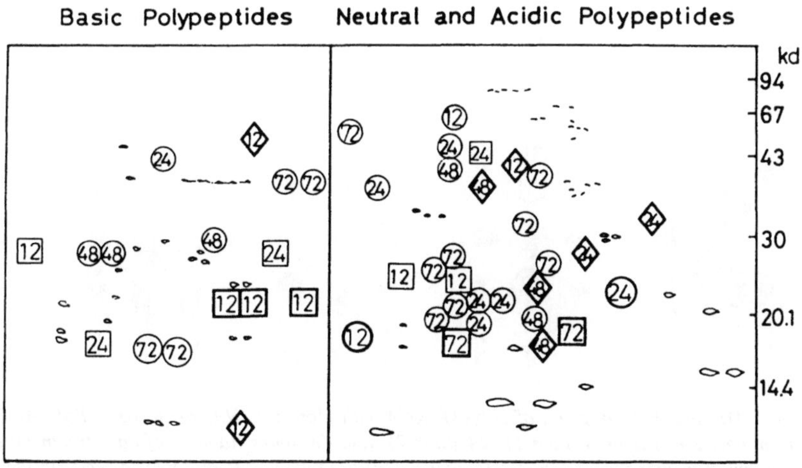

FIGURE 2. Schemetic representation of the changes in translatable mRNA during dark-induced senescence of radish cotyledons (Kawakami and Watanabe, 1988a). Circles and squares denote the polypeptides which increased and decreased, respectively in darkening on the fluorograms. The numbers denote the time when the change in labelling intensity was first observed. Diamonds denote polypeptides which increased until the time designated and decreased thereafter.

cDNA CLONING FOR mRNA ACCUMULATED IN SENESCING COTYLEDONS

Because it is difficult to quantify the level of mRNA by analysis of translation products by 2-D PAGE, we tried to measure the content of specific mRNA using cDNA complementary to the mRNA which accumulated in the cotyledonary cells during dark-induced senescence. This method also enables us to determine more precisely the timing of changes in the steady state level of mRNA and to see the effects of other environmental stimuli which may modulate the progress of senescence by influencing the expression of corresponding genes.

A cDNA library was constructed from poly(A)$^+$RNA of radish cotyledons after 24 h of dark-treatment using M13 as a cloning vector by the method of Shirras and Northcote (1984). About 6,000 recombinants were screened by differential hybridization with labelled probes of poly(A)$^+$RNA prepared from cotyledons before and after the dark-treatment. Four clones were selected that showed stronger hydridization signals with a poly(A)$^+$RNA probe from dark-treated cotyledons than with that from light-grown cotyledons. The results of dot-blot hybridization are shown in Fig. 3 in which varying amount of poly(A)$^+$RNA from dark-treated cotyledons was hybridized with

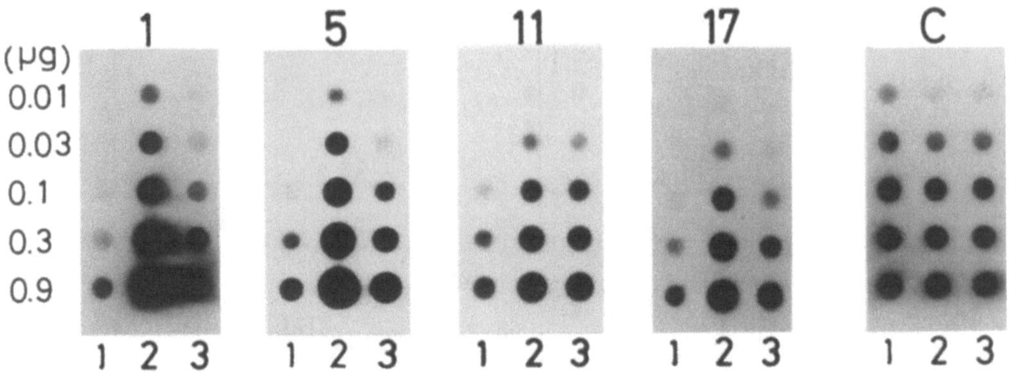

FIGURE 3. Dot-blot hybridization of poly(A)$^+$ RNA *with cloned cDNAs as probes. Poly(A)*$^+$ RNA *was isolated from cotyledons after 0 (lane 1), 24 (lane 2), and 48 hours (lane 3) of dark-treatment and spotted on a nylon membrane. Cloned cDNAs were labelled with* ^{32}P *and hybridized with the membrane-fixed RNA. The number designated above the autoradiogram is the identification number of the clone used for the probe. C denotes the result with a control clone.*

labelled probes of cloned cDNAs. It clearly shows that the content of RNA complementary to some of these cDNAs increased more than thirty-fold after 24 hours in the dark and decreased during longer dark-treatment. Increase in some of the mRNAs was detected as early as 6 hours in the dark (unpublished observation).

Similar increases were observed for the steady state level of RNA complementary to clones 1 and 5 when the seedlings were treated with ethylene at 10 µl/l in the light or heated at 35°C for 2 hours. The increase caused by dark-treatment was partially repressed when the cotyledons were sprayed with 100 µM benzyladenine (unpublished observation). The results of hydrid-select translation with these cloned cDNAs and of DNA sequencing indicated that clones 1 and 5 carried cDNA sequences for different parts of a unique mRNA which encoded a hydrophilic polypeptide of 23 kDa, the function of which is not yet known (unpublished observation). These results suggest that genes corresponding to the cloned cDNA sequences are regulated in a manner expected for those programmed to work in the cells at an early stage of senescence.

SENESCENCE-SPECIFIC ACCUMULATION OF CYTOSOLIC GLUTAMINE SYNTHETASE

One of the important events occurring in senescing leaves is the translocation of nitrogen and other nutritionally valuable cellular constituents to a storage organ and/or to a younger part of the plant. Cellular proteins in senescing leaves are hydrolyzed and their nitrogen moiety is translocated as amides, glutamine and asparagine (Zimmerman, 1960). We have examined the level of glutamine synthetase (GS) in senescing radish cotyledons as a possible regulatory step for nitrogen translocation.

We first measured total GS activity in soluble extracts of senescing radish cotyledons. Unexpectedly, the total activity decreased, to about 50% of the initial level during 72 hours in the dark. Leaf cells are known to contain two isoforms of GS, one in the cytosol (GS_1) and the other in the chloroplast (GS_2). We tried to separate these isoforms by conventional column chromatography to see if the level of these isoforms was regulated differently in senescing leaves, but failed to separate them satisfactorily. Therefore, we looked for changes in the relative content of the isoform proteins by detecting them after immunoblotting with anti-maize GS_2 antiserum. When a constant amount of soluble protein from senescing cotyledons was analyzed, the content of GS_2 protein decreased along with the progress of senescence to reach about 50% of the initial value after 72 hours of dark-treatment. On the contrary, the content of GS_1 increased two-fold during the same period (Kawakami and Watanabe, 1988c).

The differential change in the content of GS isoforms was likely to be a result of differences in the rates of synthesis of the two proteins in the cotyledonary cells. We examined this possibility by determining the level of translatable mRNAs for the two proteins by analyzing cell-free translation products from poly(A)$^+$RNA of senescing cotyledons (Fig. 4). Anti-maize GS_1 antibodies precipitated polypeptides with molecular masses of 45 kDa and 40 kDa from the translation product of GS_1 subunit. The 40 kDa polypeptide is the primary translation product for the GS1 subunit, and the 45 kDa polypeptide is the of precursor polypeptide for the GS_2 subunit. The relative content of mRNA for GS_1 stayed at the initial level for the first 12 hours, and increased linearly to a seven-fold level within 72 hours of dark-treatment. On the contrary, mRNA for GS_2 decreased quickly after transferring the cotyledons to the dark, and reached a level lower than 10%

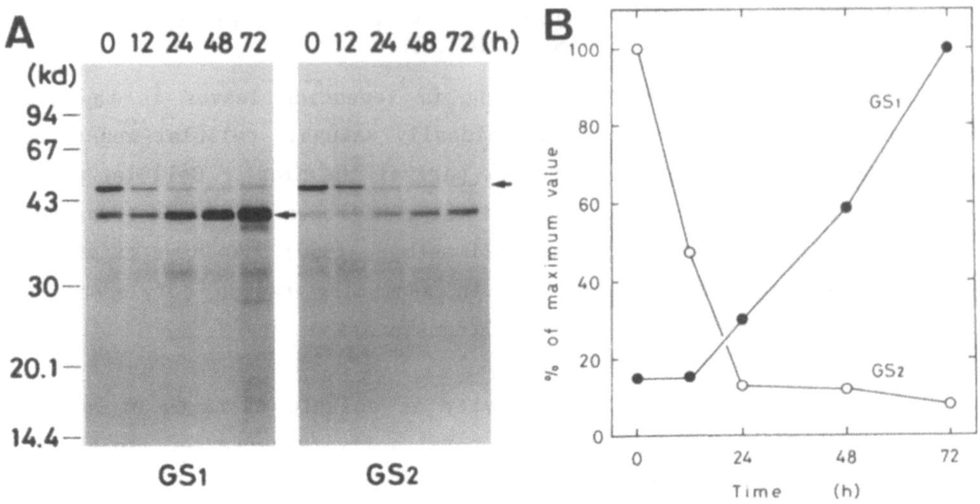

FIGURE 4. Change in translatable mRNA coding for GS_1 and GS_2 in radish cotyledons during dark-induced senescence (Kawakami and Watanabe, 1988c). (A) Poly(A)$^+$ RNA was prepared from the cotyledons at 0, 12, 24, 48 and 72 hours in the dark, and translated in a wheat germ system. The translation products were treated with anti-maize GS_1 (GS_1) or anti-maize GS_2 and analyzed by SDS-PAGE and fluorography. (B) The fluorograms in (A) were quantified by scanning in a densitometer, and the values were expressed as the percent of maximum reached during the experimental period.

after 72 hours. Similar changes in the relative content of mRNA were observed in the cotyledons senesced on seedlings grown under continuous light for longer period (Kawakami and Watanabe, 1988c).

Plant GS is known to consist of subunit isoproteins with different pIs (Hirel *et al.*, 1984). We have also examined mRNA levels of individual subunit isoproteins for GS_1 and GS_2 in senescing cotyledons. While all of the mRNAs for five different subunit isoproteins of GS_2 decreased rapidly, those of two among six isoproteins of GS_1 increased several fold in the later stage of senescence (unpublished observation). These results suggest that all of the genes for GS_2 are repressed in senescing cotyledons, whereas specific members of the gene family of GS_1 are activated and eventually alter the population of the enzyme in the cotyledonary cells. This altered cytoplasmic enzyme may facilitate the translocation of nitrogen derived from degraded cellular proteins in the senescing cotyledons to young and actively growing part of the plant.

CONCLUSION

Our results suggest that senescence is a physiological process which involves not only the turning-off of certain vital genes but also the activation of a new set of genes leading to accumulation of specific mRNAs in the cell. Furthermore, the activation seems to occur in a sequential manner which is programmed into the life cycle of the tissue and is modulated not only by plant hormones but also by such environmental stimuli as light and temperature. We don't know yet how these events signal separation processes in the petiole abscission zone, but it is very likely that some of them are prerequisites for abscission of the leaf.

The regulation of GS genes, which is activated at rather a later stage of senescence is much more complicated. Even individual members of GS_1 gene family are differentially activated in senescing leaves. Isolation of cDNA and genomic clones for these gene family members will allow more precise experimentation on the regulation of GS genes in senescing leaves.

ACKNOWLEDGEMENTS

This work is supported in part by a grant (Original and Creative Research Project on Biotechnology) from the Research Council, Ministry of Agriculture, Forestry and Fisheries, and also in part from Grants-in-Aid for Special Project Research and for Scientific Research on Priority Areas from the Ministry of Education, Science and Culture of Japan.

REFERENCES

Hirel B, Weatherley C, Cretin C, Bergounioux C, Gadal P (1984) Multiple subunit composition of chloroplast glutamine synthetase of *Nicotiana tabacum* L. Plant Physiol 74: 448-450

Kawakami N, Watanabe A (1988a) Change in gene expression in radish cotyledons during dark-induced senescence. Plant Cell Physiol 29: 33-42

Kawakami N, Watanabe A (1988b) Effects of light illumination on the population of translatable mRNA in radish cotyledons during dark-induced senescence. Plant Cell Physiol 39: 347-353

Kawakami N, Watanabe A (1988c) Senescence-specific increase in cytosolic glutamine synthetase and its mRNA in radish colyledons. Plant Physiol (in press)

Martin C, Thimann KV (1972) The role of protein synthesis in the senescence of leaves. Plant Physiol 49: 64-71

O'Farrell PZ, Goodmann HM, O'Farrell PH (1977) High resolution two-dimensional electrophoresis of basic as well as acidic proteins. Cell 12: 1133-1142

Shirras AD, Northcote DH (1984) Molecular cloning and characterisation of cDNAs complementary to mRNAs from wounded potato (*Solanum tuberosum*) tuber tissue. Planta 162: 353-360

Yoshida Y (1961) Nuclear control of chloroplast activity in *Elodea* leaf cells. Protoplasma 54: 476-492

Zimmerman MH (1960) Transport in the phloem. Ann Rev Plant Physiol 11: 167-190

REGULATION OF ETHYLENE BIOSYNTHESIS IN HIGHER PLANTS: INDUCTION AND IDENTIFICATION OF 1-AMINOCYCLOPROPANE-1-CARBOXYLATE SYNTHASE

A. K. Mattoo, A. M. Mehta, J. F. D. Dean and J. D. Anderson
Plant Molecular Biology and Plant Hormone Laboratories
USDA/ARS Beltsville Agricultural Research Center-West
Beltsville, MD. 20705, U.S.A.

INTRODUCTION

Of all the known plant growth regulators, ethylene is one of the most immediate hormonal signals involved in promoting processes that cause cell separation (abscission, fruit softening) and plant tissue senescence (see Lieberman, 1979; Mattoo and Aharoni, 1988). Cell separation occurs as a result of cell wall hydrolysis in target cells within the separation layer (Osborne, 1979). Enzymes that specifically hydrolyze cell walls, viz. cellulase and polygalacturonase, are known to be regulated positively by ethylene and negatively by auxin (Abeles, 1973; Greenberg *et al.*, 1975; Jeffrey *et al.*, 1984; Tucker *et al.*, 1988).

Endogenous levels of ethylene increase prior to the abscission process. Also, inhibitors of ethylene action retard or prevent this process and suppress cell wall hydrolytic enzyme activities (Jackson and Osborne, 1970; Aharoni *et al.*, 1979; Osborne, 1982; Baker, 1983; Sexton and Woolhouse, 1984; Sisler *et al.*, 1985; for additional references, see Mattoo and Aharoni, 1988). Thus, ethylene seems to be intimately associated with cell separation in the targeted plant tissue. However, the exact mechanisms by which ethylene brings about physiological responses in plants are largely unknown.

The biosynthetic pathway of ethylene follows the metabolic sequence: methionine $\rightarrow$ S-adenosylmethionine $\rightarrow$ 1-aminocyclopropane-1-carboxylic acid (ACC) $\rightarrow$ ethylene

NATO ASI Series, Vol. H 35
Cell Separation in Plants
Edited by D. J. Osborne and M. B. Jackson

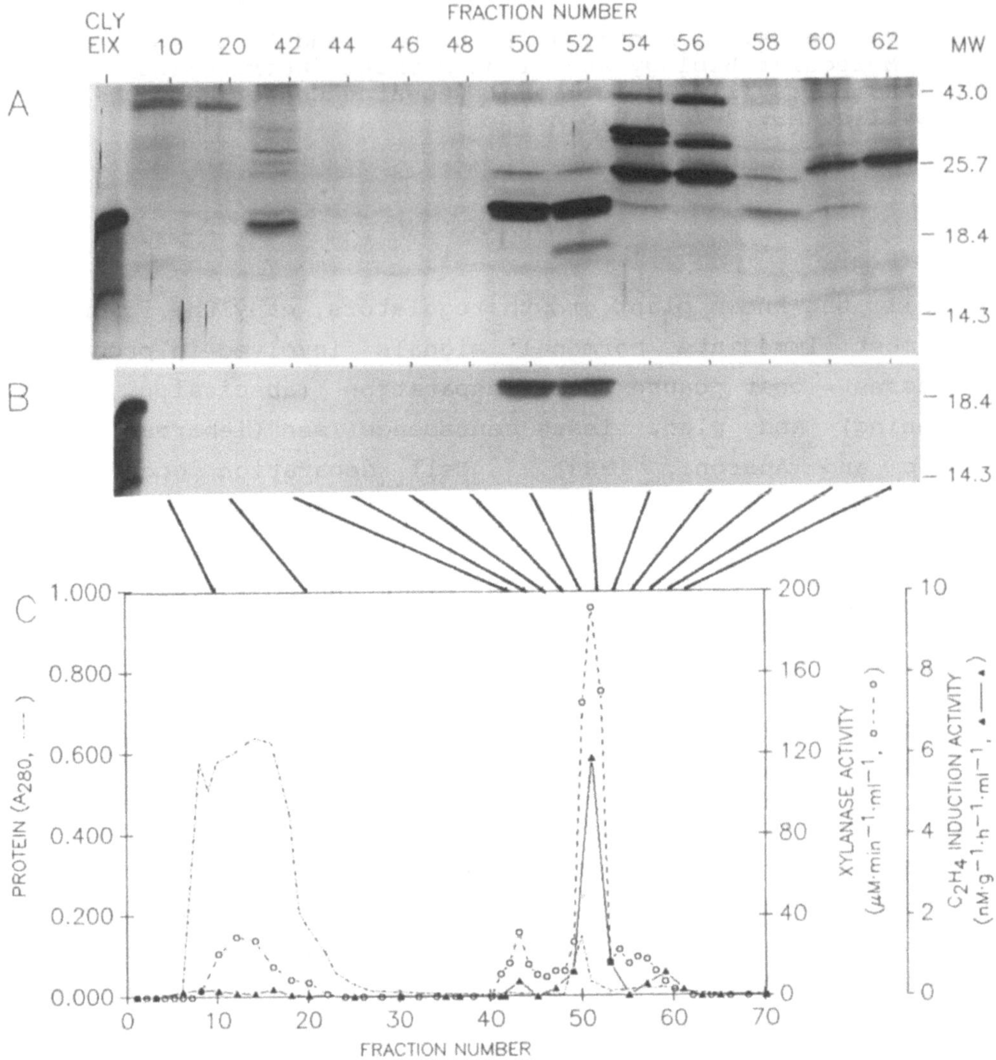

FIGURE 1. Chromatogram of T. viride culture filtrate. T. viride culture filtrate was passed through a CM-Sepharose column and adsorbed proteins were eluted by an increasing salt gradient. Proteins in equal volumes from the indicated fractions were precipitated with 10% trichloroacetic acid, and resolved in duplicate by SDS-PAGE. The first lane (CLY E1X) was loaded with E1X purified from Cellulysin. One gel was silver stained (A). The other transferred onto a nitrocellulose sheet and immunodecorated with antibody to 22kDa protein from Cellulysin (B). C depicts profiles of protein (A_{280}, ●), xylanase activity (O), and ethylene-inducing activity (▲).

(Yang and Hoffman, 1984). In many tissues, the step of conversion of S-adenosylmethionine to ACC catalyzed, by the enzyme ACC synthase, is rate limiting for ethylene production (see Yang and Hoffman, 1984; Mattoo and Aharoni, 1988). Our report here deals with factors that regulate ethylene biosynthesis, and with the identification of ACC synthase. These initial studies should add to our understanding of the cell separation process in plants.

INDUCTION OF ETHYLENE BIOSYNTHESIS BY AN ENDO-XYLANASE

Methods for isolation of protoplasts employ incubation of plant tissues with hydrolytic enzyme mixtures that degrade cell walls. During this process considerable amounts of ethylene are generated (Anderson *et al.*, 1982; Guy and Kende, 1984). It was shown that a proteinaceous factor present in a cell wall digesting enzyme mixture (Cellulysin) induces ethylene biosynthesis by means of enhanced ACC formation (Anderson *et al.*, 1985). Treatment of plant tissue with ethylene prior to incubation with Cellulysin enhances by several-fold the production of ethylene (Chalutz *et al.*, 1984). The proteinaceous factor from Cellulysin was purified and shown to possess endo-xylanase activity (Fuchs *et al.*, 1989). To study in more detail the nature of this ethylene-inducing principle, we cultured Trichoderma viride and optimized growth conditions for maximal production of the ethylene-inducing endo-xylanase (EIX) (Anderson *et al.*, 1989; Dean *et al.*, 1989). We confirmed that cultures of T. viride produce an extracellular protein that induces ethylene formation in tobacco leaf discs. The subunit molecular weight of the purified protein was determined to be 22,000 (22 kDa) and it co-purified with endo-xylanase activity (Fig. 1C). The purified protein shares the following properties with the protein isolated from Cellulysin: (1) Both have a 22 kDa protein component that co-electrophorese on sodium dodecyl sulfate polyacrylamide gels (Fig. 1 A, B); (2) both bind CM Sepharose; (3) both possess endo-xylanase activity and the ability to induce ethylene biosynthesis; and (4) antibodies raised against one cross react with the other (antigen) (see also Fig. 1B). However, the two proteins have slightly different pI's. These data suggest the two endo-xylanases to be very similar.

The relation between endo-xylanase activity and the ability to induce ethylene biosynthesis by the protein (ElX) purified from T. viride is shown in Fig. 2. As little as 20 ng (~1 pM) of ElX per 85 mg of tobacco leaf tissue was able to produce half-maximal increase in the induction of ethylene but measurable levels of ethylene were elicited at 5 ng of ElX. Although this increase in ethylene occurs via an increase in the activity of ACC synthase (Anderson *et al.*, 1982; Chalutz *et al.*, 1984), the mechanism by which the 22 kDa protein (given to tissue segments) causes this increase remains to be elucidated. Possibly, ethylene biosynthesis is regulated in plant tissue by a similar protein. Preliminary immunological data suggest the presence of antigenically similar proteins in plant tissues.

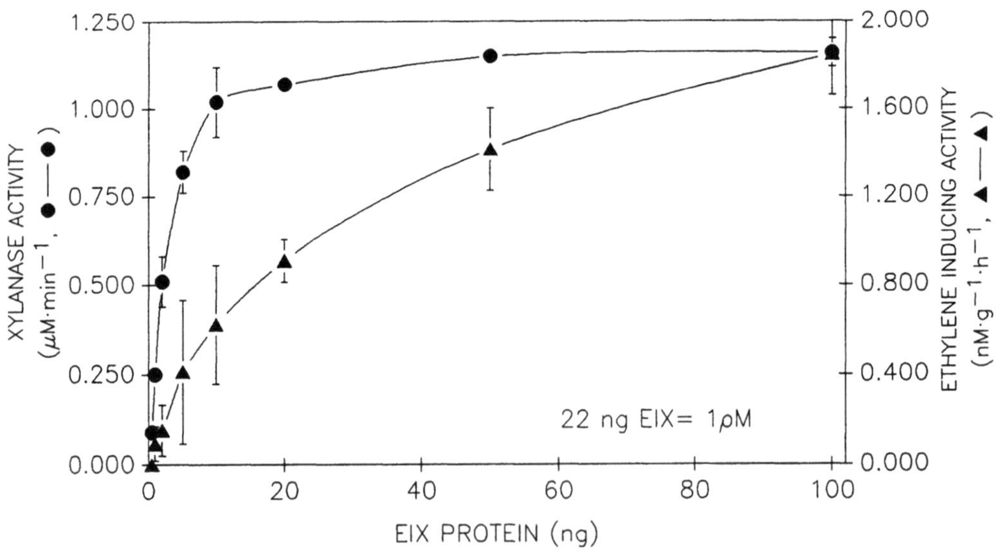

FIGURE 2. *Activity curves of ethylene-inducing protein purified from T. viride cell cultures. Xylanase (●) and ethylene inducing (▲) activities are depicted as a function of the purified protein under standard conditions (Dean et al., 1989).*

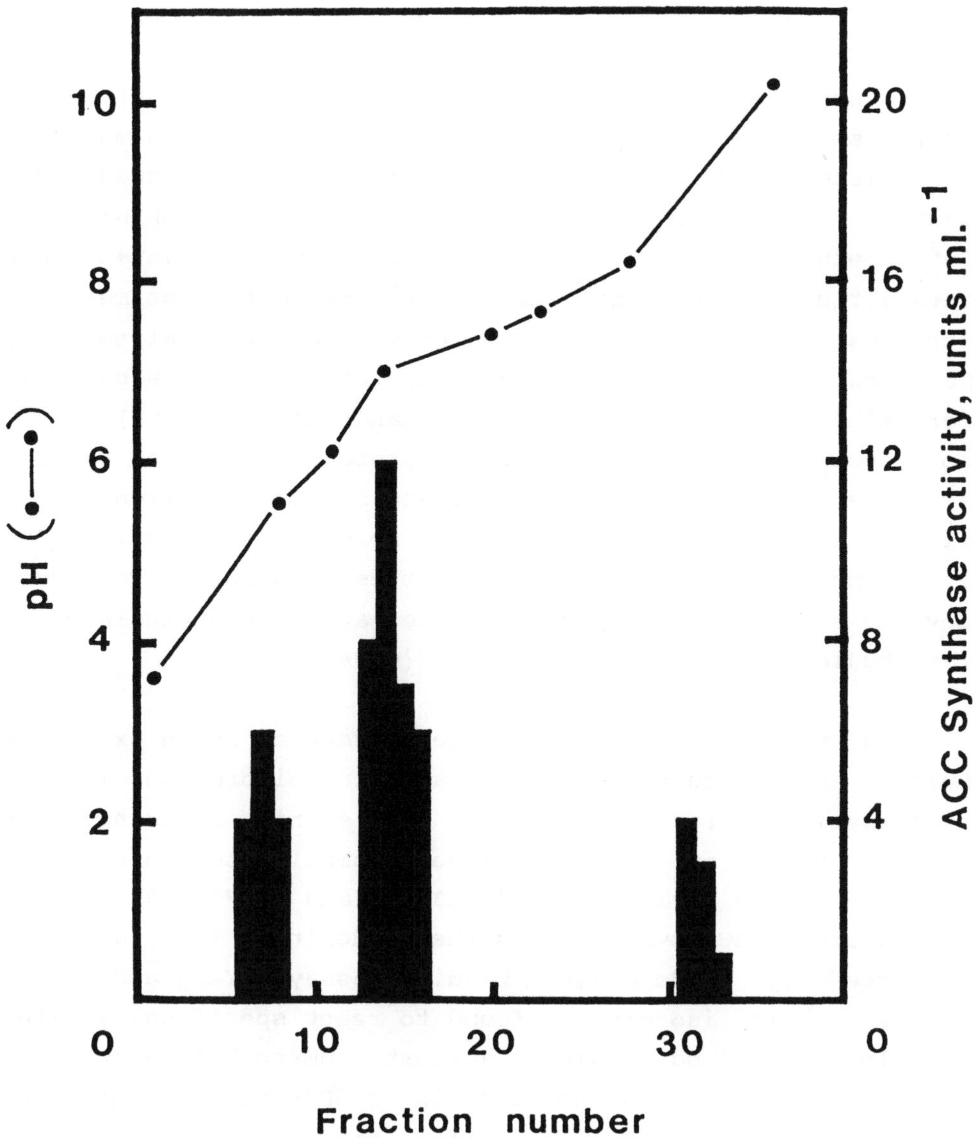

FIGURE 3. Presence of multiple isoforms of ACC synthase in tomato fruit. Fruit extract (200 ug protein) was electrophoresed under non-denaturing condtions on isoelectric focusing agarose gels with a pH gradient between 3.5 and 10. The gel was sliced into 5-mm strips and proteins were eluted overnight at 4°C in a buffer. The elements were assayed for ACC synthase activity (represented by filled bars).

IDENTIFICATION OF ISOFORMS OF ACC SYNTHASE

ACC synthase is a highly regulated protein that is of enormous interest to plant physiologists as well as molecular biologists. This enzyme is developmentally regulated and is induced or repressed by hormones (auxin, cytokinin plus Ca^{2+}, ethylene). It is also induced during plant-pathogen interactions, and by several environmental stimuli and stresses such as wounding, water stress, chemical toxicity, polluted air, thigmomorphogenesis, where it is a pacemaker for ethylene biosynthesis (Mattoo and Aharoni, 1988). It is not known if the same ACC synthase protein is induced under all these varied physiological and stress conditions, or if plant tissues elaborate different forms of this enzyme under different stimuli. To answer these questions we need to develop sensitive and specific probes at the protein level (antibodies) and at the gene level (cDNAs).

Red tomato fruit homogenates prepared in an extraction buffer that contained protease inhibitors to minimize proteolysis were resolved on isoelectric focusing (IEF) agarose gels (or on IEF-sucrose columns) into three enzyme activity peaks having pI's of 5.3, 7, and 9 (Fig. 3). The pI 7 isoform of ACC synthase was the predominant form, a finding supported by pH-dependent binding assays (Yang and Langer, 1987). This isoform was found to react specifically with a monoclonal antibody raised against tomato ACC synthase and was resolved into a 67 kDa protein on SDS-polyacrylamide gels (Mehta *et al.*, 1988b; see also Fig. 4). The monoclonal antibody did not react with any of the other isoforms of ACC synthase, suggesting that the epitope on the 67 kDa protein recognized by the antibody is probably not shared by other forms of the enzyme. In this context it is of interest that polyclonal antibodies raised against wound-induced ACC synthase do not immunoreact or immunoprecipitate IAA-induced enzyme activity (Imaseki *et al.*, see this volume). These data indicate the possibility that different isoforms of ACC synthase may represent different gene products.

STATE OF THE SIZE OF ACC SYNTHASE

A summary of various reports on the molecular mass of ACC synthase from different tissues is given in Table 1.

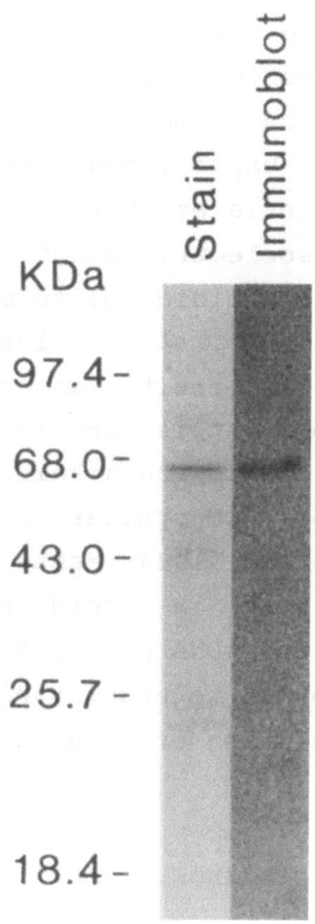

FIGURE 4. Monoclonal antibody (mAB T20C) identification of the 67 kDa ACC synthase. Five micrograms of partially purified ACC synthase were fractionated by SDS-PAGE in duplicate. One part was stained with Coomassie (Stain) and the other part transferred onto a nitrocellulose sheet and immunodecorated by the monoclonal antibody specific for ACC synthase (Immunoblot).

It is evident that two major forms of ACC synthase are present in plant tissues analyzed thus far. One of these has a molecular weight ranging between 45,000 and 50,000, and reported to be present in tomato and squash fruits. However, while the enzyme from tomato fruit is a monomer (Bleecker *et al.*, 1986), the one from squash is a dimer (Nakajima *et al.*, 1988). In both cases, the enzyme is induced upon wounding. The other form of ACC synthase was shown to have a size of 65-67 kDa, and reported to be present in tomato and mungbean hypocotyls. This form exists as a monomer in tomato (Mehta *et al.*, 1988b) and as a dimer in mungbeans (Tsai *et al.*, 1988). Moreover, the tomato enzyme was found in wounded fruit while the one from mungbeans was induced by IAA. An ACC synthase with a similar size may be present in potato (Burns and Evensen, 1986). Occasionally, a 58 kDa polypeptide is seen on immunoblots (Nakajima *et al.*, 1988; Mehta and Mattoo, unpublished observations). Another property that differentiates between these forms of ACC synthase is their isoelectric point. The 45-50 kDa protein appears to belong to acidic pI form (Privalle and Graham, 1987; van Der Straeten *et al.*, 1988) while the 67 kDa polypeptide from tomato fruit was shown to have a pI of 7 (Mehta *et al.*, 1988b). It is tempting to speculate that the 58 kDa form may belong to the alkaline pI category, since a significant amount of ACC synthase activity focuses at a pI of 9 (Mattoo and Anderson, 1984; Mehta *et al.*, 1988b). It is clear, therefore, that a need exists for developing additional antibody and cDNA probes specific to each isoform of ACC synthase in order to understand in greater detail the complex molecular regulation of this rate limiting enzyme.

ACKNOWLEDGMENTS

Results reported in this report were supported in part by a U.S.A.-Israel Binational Agricultural Research (BARD) fund (I-1165-86). We thank Mr. Dave Clark for help with illustrations.

TABLE 1. Summary of ACC synthase molecular weight and pI in higher plants

Source	Special Conditions	Subunit Mol. Wt. (kDa)	pI	Detection Method	References
Tomato fruit	Wounded, pink	50	?	Monoclonal antibody	Bleecker *et al.*, (1986)
Tomato fruit	Wounded	50	4.5	Reduction with NaB^3H_4	Privalle & Graham (1987)
Tomato fruit	LiCl-treatment	45	5.8	^{14}C-SAM labeling	Van Der Straeten *et al.*,(1988)
Tomato fruit	Wounded, red	67	7.0	Monoclonal antibody, reduction with NaB^3H_4	Mehta *et al.*, (1988 a,b)
Tomato fruit	Wounded, red	58(?) N.D.	9.0 5.3	IEF, activity	Mehta, *et al.*, (1988b)
Potato	Wounded	72	?	Gel filtration, activity	Burns and Evensen (1986)
Mungbean hypocotyl	IAA-induced	65	?	Gel filtration, SDS-PAGE	Tsai *et al.*, (1988)
Squash fruit	Wounded	67(*) 58 50	?	Polyclonal antibody	Nakajima *et al.*, (1988)
Squash fruit	Wounded, LiCl, IAA	48	?	Polyclonal antibody, expression in E. coli	Theologies (personal communication)

N.D. = not determined; ? = not known; * = polypeptide of this molecular weight is seen on the immunoblot presented in the paper.

REFERENCES

Abeles FB (1973) Ethylene in plant biology. Academic Press, New York, London

Aharoni N, Lieberman M, Sisler HD (1979) Patterns of ethylene production in senescing leaves. Plant Physiol 64:796-800

Anderson J, Dean JFD, Gamble RH, Mattoo AK (1989) Induction and characterization of the ethylene biosynthesis-inducing xylanase produced by the fungus, Trichoderma viride. In: Clijsters H , Van Poucke M (eds). Biochemical and Physiological Aspects of Ethylene Production in Lower and Higher Plants. Kluwer Dordrecht, (In press)

Anderson JD, Mattoo AK, Lieberman M (1982) Induction of ethylene biosynthesis in tobacco leaf discs by cell wall digesting enzymes. Biochem Biophys Res Commun 107:588-596

Anderson JD, Chalutz E, Mattoo AK (1985) Induction of ethylene biosynthesis by cell wall digesting enzymes. In: Key JL, Kosuge K (eds) Cellular and Molecular Biology of Plant Stress, Vol 22, New Series, Alan R. Liss Inc., New York, pp 263-273

Baker JE (1983) Preservation of cut flowers. In: Nickell L (ed) Plant Growth Regulating Chemicals Vol. 2, CRC Press, Boca Raton, pp 177-191

Bleecker AB, Kenyon WH, Sommerville SC, Kende H (1986) Use of monoclonal antibodies in the purification and characterization of 1-aminocyclopropane-1-carboxylate synthase, an enzyme in ethylene biosynthesis. Proc Natl Acad Sci (USA) 83:7755-7759

Burns JK, Evensen KB (1986) Ca^{2+} effects on ethylene, carbon dioxide and 1-aminocyclopropane-1-carboxylic acid synthase activity. Physiol Plant 66:609-615

Chalutz E, Mattoo AK, Solomos T, Anderson JD (1984) Enhancement by ethylene of Cellulysin-induced ethylene production by tobacco leaf discs. Plant Physiol 74:99-103

Dean JFD, Gamble HR, Anderson JD (1989) The ethylene biosynthesis-inducing xylanase: Its induction in Trichoderma viride and certain plant pathogens. (Submitted)

Fuchs Y, Saxena A, Gamble HR, Anderson JD (1989) Ethylene biosynthesis-inducing protein from Cellulysin is an endoxylanase. Plant Physiol 89:138-143

Greenberg J, Goren R, Riov J (1975) The role of cellulase and polygalacturonase in abscission of young and mature Shamouti orange fruits. Physiol Plant 34:1-7

Guy M, Kende H (1984) Conversion of 1-aminocyclopropane-1-carboxylic acid to ethylene by isolated vacuoles of Pisum sativum L. Planta 160:281-287

Jackson MB, Osborne DJ (1970) Ethylene, the natural regulator of leaf abscission. Nature 225:1019-1022

Jeffrey D, Smith C, Goodenough P, Prosser I, Grierson D (1984) Ethylene-independent and ethylene-dependent biochemical changes in ripening tomatoes. Plant Physiol 74:32-38

Lieberman M (1979) Biosynthesis and action of ethylene. Ann Rev Plant Physiol 30:533-591

Mattoo AK, Aharoni N (1988) Ethylene and plant senescence. In: Leopold AC and Nooden L (eds). Senescence and Aging in Plants. Academic Press, New York, pp 241-280

Mattoo AK, Anderson JD (1984) Wound-induced increase in 1-aminocyclopropane-1-carboxylate synthase activity: regulatory aspects and membrane association of the enzyme. In: Fuchs Y and Chalutz E (eds), Ethylene: Biochemical, Physiological, and Applied Aspects, Martinus Nijhoff/Dr. W. Junk, The Hague, pp 139-147

Mehta AM, Anderson JD, Mattoo AK (1988a) Sodium boro[^{3}H]hydride reduction and monoclonal antibody decoration of tomato fruit 1-aminocyclopropane-1-carboxylic acid synthase. FASEB J 2:423

Mehta AM, Jordan R, Anderson JD, Mattoo AK (1988b) Identification of a unique isoform of 1-aminocyclopropane-1-carboxylic acid synthase by monoclonal antibody. Proc Natl Acad Sci (USA) 85:8810-8814

Nakajima N, Nakagawa N, Imaseki H (1988) Molecular size of wound-induced 1-aminocyclopropane-1-carboxylate synthase from Cucurbita maxima Duch. cv. Ebisu and change of translatable mRNA of the enzyme after wounding. Plant Cell Physiol 29:989-998

Osborne DJ (1979) Target cells - new concepts for plant regulation in horticulture. Scientific Horticulture 30:1-13

Osborne DJ (1982) The ethylene regulation of cell growth in specific target tissues of plants. In: Wareing PF (ed), Plant Growth Substances 1982, Academic Press, London pp 279-290

Privalle LS, Graham JS (1987) Radiolabeling of a wound-inducible pyridoxal phosphate-utilizing enzyme. Evidence for its identification as ACC synthase. Arch Biochem Biophys 253:333-340

Sexton R, Woolhouse HW (1984) Senescence and abscission. In: Wilkins MB (ed), Advanced Plant Physiology, Pitman, London, pp 469-497

Sisler EC, Goren R, Huberman M (1985) Effect of 2,5-norbornadiene on abscission and ethylene production in citrus leaf explants. Physiol Plant 63:114-120

Tsai D-S, Arteca RN, Bachman JM, Phillips AT (1988) Purification and characterization of 1-aminocyclopropane-1-carboxylate synthase from etiolated mung bean hypocotyls. Arch Biochem Biophys 264:632-640

Tucker ML, Sexton R, Campillo ED, Lewis LN (1988) Bean abscission cellulase. Characterization of a cDNA clone and regulation of gene expression by ethylene and auxin. Plant Physiol 88:1257-1262

Van Der Straeten D, Van Wiemeersch L, Van Damme J, Goodman H, Van Montagu M (1988) Purification, characterization, and amino-acid sequence analysis of ACC synthase from tomato pericarp. Proc 2nd Intl Cong Plant Mol Biol (November 1988), p 210 (Abstract)

Yang SF, Hoffman NE (1984) Ethylene biosynthesis and its regulation in higher plants. Ann Rev Plant Physiol 35:155-189

Yang VC, Langer R (1987) A simple and economical technique for pI measurement. BioTechniques 5:138-144

AUXIN- AND WOUND-INDUCED EXPRESSION OF ACC SYNTHASE

H. Imaseki, N. Nakajima and N. Nakagawa
Research Institute for Biochemical Regulation
School of Agriculture
Nagoya University
Chikusa, Nagoya 464
Japan

INTRODUCTION

Ethylene plays a vital role in the cell separation processes involved in leaf abscission, fruit softening and probably in the formation of large intercellular air spaces (Abeles, 1973). Increases in the rate of ethylene production precede abscission and tissue softening. Unscheduled abscission evoked by pathogenic invasion or wounding is also preceded by production of a massive amount of ethylene. An elevated concentration of ethylene in abscission zones or fruits has been shown to induce formation of enzymes which hydrolyze cell wall polysaccharides, leading to loss of cell-to-cell adhesion (Sexton and Roberts, 1982). Thus, ethylene production, on the one hand, is regulated developmentally and on the other hand, controlled by environmental stimuli imposed irregularly upon plants.

Ethylene biosynthesis in higher plants may be physiologically classified into three types; a) auxin-regulated, b) stress-regulated and c) ripening-associated biosynthesis. Auxin-regulated biosynthesis may constitute the basic ethylene production in vegetative tissues and so is regulated developmentally, but stress-induced biosynthesis is developmentally unscheduled and is not affected by auxin. Ripening-associated biosynthesis is also developmentally regulated but with little control by plant hormones and a suppression by auxin. However, all of these ethylene biosyntheses proceed through the same biosynthetic pathway starting from methionine via S-adenosylmethionine (SAM) and 1-aminocyclopropane-1-carboxylic acid (ACC), which is oxidized to ethylene by the so-called ethylene forming enzyme (EFE) (Yang and Hoffman, 1984). In most cases, the endogenous activity of ACC synthase (ACC methylthioadenosine lyase) producing ACC from SAM determines the rate of ethylene production in tissues. Thus, ACC synthase serves as the rate-limiting enzyme in ethylene biosynthesis (Yu *et al.*, 1979, Yoshii and Imaseki, 1981). Auxin, tissue wounding or ripening provide essentially different stimuli to plant

NATO ASI Series, Vol. H 35
Cell Separation in Plants
Edited by D. J. Osborne and M. B. Jackson

tissues, although each induces formation of ACC synthase in the same plant species and in different organs. However, it is not known whether ACC synthase induced by the different stimuli in each species is the product of the same gene.

In this paper, evidence is summarized to show that auxin- and wound-induced increases in ACC synthase result from the expression of two different genes.

AUXIN-INDUCED ACC SYNTHASE

It has been well documented that stem sections of etiolated mung bean (Yoshii and Imaseki, 1981, 1982) and pea (Jones and Kende, 1979) seedlings increase the endogenous activity of ACC synthase in response to exogenously applied indole-3-acetic acid (IAA). As shown in Table 1, the auxin-induced increase of enzyme activity in mung bean hypocotyls is effectively suppressed by inhibitors of protein synthesis (cycloheximide and 2-(4-methyl-2,6-dinitroanilino)-N-methylpropionamide) and of RNA (actinomycin D and α -amanitin)(Yoshii and Imaseki, 1982). Although the evidence is indirect, the results suggest that auxin-induced development of ACC synthase activity is dependent upon new synthesis of mRNA followed by *de novo* synthesis of ACC synthase protein.

TABLE 1. *Inhibition of IAA-induced increase in ACC synthase activiy in etiolated mung bean hypocotyls by inhibitors of protein or RNA synthesis [after Yoshii and Imaseki, 1982].*

Inhibitor added (μg/ml)	ACC synthase activity (nmol ACC/h.mg protein)	% Activity
None	0.400	100
Cycloheximide (28)	0.016	4
Cycloheximide (14)	0.022	5.5
MDMP (5)	0.010	2.5
Actinomycin D (25)	0.070	17.6
α-Amanitin (5)	0.042	10.6

Hypocotyl sections were incubated with 50 mM phosphate buffer, pH 6.8 containing 0.5 mM IAA and 5 μM benzyladenine with or without inhibitors for 4 hours.

MDMP; 2-(4-methyl-2,6-dinitroanilino)-N-methylpropionamide.

Similar auxin-induced increases ACC synthase were observed in stem sections of etiolated squash (*Cucurbita maxima* Duch. cv. Ebisu) and tomato (*Lycopersicon esculentum* Mill.) seedlings. Tissues so far examined for auxin-induced ACC synthase are vegetative tissues of dicot seedlings. In order to find if auxin-induced increases of enzyme activity are limited to vegetative tissues, we examined several fruit tissues and found that IAA treatment to immature cucumber (*Cucumis sativa* L.) fruit slices also increased ACC synthase activity, although no enzyme activity was induced by wounding (Fig. 1A).

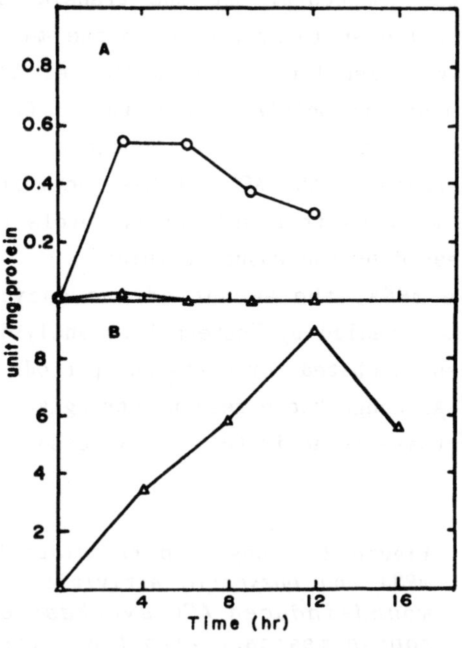

Figure 1. *Induction of ACC synthase by auxin (panel A) or wounding (panel B) in slices of cucumber fruits. Outer skin of cucumber fruits was removed and white mesocarp tissues were cut into 2 mm slices. The slices were incubated with 50 mM phosphate buffer, pH 6.8 with (circle) or without (triangle) 0.5 mM IAA, and at intervals, the slices were extracted and assayed for activity of ACC synthase. Panel A, with immature fruits; panel B, with mature, fully grown fruits.*

WOUND-INDUCED ACC SYNTHASE

A simple cutting and incubation of slices of tomato fruits (Boller *et al.*, 1979, Yu and Yang, 1980) and squash fruits (Hyodo *et al.*, 1985) cause a dramatic increase in ACC synthase activity. Acaster and Kende (1983) showed by a density labeling experiment that wound-induced increases tomato ACC synthase were due to *de novo* synthesis of the enzyme. Inhibition by cycloheximide of wound-induced increase in the enzyme of squash fruit slices also suggested *de novo* synthesis of the enzyme (Hyodo *et al.*, 1985).

As described above, immature cucumber fruit tissues produced auxin-induced ethylene but not wound-induced ethylene. However, fully grown, partially yellowed mature fruit tissue produced large amounts of wound-induced ethylene (Fig. 1B), and the wound-induced production was not affected by auxin. Thus, cucumber fruit responds to two different stimuli in ways that depend on the growth stage. This provides a useful comparison of auxin- and wound-induced enzymes.

We purified the wound-induced enzyme from squash fruit mesocarp and raised antibodies against the purified enzyme (Imaseki *et al.*, 1988, Nakajima *et al.*, 1988). The antibody was used to measure changes in translatable mRNA coded for ACC synthase after tissue wounding. *In vitro* translation in a wheat germ extract was directed by poly(A)$^+$RNA isolated from fresh and wounded tissues of squash mesocarp, and the translation products were treated with the antibody. The antibody picked up one major translation product of 55-58 kDa relative molecular size (Nakajima *et al.* 1988). mRNA for ACC synthase was contained in poly(A)$^+$RNA extracted from wounded tissue but not from fresh tissue, and the relative amount of the mRNA increased corresponding to the increase in the ACC synthase activity after wounding (Fig. 2). The result indicates that endogenous levels of translatable mRNA for ACC synthase increased by the wound stimulus.

Corresponding to this increase in mRNA, the amount of the enzyme protein also increased after wounding as revealed by Western blot analysis of partially purified enzyme from tissues incubated for different periods after wounding (Nakajima *et al.*, 1988). Although these results strongly suggest that the gene coding for ACC synthase is activated and expressed by

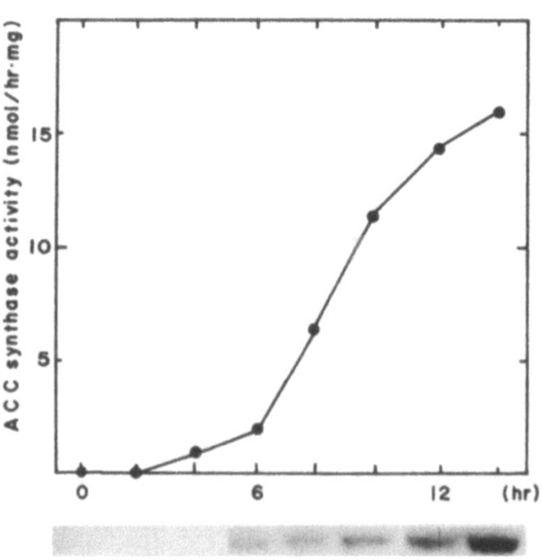

Figure 2. *Changes in translatable mRNA and enzymatic activities of wound-induced ACC synthase of squash mesocarp with time after wounding. Mesocarp slices were incubated in a humid chamber, and at intervals, the tissues were extracted for assay of ACC synthase activity and for preparation of poly(A)$^+$RNA. RNA was translated in a wheat germ extract in the presence of [^{35}S]methionine, and the translation products bound to the antibody against wound-induced ACC synthase were detected by fluorography after SDS-PAGE. Flurograms of the translation product at respective times are shown in the bottom [from Imaseki et al., 1988].*

a wound stimulus, it could be that an inactive form of mRNA is changed to a translatable form by wounding.

To obtain conclusive evidence for wound-induced gene expression, a cDNA library was constructed in an expression vector, gt11, from poly(A)$^+$RNA extracted from wounded tissue of squash mesocarp, and the library was screened by the antibody. A cDNA clone containing an insert DNA of 0.85 kbp was obtained and a fusion protein with β-galactosidase produced by the phage-infected *E. coli* reacted with the antibody. The insert DNA of the clone was used as a probe to determine changes in the content of mRNA for ACC synthase by Northern blot analysis. No detectable amount of this mRNA was initially present in fresh tissue but the mRNA appeared increasingly after wounding (Fig. 3). The size of mRNA for the enzyme was found to be about 2.5 kb. The result, together with the previous results, clearly indicate that mRNA for ACC synthase increased in an amount corresponding to its translatability. That there was no RNA in fresh tissue which hybridized with the cDNA, indicates that the mRNA was newly synthesized and accumulated after wounding.

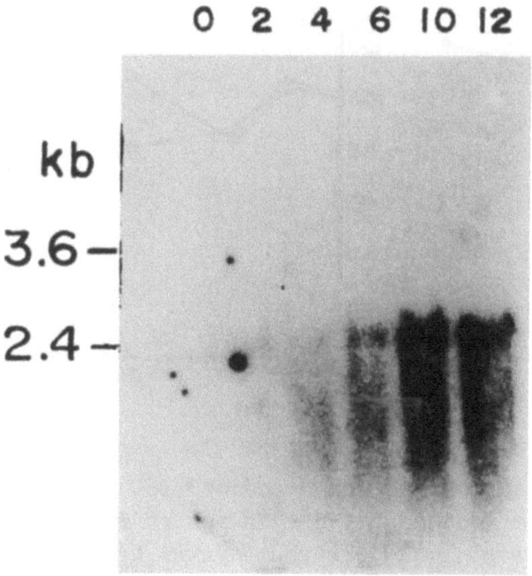

Figure 3. *Changes in the content of mRNA for wound-induced ACC synthase of squash mesocarp after wounding. Total RNA of mesocarp extracted at different time after wounding was electrophoresed on agarose and probed with cloned cDNA of ACC synthase.*

IMMUNOCHEMICAL COMPARISON OF AUXIN- AND WOUND-INDUCED ACC SYNTHASE

Several studies on ACC synthase induced by auxin or wounding report enzymological properties of partially purified or crude preparations. The data show that kinetic parameters of ACC synthase induced by wounding and auxin are very similar (Nakagawa *et al.*, 1988). As we obtained the antibody against wound-induced ACC synthase purified from squash mesocarp, we examined immunochemical cross-reactivity of the antibody to ACC synthases induced by different stimuli.

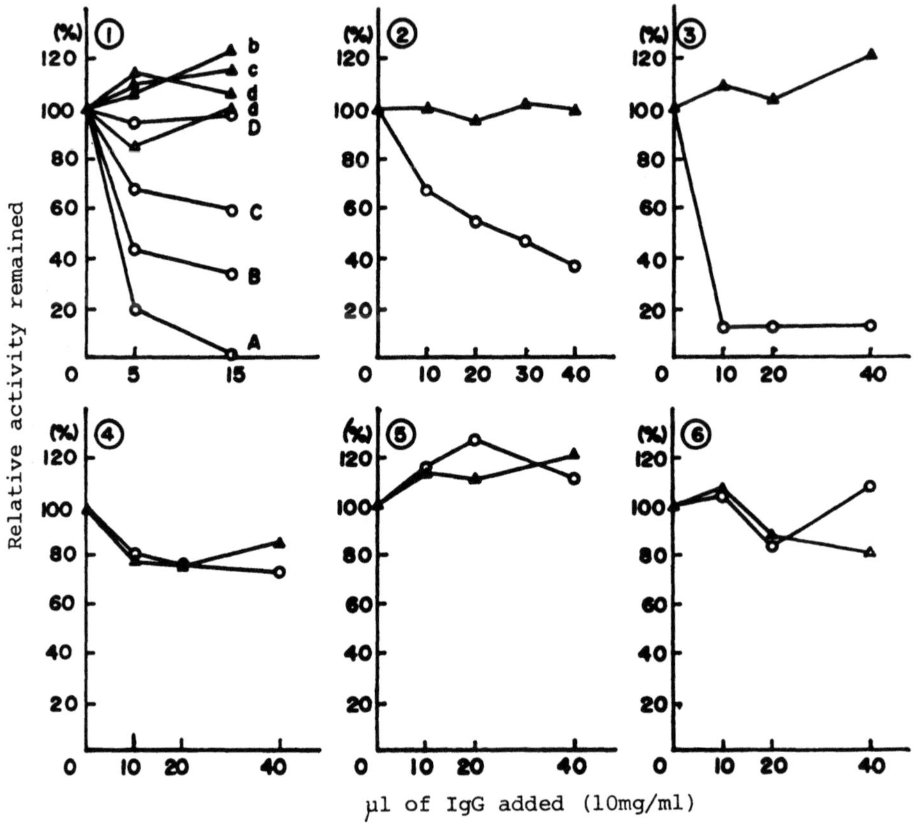

Figure 4. *Immunochemical difference of auxin- and wound-induced ACC synthases. ACC synthase preparations extracted from wounded tissues of squash mesocarp (panel 1, line A and a), tomato pericarp (panel 2) and squash hypocotyl (panel 3), and IAA-treated hypocotyls of squash (panel 4), tomato (panel 5) and mung bean (panel 6) seedlings were precipitated with the antibody against squash wound-induced ACC synthase (circles) or non-immune serum (triangles), and the enzymatic activities in the supernatant were determined. Results with mixtures of the wound-induced enzyme of squash mesocarp and the auxin-induced enzyme of squash hypocotyl at the activity ratio of 2:1 and 1:2 are shown in line B and b, and line C and c of panel 1, respectively [from Nakagawa et al. 1988]*

The antibody efficiently bound the acitivity of ACC synthase extracted from wounded squash mesocarp, but not from IAA-treated hypocotyls of squash seedlings (Fig. 4, panel 1 and 4). As there was a possibility that the auxin-induced enzyme preparation from squash hypocotyls contained substances which interfered with the immunochemical reaction, the auxin- and wound-induced enzyme preparations were mixed at different activity ratios and the mixtures were precipitated with the antibody. The residual activities can be accounted for by the auxin-induced enzyme (Fig. 4, panel 1) indicating the lack of immunochemical reaction of this enzyme with the antibody to the woiund-induced enzyme. To compare the auxin- and wound-induced enzymes from the same tissue, ACC synthase was induced in squash hypocotyls by wounding or IAA treatment. As shown in Fig. 4, the wound-induced enzyme of hypocotyls was recognized by the antibody (panel 3), but the auxin-induced enzyme of the same tissue was not (panel 4). The results clearly show that in squash plant, the auxin-induced enzyme is immunochemically different from the wound-induced enzyme.

The wound antibody of squash also reacted with the wound-induced enzyme of tomato fruits, although the immunochemical reaction was weaker than with the wound-induced enzyme of squash fruits, but the auxin-induced enzymes of tomato hypocotyls (Fig. 4, panel 2) and mung bean hypocotyls (Fig. 4, panel 6) were not bound at all. The same trend was also observed with auxin- and wound-induced ACC synthase of cucumber fruits (data not shown).

The above results indicate that ACC synthase induced by wounding has a common immunoreactive region which is lacking in the auxin-induced ACC synthase regardless of plant species and of tissues used. Therefore, the primary structure of the wound-induced enzyme must be different from that of the auxin-induced enzyme, and the enzymes of the two types originate from two different genes.

SUMMARY AND CONCLUSION

ACC synthase activity which regulates the rate of ethylene production in higher plant tissues is induced by auxin, a chemical signal, and wounding, a physical signal. For the auxin-induced enzyme, indirect evidence is presented to show that the auxin-induced enzyme is synthesized coupled with RNA and protein synthesis. For the wound-induced enzyme, it is shown that the amount, as well as translatability, of mRNA coding for the enzyme and the amount of the enzyme protein are increased by wound stimulus. The results indicate that both auxin- and wound-induced increases in ACC synthase activity resulted from activation and expression of the gene by either auxin or wounding. A question is whether or not the auxin- and wound-induced enzyme is a product of a single gene. If the same gene is activated by auxin or wounding, the cell must have a common sequence of reactions leading to expression of the gene after perception of the two different stimuli, or the gene must contain two different regulatory regions, corresponding to auxin and wound stimuli. On the other hand, a possibility exists that there are two different genes for ACC synthase in the cell. Immunochemical studies with the antibody against the purified wound-induced enzyme clearly show that the wound-induced enzymes are immunochemically different from the auxin-induced enzymes regardless of plant species and tissues within a range where cross-reactivity of the antibody was observed. The result indicates that the primary structures of ACC synthases of the two types are different and we conclude that there must be at least two different genes in a cell, one for auxin activation and the other for wound activation. Ripening-associated ACC synthase could be coded by the third gene.

Auxin-induced synthesis of ACC synthase is developmentally regulated, whereas wound-induced synthesis occurs only when unscheduled wounding is given. It may be that plants have evolved two different genes, one for an enzyme which plays a vital role in normal growth and development and another for the wound responses. The latter may hold survival value for individual plants and for the species as well.

REFERENCES

Abeles FB (1973) Ethylene in plant biology, Academic Press, New York London

Acaster MA, Kende H (1983) Properties and partial purification of 1-aminocyclopropane-1-carboxylic acid synthase, Plant Physiol 72:139-145

Boller T, Herner RC, Kende H (1979) Assay for and enzymatic formation of an ethylene precursor, 1-aminocyclopropane-1-carboxylic acid, Planta 145: 293-303

Hyodo H, Tanaka K, Yoshisaka J (1985) Induction of 1-aminocyclopropane-1-carboxylic acid synthase in wounded mesocarp tissue of winter squash fruit and effects of ethylene, Plant Cell Physiol 26:161-167

Imaseki H, Nakajima N, Nakagawa N (1988) Biosynthesis of ethylene and its regulation in plants, In:Steffens GL, Rumsey TS (eds) Biomechanisms regulating growth and development. Kluwer Academic, Dordrecht Boston London, p205

Jones JF, Kende H (1979) Auxin-induced ethylene biosynthesis in subapical stem sections of etiolated seedlings of *Pisum sativum* L, Planta 146: 649-656

Nakagawa N, Nakajima N, Imaseki H (1988) Immunochemical difference of wound-induced 1-aminocyclopropane-1-carboxylate synthase from the auxin-induced enzyme, Plant Cell Physiol 29:1255-1259

Nakajima N, Nakagawa N, Imaseki H (1988) Molecular size of wound-induced 1-aminocyclopropane-1-carboxylate synthase from *Cucurbita maxima* Duch. cv. Ebisu and change of translatable mRNA of the enzyme after wounding, Plant Cell Physiol 29:989-998

Sexton R, Roberts JA (1982) Cell biology of abscission, Annu Rev Plant Physiol 33:133-162

Yang SF, Hoffman NE (1984) Ethylene biosynthesis and its regulation in higher plants, Annu Rev Plant Physiol 35:155-189

Yoshii H, Imaseki H (1981) Biosynthesis of auxin-induced ethylene. Effects of IAA, benzyladenine and ABA on endogenous levels of 1-aminocyclopropane-1-carboxylic acid (ACC) and ACC synthase, Plant Cell Physiol 22:369-379

Yoshii H, Imaseki H (1982) Regulation of auxin-induced ethylene biosynthesis. Repression of inductive formation of ACC synthase by ethylene, Plant Cell Physiol 23:639-649

Yu Y-B, Yang SF (1980) Biosynthesis of wound ethylene, Plant Physiol 66:281-285

Yu Y-B, Adams DO, Yang SF (1979) 1-Aminocyclopropane-1-carboxylate synthase, a key enzyme in ethylene production, Arch Biochem Biophys 198:280-286

CHANGES IN GENE EXPRESSION DURING ETHYLENE-INDUCED LEAF ABSCISSION

J.A. Roberts, J.E. Taylor, Y.V. Lasslett and G.A. Tucker
Department of Physiology and Environmental Science
University of Nottingham
Faculty of Agricultural and Food Science
Sutton Bonington
Loughborough
Leicestershire
U.K.

INTRODUCTION

The natural shedding of leaves takes place at predictable locations on a plant. It is the culmination of a spectrum of physiological and biochemical events which are restricted to precise cellular locations known as abscission zones (Sexton and Roberts, 1982). It is possible to induce leaf abscission prematurely by exposure of the tissue to the plant growth regulator ethylene (Abeles, 1973). Ethylene-induced shedding shares many features in common with the natural process, and there is good reason to believe that this gaseous regulator may influence the timing of leaf fall in vivo (Morgan, 1986).

ANATOMY OF LEAF ABSCISSION

The leaf abscission zone is often morphologically distinguishable prior to the onset of separation. It consists of one or more rows of flattened cells which, on treatment with ethylene, exhibit cell wall breakdown (Sexton and Roberts, 1982). Associated with wall degradation is an increase in the activity of several hydrolytic enzymes (Reid *et al.*, 1974). These ethylene-induced changes do not take place in adjacent non-separating cells (Osborne and Sargent, 1976a). In order to ascertain the significance of these changes it is necessary to identify the sequence of events which occur in the leaf abscission zone cells after they have been induced to separate. Such a study can only be readily carried out on a zone which is composed of sufficient rows of cells for it to be easily

NATO ASI Series, Vol. H 35
Cell Separation in Plants
Edited by D.J. Osborne and M.B. Jackson

excised in the absence of contamination by non-abscission zone tissue. For this reason we have chosen to study the leaf abscission zones of tomato (Lycopersicon esculentum) and elder (Sambucus nigra).

PHYSIOLOGY AND BIOCHEMISTRY OF TOMATO LEAF ABSCISSION

Tomato leaf abscission takes place at the junction between the leaf petiole and the stem and the zone comprises approximately 10 rows of cells. Our studies have been carried out using abscission zone explants which consist of a 2cm length of stem tissue containing a subtending petiole segment. Explants are routinely employed in abscission studies since their excision from the parent plant initiates separation and synchronises its rate of progress (Sexton *et al.*, 1985). Exposure of explants to ethylene accelerates abscission and 100% separation can be achieved within 60 hours of treatment.

Associated with abscission is an increase in the activity of the cell wall degrading enzyme polygalacturonase E.C.3.2.1.15. (PG). This increase in enzyme activity is detectable within 42 hours of ethylene treatment and is primarily restricted to the abscission zone tissue (Fig. 1).

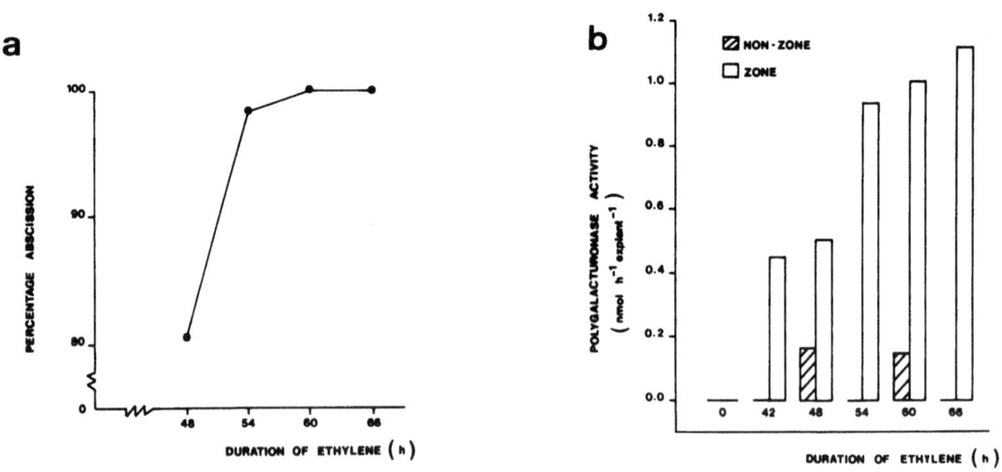

FIGURE 1. Time course of (a) tomato leaf explant abscission and (b) polygalacturonase activity in the presence of 10 μl l^{-1} ethylene.

An increase in the activity of β1,4-glucanase E.C.3.2.1.4. (cellulase) can also be detected, but this rise is not restricted to the abscission zone tissue.

Leaves of the woody shrub Sambucus nigra comprise two pairs of leaflets and a terminal leaflet (Osborne and Sargent, 1976b). During the natural shedding process, abscission takes place at the base of each leaflet and between the leaflet insertions. The zone is made up of approximately three times as many rows of cells as tomato, which makes it a better system to study the biochemical and molecular changes which accompany ethylene-promoted cell separation.

An increase in the activity of PG also accompanies S. nigra leaflet abscission (Fig. 2).

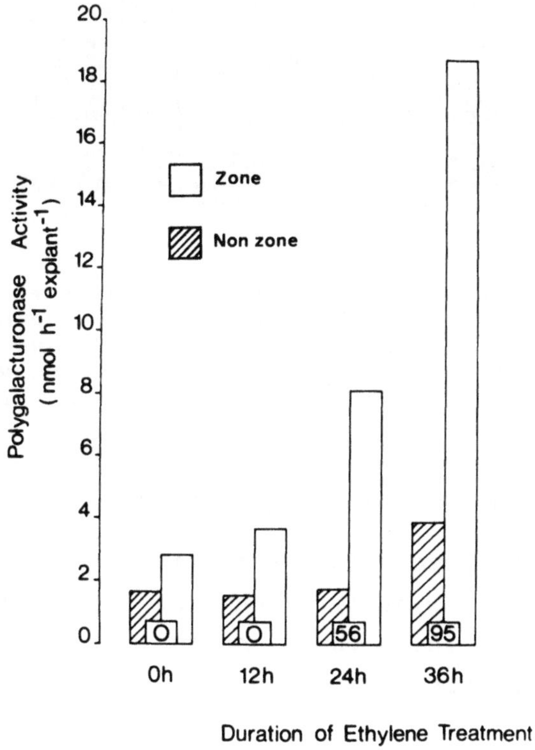

FIGURE 2. Time course of PG activity in S. nigra leaflet zone or non zone tissue after exposure to 10 μl l^{-1} ethylene for 0h, 12h, 24h or 36h. Numbers in the boxes at the base of each pair of histograms represents percentage of abscinded explants at that time.

If enzyme extracts from zone or non-zone tissues are run on a non-denaturing acrylamide gel and stained for enzyme activity it is possible to resolve at least two isoenzymes of PG (Fig. 3). Tomato fruit softening is also associated with an increase in the activity of PG and this enzyme has been shown to consist of a mixture of two isoenzymes - namely PG_1 and PG_2 (Tucker, Robertson and Grierson, 1980). It is clear however, that the PG isoenzymes from S. nigra run at different locations from PG_1 and PG_2.

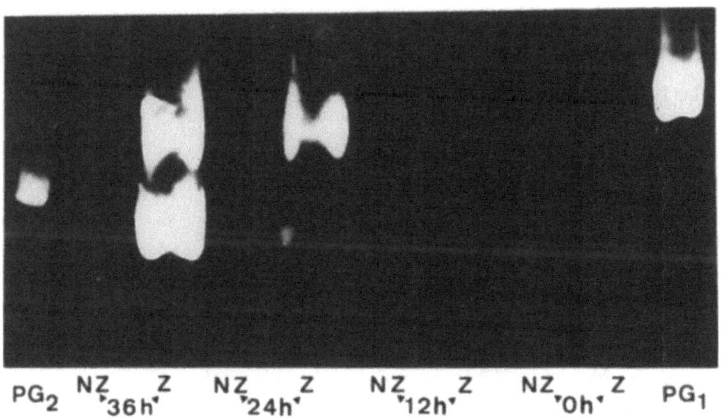

FIGURE 3. Natured gel electrophoresis of proteins extracted from zone (Z) or non-zone (NZ) S. nigra tissue exposed to $10\ \mu l\ l^{-1}$ ethylene for 0h, 12h 24h or 36h. PG_1 and PG_2 were extracted from ripe tomato fruits. The gel was incubated in polygalacturonic acid substrate and then stained with methylene blue.

ABSCISSION AND GENE EXPRESSION

If ethylene-induced abscission is mediated by an increase in the activity of several hydrolytic enzymes, it is important to ascertain the level at which these enzymes are regulated. One approach to this problem has been to pulse abscission zones with ^{35}S-methionine and examine the profile of proteins which are synthesised de novo during abscission. This in vivo approach has been carried out on both tomato and S. nigra tissue. Explants complete with subtending leaf were labelled for 12h after which time the leaf lamina was excised and the remaining explant was exposed to ethylene for 36h. Our studies, so far, have failed to detect any major changes in protein profile during the course of leaf abscission in either tomato or S. nigra. Some differences, however, are apparent in the banding patterns between zone and non-zone tissues (Fig 4).

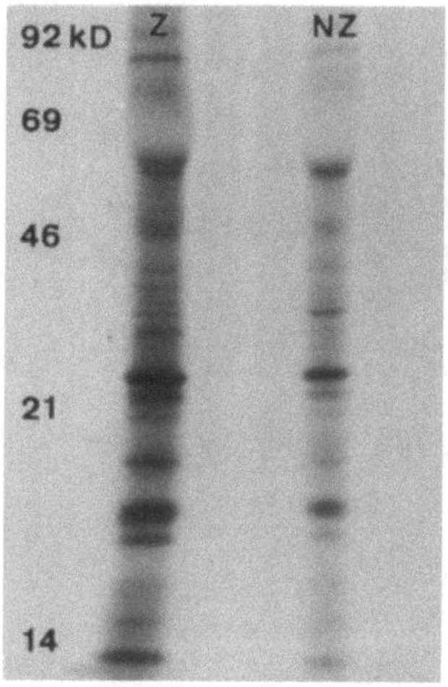

FIGURE 4. In vivo labelling of abscission zone (Z) or non-zone (NZ) proteins from tomato leaves. Tissue was labelled for 12h with ^{35}S methionine and then incubated for 36h in an atmosphere containing 10 µl l^{-1} ethylene.

An alternative approach has been to generate a cDNA library from messages present during a particular developmental event and analyse how the levels of these messages change (Grierson *et al.*, 1985). Since abscission zones are composed of such a small population of cells, this approach is particularly difficult. We have attempted to 'short circuit' the procedure by using cDNA probes for messages from ripe tomato fruit since softening shares many features in common with abscission. In particular, we have employed a cDNA which codes for polygalacturonase (Grierson *et al.*, 1986). mRNA has been extracted from petiole tissue of tomato or S. nigra during abscission and this has been probed with the PG cDNA. In all the tissues that we have studied, both zone and non-zone, it has been possible to detect, by dot blot analysis, hybridization of mRNA to the PG cDNA. Whilst the signal from the tomato tissue has been much weaker than that obtained from the <u>S. nigra</u> tissue, it is clear that there is no difference between zone and non-zone mRNA (see Fig. 5).

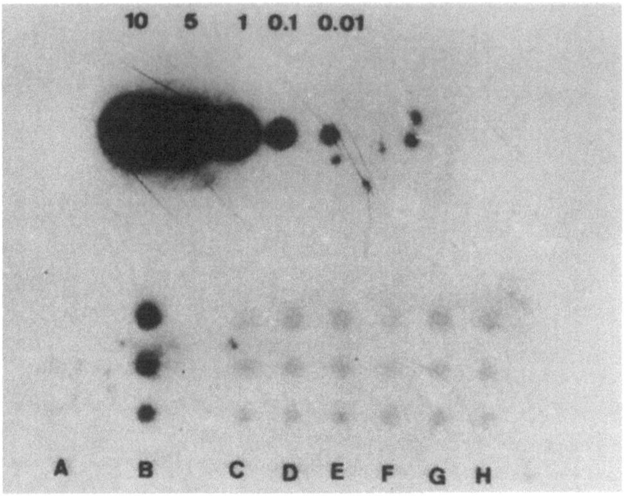

FIGURE 5. Dot blot analysis of total RNA from tomato or <u>S. nigra</u> tissues probed with a cDNA coding for PG (pTom6).

The upper set of dots represent a range of pTom6 DNA standards of 10, 5, 1, 0.1 and 0.01ng. The blot represents 10, 5 and 1μg total RNA from A. tomato leaf; B. orange tomato fruit. Tracks C-H are dots of total RNA extracted from <u>S. nigra</u>; C. zone prior to treatment; D. zone 40h ethylene (10μl l^{-1}) treatment; E. zone 40h minus ethylene; F. non-zone prior to treatment; G. non-zone 40h ethylene (10 μl l^{-1}); H. non-zone 40h minus ethylene.

The nature of this hybridization has been further studied by running the mRNA from <u>S. nigra</u> on an agarose gel and probing it with PG cDNA. In order to size the mRNA which hybridizes to the cDNA, mRNA from ripe tomato fruit was also run on the Northern gel. Although the PG mRNA from tomato fruit has a molecular size of 1.7kb, it is clear that the cDNA hybridizes to mRNA from <u>S. nigra</u> tissue of a different size class. The markers indicate that this is approximately 2.5kb (see Fig. 6).

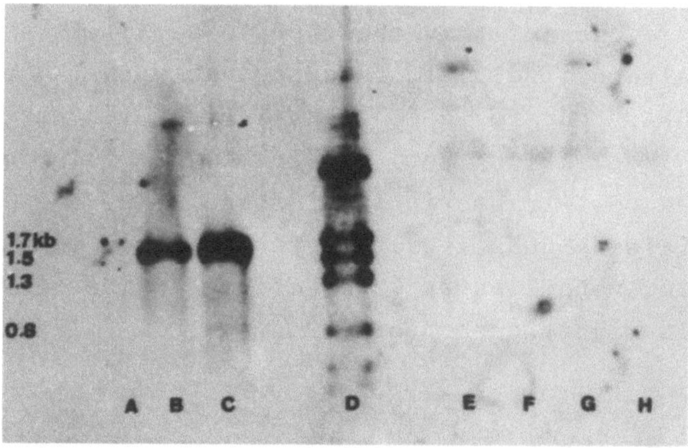

FIGURE 6. Northern analysis of tomato and S. nigra total RNA probed with PG cDNA (pTom6). Track A. 10 µg tomato leaf RNA; B. 10 µg orange tomato fruit RNA; C. 4 µg orange tomato fruit Poly A$^+$ RNA; D. RNA markers; E/F. 10 µg RNA from S. nigra tissue prior to and after 40h ethylene (10 µl l^{-1}) respectively; G/H. 10 µg RNA from S. nigra non-zone tissue prior to and after 40h ethylene (10 µl l^{-1}) respectively. Film exposed for 48h.

CONCLUSIONS

It is evident that if we are to study abscission in detail, we must choose a system where the zone is composed of many rows of separating cells. Leaf abscission zones are ideally suited for this purpose. During ethylene-induced tomato and S. nigra leaf abscission, the enzyme polygalacturonase increases markedly, and this rise is restricted to the zone tissue. In an effort to ascertain how the level of PG is regulated during abscission, we have utilised a cDNA for PG from tomato fruit with which to probe abscission zone mRNA. Although hybridization can be detected, in general the signal is weak. Furthermore, no differential hybridization is apparent between abscission zone and non-zone tissue during cell separation. This observation raises a number of possibilities. Firstly, since the molecular size of the mRNA from S. nigra differs from that from tomato fruit which hybridizes to the PG cDNA, it may represent a message with some homology to fruit PG but with a different function. Secondly, the mRNA might code for a PG with some sequence similarity to tomato fruit PG. In both these instances there is no evidence that these proteins play a role in abscission. Alternatively, it is conceivable that the S. nigra and tomato mRNAs code for a PG which is involved in ethylene-induced cell separation but the level of this enzyme is regulated at a post-transcriptional stage. We are presently engaged in raising cDNA libraries to mRNA from tomato and S. nigra abscission zone tissues in an effort to identify which of these proposals might be correct.

REFERENCES

Abeles FB (1973) Ethylene in plant biology. Academic Press, New York/London

Grierson D, Tucker GA, Keen J, Ray J, Bird, CR, Schuch, W (1986) Sequencing and identification of a cDNA clone for tomato polygalacturonase. Nucleic Acid Res. 14:8595-8603

Morgan PW (1984) Is ethylene the natural regulator of abscission? In: Fuchs Y, Chalutz E (eds) Ethylene: biochemical, physiological and applied aspects. Martinus Nijhoff/Dr W Junt, The Hague, pp 231-240

Osborne DJ, Sargent JA (1976a) The positional differentiation of ethylene-responsive cells in rachis abscission zones in leaves of Sambucus nigra and their growth and ultrastructural changes at senescence and separation. Planta 130:203-210

Osborne DJ, Sargent JA (1976b) The positional differentiation of abscission zones during the development of leaves of Sambucus nigra and the response of the cells to auxin and ethylene. Planta 132:197-204

Reid PD, Strong HG, Lew F, Lewis LN (1974) Cellulase and abscission in the red kidney bean (Phaseolus vulgaris). Plant Physiol. 53:312-316

Sexton R, Lewis LN, Trewavas AJ, Kelly P (1985) Ethylene and abscission. In: Roberts JA, Tucker GA (eds) Ethylene and plant development. Butterworths, London, p 173-196

Sexton R, Roberts JA (1982) Cell biology of abscission. Ann Rev Plant Physiol 33:133-162

Tucker GA, Robertson NG, Grierson D (1980) Changes in PG isoenzymes during the ripening of normal and mutant tomato fruit. Eur. J. Biochem. 112:119-24

THE CELL BIOLOGY OF BEAN LEAF ABSCISSION

R. Sexton, M.L. Tucker[1], E. del Campillo[1], L.N. Lewis[1]
Department of Biological Science,
Stirling University, Stirling,
FK9 4LA
Scotland

INTRODUCTION

Over the last 25 years the primary leaves of bean have emerged as the favoured system to study the cell biology of abscission. The petioles of these leaves contain two genetically determined regions known as *abscission zones* (AZ) where fracture can occur. The upper AZ is found just below the lamina and its weakening results in the shedding of the leaf blade. The basal zone resides at the petiole-stem junction and its activation leads to the loss of the petiole. Weakening and subsequent fracture of these zones can be induced experimentally by removing the leaf blade.

The progressive weakening of the AZs after delamination can be followed with a strain gauge, giving a value called the *breakstrength* (Fig. 1). After 3-4 days in air the zone starts to weaken (Fig. 1). As weakening progresses a white line becomes visible round the circumference of the petiole. Initially it is only present on the upper side of the petiole, then slowly extends to the lower side. When it is examined under the SEM the line is seen to be a break in the epidermis involving 2 or 3 rows of cells (Fig. 2A).

THE MORPHOLOGICAL FEATURES OF ABSCISSION

The general AZ regions of the petiole where fracture will eventually occur are characterized by a number of subtle features (Webster, 1968). The stele branches, the tracheids are shorter and the pith cells are less lignified. Although the general region where abscission will take place can be roughly defined, anatomists have been unable to identify the actual band of cells within this AZ that will ultimately be involved in wall breakdown and fracture. It is assumed that the cells that will ultimately compose this *separation layer* are biochemically differentiated to respond to the abscission signal. In other abscission systems there is both morphological

[1]Plant Molecular Biology, University of California, Berkeley, CA 94720, USA

NATO ASI Series, Vol. H 35
Cell Separation in Plants
Edited by D. J. Osborne and M. B. Jackson

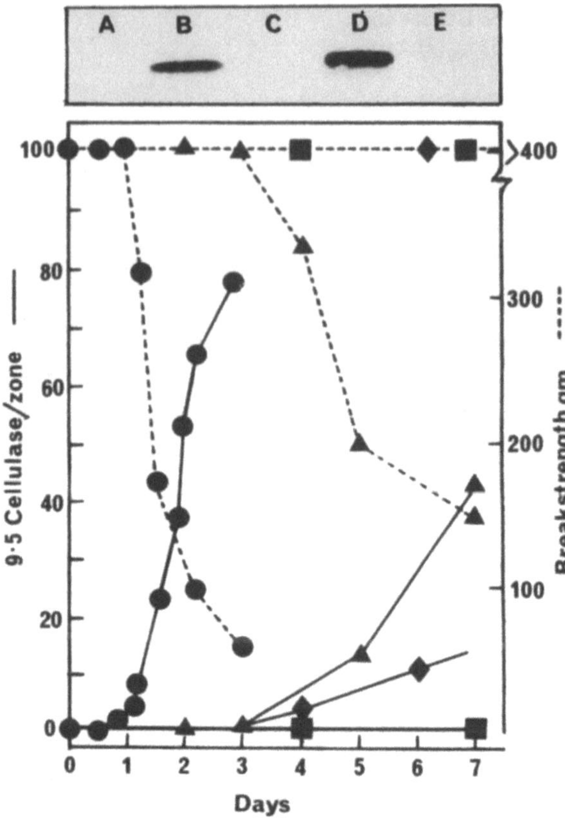

FIGURE 1. (Below) A time-course of the changes in breakstrength (dashed line) and 9.5 cellulase (solid line) in the basal abscission zones of bean. The debladed petioles were incubated in air (triangles); 50 $\mu l\ l^{-1}$ ethylene (circles); 10^{-3} M IAA added to the end of the debladed petiole at 0 h and incubated in air (squares); 10^{-3} M IAA added to the end of the petioles at 0 h and incubated in 50 $\mu l\ l^{-1}$ ethylene (diamonds). (Above) An immunoblot of SDS PAGE- separated total proteins extracted from basal zones after A, 0 h; B,48 h; D,72 h in 25 $\mu l\ l^{-1}$ ethylene; C,48 h; E, 72h in 25 $\mu l\ l^{-1}$ ethylene, 10^{-4} M IAA having been added to the cut petiole surface at 0 h. Only the area of the blot to which 9.5 cellulase antiserum binds is shown.

and immunological data to support this view (Sexton and Roberts, 1982; Osborne and McManus, 1984).

There is general agreement that leaf abscission involves breakdown of the cell walls in the separation layer. Fracture predominantly follows the line of the middle lamella between 2 or 3 layers of cells leaving the majority of cells intact. When the exposed scar surfaces are viewed with the SEM they are covered with rounded cells (Fig. 2B, Osborne, 1984). Under the TEM these surface cells contain undegraded cytoplasm with large numbers of active Golgi (Sexton and Hall, 1974; Osborne and McManus, 1984).

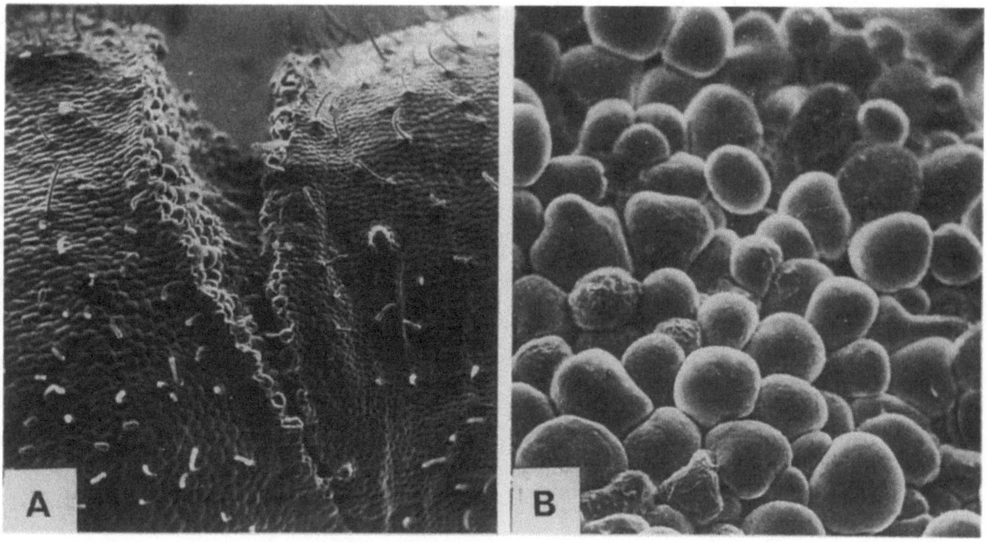

FIGURE 2. (A) An SEM view of the developing fracture line in the apical abscission zone of bean leaves. (B) An SEM view of the proximal fracture surface. Note the apparently intact cells.

Thus the idea that abscission is a degenerative process accompanied by cellular collapse is questionable.

Rasmussen and Bukovac's (1969) histochemical observations showed extensive swelling of the wall pectin during separation layer weakening. Later EM work revealed a loss of cellular adhesion due to dissolution of the middle lamella and disruption of adjacent regions of the primary wall matrix (Sexton and Hall, 1974). 11% of the wall material is lost during weakening (Morré, 1968), water soluble pectins increase while acid soluble pectins decline. X-ray microprobe analysis also showed Ca^{2+} is removed (Poovaiah and Rasmussen, 1973). The cells on the proximal side of the fracture line swell as the walls degrade. This enlargement produces a force which facilitates cell separation and rupture of the non-living cells of the stele (Wright and Osborne, 1974).

THE EFFECTS OF ADDING IAA AND ETHYLENE

Ethylene is a potent accelerator of abscission in many different abscission systems. Concentrations of 0.1 ul l^{-1} will speed up bean explant weakening (Abeles *et al.*, 1971). In the case of the basal zone, ethylene treatment reduces the lag before weakening commences and increases the rate of loss of structural integrity (Fig. 1). The continued presence of ethylene is necessary if this acceleration is to be maintained (Abeles *et al.*, 1971).

There is strong evidence that ethylene produced naturally by ageing explants is also implicated in the abscission of air-held zones. Jackson and Osborne (1970) showed that ethylene production rates of bean explant pulvinii increase as they senesce. Kushad and Poovaiah (1984) have used the ethylene synthesis inhibitor aminoethoxyvinylglycine to inhibit this increase and delay abscission. ABA will also stimulate bean explant abscission. However, it seems likely that this effect is mediated by ABA through an enhanced senescence-induced ethylene production (Jackson and Osborne, 1972).

Auxin has long been known to be a potent inhibitor of abscission when applied to the end of a debladed petiole. In the intact plant it is thought that IAA produced in the healthy lamina prevents abscission. Both IAA and its analogues are effective even in the present of ethylene (Fig. 1).

THE ROLE OF RNA AND PROTEIN SYNTHESIS

The inhibition of bean abscission by both transcriptional and translational protein synthesis inhibitors (Abeles *et al.*, 1971) suggested the *de novo* synthesis of proteins was a vital component of the process. Analysis of the protein (Reddy *et al.*, 1988) and mRNA populations in the abscission zone (Kelly *et al.*, 1987) demonstrated that new species appeared during abscission while others declined. Four translatable mRNA species increased dramatically, one of which has been subsequently identified as chitinase (Gomez Lim *et al.*, 1987). Chitinase may well be involved in protecting the exposed scar surface against fungal attack.

WALL DEGRADING ENZYMES

The anatomical evidence clearly suggested the involvement of wall degrading enzymes. Using a cucumber pericarp bioassay, Morré (1968) reported an increase in *pectinase* during bean abscission. However Berger and Reid (1979) employing more conventional assays found no increase in polygalacturonase.

In 1967, Horton and Osborne observed that cellulase (β,1:4-glucan glucanohydrolase increases in the bean abscission zone during fracture. On the basis of salt solubility, Lewis and Varner (1970) showed two forms of cellulase were present. The high salt soluble form increases as the tensile strength of the zone diminishes. ^2H_2O labelling showed this abscission induced form is synthesised *de novo*. Isoelectric focussing was subsequently used to separate the two types of cellulase. The form with the acidic pI (4.5 cellulase) decreases during abscission while the form with the basic pI (9.5 cellulase) appears (Fig. 1). Initially, it was thought that 4.5 cellulase could be converted into 9.5 cellulase, but recent work suggests the two forms are very unlikely to be related (del Campillo *et al.*, 1988).

THE REGULATION OF 9.5 CELLULASE

9.5 cellulase has been purified and antibodies raised against it. The protein has a MW of 51 kD and appears as a very minor band in SDS PAGE separations of total AZ proteins. The antibodies only recognise 9.5 cellulase and thus it has proved possible to quantify both forms of the enzyme. These data confirm that 4.5 cellulase decreases and 9.5 cellulase increases in both leaf zones as weakening commences (Fig. 1; Durbin *et al.*, 1981). The increase in 9.5 cellulase is predominantly localized in the separation layer, though there are also smaller increases in adjacent tissues (Sexton *et al.*, 1981). Radio-immunoassay and immunoblots have confirmed the enzyme is synthesised *de novo* (Fig. 1). If abscission is speeded up with ethylene, the appearance of 9.5 cellulase is also accelerated (Fig. 1, 3). Both abscission and 9.5 cellulase formation in air held apical zones is inhibited by 2,5 norbornadiene (NBD), a competitive inhibitor of ethylene action (Fig. 3). The addition of high ethylene concentrations to overcome the competitive inhibition of NBD reinstates 9.5 cellulase formation (Fig. 3). These data suggest naturally produced ethylene is an important regulator of cellulase production in air-held explants. If 10^{-3}M IAA is added to the apical petiolar surface in the presence of ethylene, the appearance of 9.5 cellulase is virtually abolished (Fig. 1).

Recently a cDNA clone pBAC1 complementary to bean 9.5 cellulase mRNA has been identified (Tucker *et al.*, 1988). pBAC1 showed no hybridization to RNA extracted from abscission zones immediately after deblading (Fig. 4). During ethylene induced abscission of the basal zone the amount of 2.0 kb polysomal RNA that hybridizes to pBAC1 (9.5C mRNA) increases dramatically

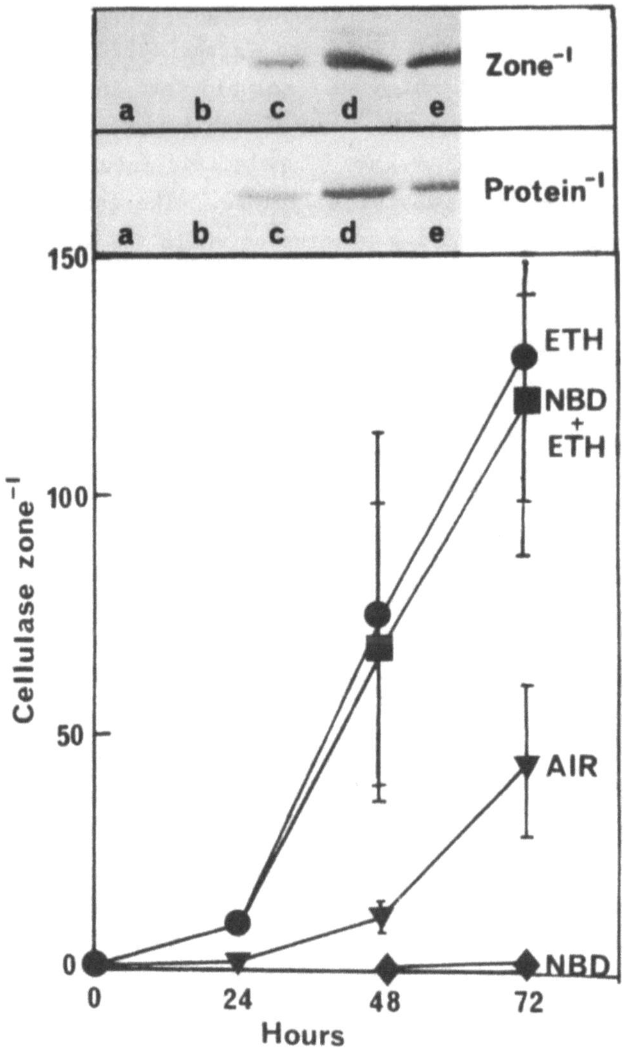

FIGURE 3. (Below) a time-course of the cellulase activity that develops in the excised apical zones incubated in an atmosphere containing air (triangles); 7.5 µl l^{-1} ethylene (circles); 2,5-norbornadiene (NBD) 1100 µl l^{-1} (diamonds); NBD 1100 µl l^{-1} + 100 µl l^{-1} ethylene (squares). Data from three separate experiments.
(Above) An immunoblot of SDS PAGE separated total proteins extracted from the 72 hour time points of the experiment shown in the accompanying graph. A = 0h; B = 72h NBD; C = 72h air; D = 72h ethylene; E = 72h NBD+ ethylene. The area of the two blots where 9.5 cellulase antiserum binds is shown. Equal numbers of zones and amounts of protein were loaded onto the two gels.

(Fig. 4). The temporal characteristics of this accumulation correlate very closely with the observed levels of cellulase activity and immunodetectable cellulase protein in the same extracts (Tucker *et al.*, 1988). There is a long 24h lag proceeding a progressive increase in the hybridizable message over the subsequent 36h. In air held tissue, neither 9.5C mRNA nor 9.5 cellulase protein increase over this period. The spatial distribution of the 9.5C mRNA has been determined. When identical amounts of polysomal RNA from the stem, petiole and AZ of abscising plants are probed with pBAC1, 9.5C mRNA is ten times more abundant in the AZ.

Experiments were undertaken to investigate if continual ethylene presence is necessary for cellulase accumulation after the rise in 9.5C mRNA had commenced (Tucker *et al.*, 1988). Explants were placed in ethylene for 31 h to initiate the 9.5 cellulase increase, then removed into air containing NBD. The NBD was added to overcome the effects of any endogenously generated ethylene. Over the next 24 h, the mRNA hybridizing to pBAC1 drops to undetectable levels (Fig. 4) and the rate of 9.5 cellulase accumulation decreases (Tucker *et al.*, 1988). Controls where ethylene was added to overcome the competitive inhibition of NBD showed the normal 9.5 cellulase increase (Fig. 4). Therefore ethylene appears necessary both to accelerate and maintain expression of 9.5 cellulase.

These data partially substantiate the proposal of Abeles *et al.* (1971) that ethylene induced cellulase production is transcriptionally controlled. However, we have not distinguished between transcriptional regulation and increased mRNA stability. The appearance of both 9.5 cellulase and the 9.5C mRNA was inhibited by 5 x 10^{-6}M IAA in lanolin applied to the cut petioles 4h prior to the addition of ethylene (Tucker *et al.*, 1988). Controls, where lanolin alone was applied behaved normally.

Those special features of 9.5 cellulase expression which must be ultimately explained, are brought into sharp focus by comparing it with another ethylene evoked, unidentified mRNA in bean which hybridizes to cDNA clone pBZ30 (Tucker *et al.*, 1987). Both this gene and also chitinase (Gomez Lim *et al.*, 1988) are expressed in the same AZ tissue within 6 to 12h exposure to ethylene and peak from 12h to 24h. 9.5C mRNA only appears in trace amounts at 24h and is still increasing at 60h. The ethylene evoked expression of 9.5C mRNA is inhibited by IAA while the accumulation of pBZ30 mRNA is not (see below). As a result, it is possible to attribute the long lag in the appearance of 9.5C mRNA after deblading to the time it takes the IAA concentration naturally present in the tissue to decrease to a level where it no longer inhibits 9.5C mRNA accumulation. Although this is an attractive idea, there are no measurements of IAA levels in the AZ to corroborate the hypothesis. The protracted increase in 9.5C mRNA is probably attributable both to an increase per cell and a gradual recruitment of more cellulase producing cells into the expanding separation layer.

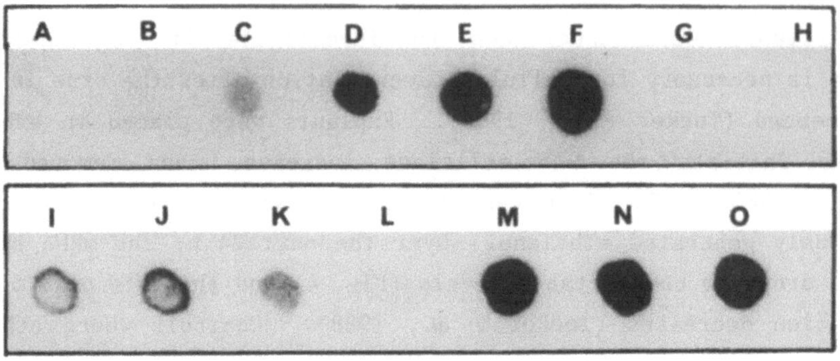

FIGURE 4. (Top row) A dot blot of polysomal RNA hybridizing to pBAC1, 5 µg per spot. RNA extracted from the basal abscission zone after incubation in 5 ul l^{-1} ethylene for A = 0h; B = 12h; C = 24h; D = 36h; E = 48h; F = 60h; or in air for G = 24h; H = 48h.
(Bottom row) A similar dot blot to that above. RNA extracted from basal abscission zones held in; I = 5 µl l^{-1} ethylene for 31h; J = 31h ethylene+8h 1000 µl l^{-1} NBD; K = 31h ethylene+16h NBD; L = 31h ethylene+24h NBD; M = 31h ethylene+8h 1000 µl l^{-1} NBD+100 µl l^{-1} ethylene; N = 16h NBD+ ethylene; 0=24h NBD+ethylene. (For quantitative data see Tucker et al., 1988).

Both chitinase and pBZ30 mRNA constitute a similar proportion of the RNA population in all the ethylene exposed tissues investigated. In contrast 9.5C mRNA is predominantly expressed in the AZ. Two different hypotheses have been proposed to account for this tissue specificity (Sexton and Roberts, 1982). The first envisages that the separation layer cells are actually different from their neighbours, without any morphological manifestation of this biochemical distinction. As discussed above most authors appear to assume such *cryptic differentiation.* The alternative hypothesis suggests that potentially any cell can produce 9.5 cellulase but the inductive conditions required for its production are normally restricted to the separation layer after deblading. Although this hypothesis of *precise positional induction* might seem unlikely it would account for the *adventitious abscission zones* which can be experimental induced to form at unnatural position in the bean plant (Sexton and Roberts, 1982).

Another important difference between ethylene evoked increase in pBZ30 mRNA and 9.5C mRNA is that only the expression in the latter is inhibited by IAA (Tucker *et al.*, 1988). Contrary to earlier belief, it appears that the increase in cellulase can be arrested by IAA application at any point in the time course (Osborne and McManus, 1984). How IAA inhibits the ethylene response is not understood though two clues may have been recently provided by Poovaiah *et al.* (1988). They have shown using *in vivo* phosphorylation that auxin treated explants contain different phosphoproteins to those in the abscising controls. Ca^{2+} channel blockers will also reverse IAA inhibition implicating Ca^{2+} levels in the response. A link between these apparently disparate observations may be provided by Ca^{2+} regulated protein phosphorylations/dephosphorylations.

THE CONTROL OF BEAN ABSCISSION

It is now 47 years since the first attempts were made to understand the role of growth substances in the control of bean abscission. Over this period many measurements of endogenous levels of growth substances have been made and the effects of experimentally increasing and decreasing their concentrations monitored in dozens of papers. These data provide clear evidence that IAA, ethylene and *tissue responsiveness* are likely to be major controlling elements and that ABA, cytokinin and Ca^{2+} may also be involved. Unfortunately these type of data have major short comings which make it difficult to draw definitive conclusions about the mechanism of abscission induction. For example ethylene responses are usually slow, suggesting the effect could be indirect perhaps mediated via a reduction in IAA levels. However, we can confidently expect the molecular biologists to provide more concrete information and it should not be long before we have a detailed understanding of some components of the abscission programme such as 9.5 cellulase expression.

REFERENCES

Abeles FB, Leather GR, Forrence LE, Craker LE (1971) Abscission: Regulation of senescence protein synthesis and enzyme secretion by ethylene. HortSci 6: 371-376.

Berger RK, Reid PD (1979) Role of polygalacturonase in bean leaf abscission. Plant Physiol 63: 1133-1137

del Campillo E, Durbin M, Lewis LN (to be published) Changes in two forms of membrane associated cellulase during ethylene induced abscission. Plant Physiol (in press)

Durbin ML, Sexton R, Lewis, LN (1981) The use of immunological methods to study the activity of cellulase isoenzymes in bean leaf abscission. Plant Cell Environ 4: 67-73

Gomez Lim MA, Kelly P, Sexton R, Trewavas AJ (1987) Identification of chitinase mRNA in abscission zones from bean during ethylene induced abscission. Plant Cell Environ 10: 741-756

Horton RF, Osborne DJ (1967) Senescence abscission and cellulase activity in *Phaseolus vulgaris.* Nature 214: 1086-1088

Jackson MB, Osborne DJ (1970) Ethylene the natural regulator of bean abscission. Nature 225: 1019-1022

Jackson MB, Osborne DJ (1972) Abscisic acid, auxin and ethylene in explant abscission. J Exp Bot 23: 849-862

Kelly P, Trewavas AJ, Lewis LN, Durbin ML, Sexton R (1987) Translatable mRNA changes in ethylene induced abscission zones of *Phaseolus vulgaris*. Plant Cell Environ 10: 11-16

Kushad MM, Poovaiah BW (1984) Deferral of senescence and abscission by chemical inhibition of ethylene synthesis and action in bean explants. Plant Physiol 76: 293-296

Lewis LN, Varner JE (1970) Synthesis of cellulase during abscission of *Phaseolus vulgaris* leaf explants. Plant Physiol 46: 194-199

Morré DJ (1968) Cell wall dissolution and enzyme secretion during leaf abscission. Plant Physiol 43: 1545-1559

Osborne DJ (1984) Abscission in agriculture. Outlook on Agriculture 13: 97-103

Osborne DJ, McManus MT (1984) Abscission and the recognition of zone specific target cells. In: Fuchs Y, Chalutz E (eds) Ethylene: Biochemical, Physiological and Applied Aspects. Martinus Nijhoff/Dr W Junk, The Hague, p 221

Poovaiah BW, Friedmann M, Reddy ASN, Rhee JK (1988) Auxin-induced delay of abscission: The involvement of calcium ions and protein phosphorylation in bean explants. Physiol Plant 73: 354-359

Poovaiah BW, Rasmussen HP (1973) Calcium distribution in the abscission zones of bean leaves. Plant Physiol 52: 683-684

Rasmussen HP, Bukovac MJ (1969) A histochemical study of abscission layer formation in the bean. Amer J Bot 56: 69-76

Reddy ASN, Friedmann M, Poovaiah BW (1988) Auxin induced changes in protein synthesis in the abscission zone of bean explants. Plant Cell physiol 29: 179-183

Sexton R, Durbin ML, Lewis LN, Thomson WW (1981) The immunological localization of 9.5 cellulase in abscission zones of bean. Protoplasma 109: 335-347

Sexton R, Hall JL (1974) Fine structure and cytochemistry of abscission zone cells of *Phaseolus* leaves. Ann Bot 38: 849-854

Sexton R, Roberts JA (1982) Cell biology of abscission. Ann Rev Plant Physiol 33: 133-162

Tucker ML, Sexton R, del Campillo E, Lewis, LN (to be published) Bean abscission cellulase: characterization of a cDNA clone and regulation of gene expression by ethylene and auxin. Plant Physiol (in press)

Webster, BD (1968) Anatomical aspects of abscission. Plant Physiol 43: 1512-1544

Wright M, Osborne DJ (1974) Abscission in *Phaseolus vulgaris*. The positional differentiation and ethylene-induced expansion of specialized cells. Planta 120: 163-170

SECTION II

Cell Wall and Enzyme Changes

SORTING SIGNALS AND TRAFFICKING OF LYSOSOMAL AND EXTRACELLULAR HYDROLASES OF CELL SEPARATION

D. James Morré
Department of Medicinal Chemistry
Purdue University
West Lafayette, IN 47907
USA

INTRODUCTION

The induction, synthesis and delivery of cell wall hydrolases to specific tissue regions where separation will occur are processes central to cell separation. All are amenable to detailed dissection using modern molecular techniques. Perhaps the next plateau, where molecular probes will be used to completely elucidate the topological and functional interaction of both large and small molecules at the limits of fine structural resolution, will bring even greater levels of understanding to the problem. In this account, the current status of knowledge concerning trafficking of intracellular hydrolases in plants and animal cells will be summarized together with available, albeit limited, evidence for sorting signals. Emphasis will be on hydrolases and secretion pathways followed by hydrolases either during movement to lysosomes in animals or to cell walls in plants.

PATHWAYS FOR TRAFFICKING OF INTRACELLULAR HYDROLASES IN MAMMALIAN CELLS

Trafficking pathways and sorting signals that ensure correct delivery of hydrolases to their sites of action in cell wall dissolution in plants are only incompletely understood. More thoroughly studied are those connected with the transport of lysosomal enzymes. Because the sorting mechanisms for lysosomal hydrolases are the only ones where details are known, information on targeting of lysosomal enzymes will be summarized first as a valuable model to provide insight into how targeting may be directed and controlled and into the membrane pathways involved.

Lysosomal enzymes are synthesized, for the most part, as typical N-linked glycoproteins (Rosenfeld *et al.*, 1982). The peptide portions are translated from m-RNAs associated with polyribosomes of the rough

NATO ASI Series, Vol. H 35
Cell Separation in Plants
Edited by D. J. Osborne and M. B. Jackson

endoplasmic reticulum and, after initial co-translational glycosylation, enter the lumens of the rough endoplasmic reticulum mixed together with other proteins destined for secretion (Fig. 1). After partial trimming and transport out of the endoplasmic reticulum to the Golgi apparatus, sorting occurs.

The work particularly of Neufeld and co-workers (1977) showed that the intracellular targeting of lysosomal enzymes is receptor mediated. Eventually, the recognition marker, mannose 6-phosphate, was elucidated.

Certain hydrolases, specifically targeted for entry into lysosomes, receive the mannose 6-phosphate recognition signal via a two-step reaction. First N-acetylglucosamine 1-phosphate is transferred from UDP-N-acetylglucosamine to certain mannose residues by a UDP-GlcNAc:lysosomal enzyme N-acetylglucosaminyl-1-phosphotransferase (phosphotransferase) (Hasilik *et al.*, 1981; Reitman and Kornfeld, 1981a) which selectively phosphorylates lysosomal enzymes (Lang *et al.*, 1984; Reitman and Kornfeld, 1981b; Waheed *et al.*, 1982). Subsequently, the N-acetylglucosamine is removed by an N-acetylglucosamine 1-phosphodiester α-N-acetylglucosaminidase (phosphodiesterase) (Varki and Kornfeld, 1980, 1981; Waheed *et al.*, 1981a), generating the phosphomannosyl recognition marker which is capable of binding to the phosphomannosyl receptor (Gabel *et al.*, 1982). Both of these enzymes are membrane bound and have been localized to the Golgi apparatus (Deutscher *et al.*, 1983; Pohlmann *et al.*, 1982; Varki and Kornfeld, 1980, 1981; Waheed *et al.*, 1981b, 1982). These activities are concentrated in cis Golgi apparatus (Minnifield *et al.*, 1986), but are found, as well, in subcellular fractions corresponding to membranes derived from the trans Golgi apparatus face. Although low levels of activity in endoplasmic reticulum have not been excluded, it is generally assumed that the recognition marker is acquired at or near the cis Golgi apparatus compartment.

Recognition of the marker is via mannose 6-phosphate receptors. Two are currently well studied. One, a high molecular weight membrane glycoprotein (215 kD), binds ligand in a cation independent manner, whereas the other, a lower molecular weight membrane glycoprotein (46 kD), is cation dependent for ligand binding (Hoflack *et al.*, 1987, Pohlmann *et al.*, 1984). According to recent evidence, the cation dependent mannose phosphate receptor is involved largely with the biosynthetic route whereas

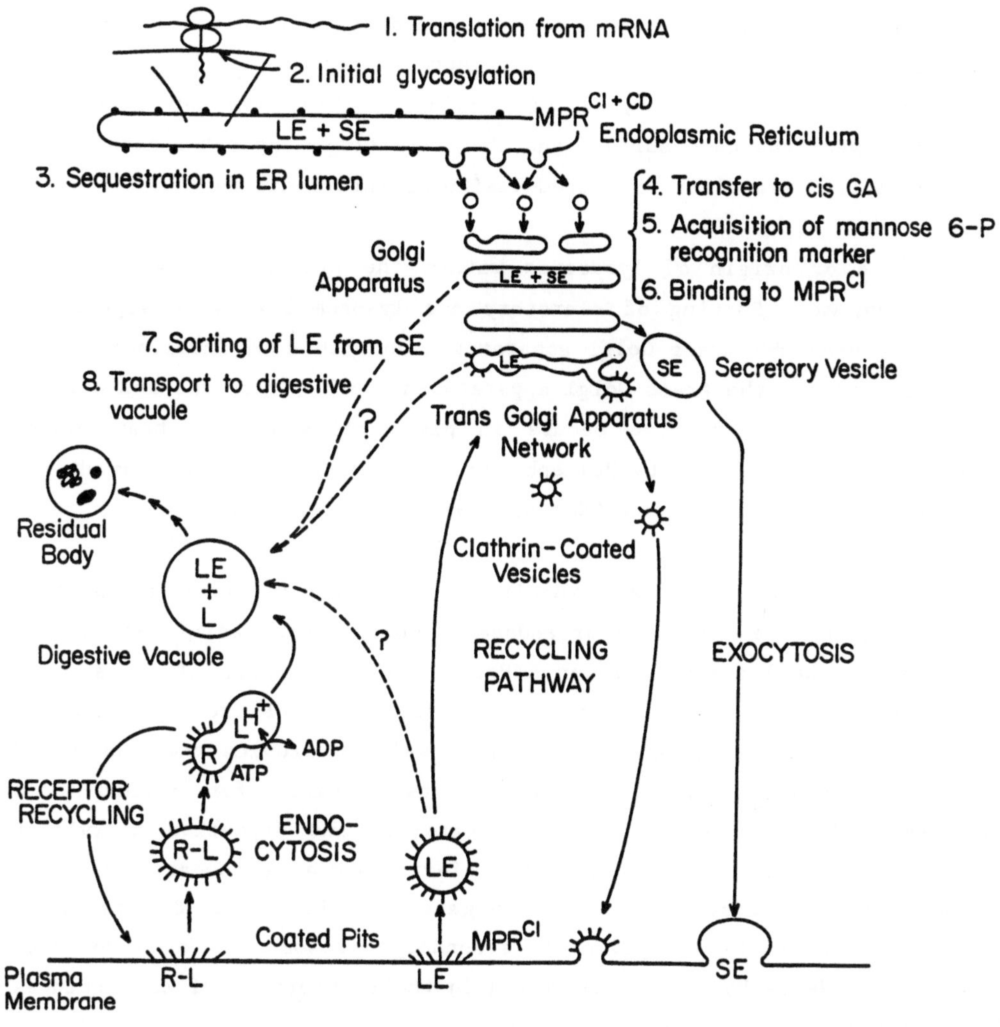

FIGURE 1. Summary diagram illustrating subcellular events in life history of lysosomal enzyme biosynthesis, targeting, trafficking, recycling and delivery to endosomes. The physiological significance of the exocytotic (recycling pathway) is unknown. It might be involved in clearance of lysosomal enzymes or other ligands (e.g. IGF-II) from the circulation or in signal mediation. The steps that involve movement from the Golgi apparatus to digestive vacuoles also are largely unknown. Involvement of a so-called "primary lysosome" is omitted from the scheme since this component has no clear structural counterpart known from electron microscopy. In keeping with this suggestion would be evidence from cultured hepatocytes showing that newly synthesized lysosomal enzymes bound to the 215 kD mannose 6-phosphate receptor are targeted to endosomes and not to lysosomes (Brown et al., 1986). The trans Golgi apparatus network may be synonymous with the GERL structure first described by Novikoff (1976). mRNA - messenger RNA; LE - lysosomal enzyme; SE - secretory enzyme; R-L - receptor-ligand; GA - Golgi apparatus; MPR^{CI} - mannose phosphate receptor, cation independent; MPR^{CD} - mannose phosphate receptor, cation dependent.

the cation independent mannose phosphate receptor is involved in both the biosynthetic and endocytic routes. The latter receptor also binds insulin-like growth factor II (Tong *et al.*, 1988). This receptor apparently recycles between the cell surface and the trans Golgi apparatus network, at least partially, via clathrin coated vesicular intermediates.

The site of origin of lysosomes within the present scheme is still largely unknown. Sorting of secretory and lysosomal enzymes appears to occur at or near the trans Golgi apparatus face with some involvement of a special structure, the trans Golgi apparatus network (Griffiths and Simons, 1986). A similar structure occurs in plant cells at the trans Golgi apparatus recognized first by Mollenhauer (1971) as a cisternal remnant and later by Pesacreta and Lucas (1984) as a partially-coated reticulum.

Eventually, the lysosomal hydrolases, by whatever route or sorting mechanism, reach the interiors of endosomes which become acidified through the action of the lysosomal proton ATPase (Geisow, 1982). The resultant acidification causes dissociation of endocytosed ligands from receptors to permit receptor retrieval and recycling as well as provide the reduced pH of the interiors of digestive compartments required for optimum acid hydrolase activity. Once endocytosed material and digestive activities become involved, the compartments are referred to as secondary lysosomes (or secondary endosomes). The organelles isolated during cell fractionation normally are primarily secondary lysosomes or even residual granules made dense by the accumulation in their interiors of the residues of successive rounds of digestion.

SUBCELLULAR DISTRIBUTION OF LYSOSOMAL ENZYMES

The subcellular distribution of lysosomal enzymes has been determined by several different approaches. These include enzyme cytochemistry, immunolocalization and cell fractionation. All three approaches provide results that are in basic agreement and show the occurrence of lysosomal enzymes in the endoplasmic reticulum, within the Golgi apparatus, within acidified endosomes (lysosomes) and within clathrin-coated vesicles. These are the major components outlined in Fig. 1 that comprise the biosynthetic and endocytic routes of entry of lysosomal enzymes into endosomes.

A variety of lysosomal enzymes have been localized from cytochemistry but chiefly the aryl sulfatases and acid phosphatase. The latter is present throughout the Golgi apparatus but often may be concentrated at the trans Golgi apparatus network (Pavelka, 1987). Some like NADPase appear most concentrated in the median Golgi apparatus (Fig. 2). Immunocytochemical localizations have been carried out for the mannose 6-phosphate receptor, β-hexosaminidase and cathepsin b, as examples, where the activity is seen, as well, in the trans Golgi apparatus network (Geuze *et al.*, 1984, 1985). Golgi apparatus subfractionation by preparative free-flow electrophoresis shows a distribution of lysosomal enzymes throughout all the different Golgi apparatus cisternae including cis, medial and trans compartments of rat liver (Minnifield *et al.*, 1986). Exceptions for Golgi apparatus of rat liver include NADPase which shows concentration in medial and lysosomal compartments (Navas *et al.*, 1986) agreeing with cytochemical analyses (Fig. 2), and acid phosphatase which is concentrated only in lysosomes. Although used as a classic marker for lysosomes, the actual passage of acid phosphatase through the Golgi apparatus en route to lysosomes must be questioned for rodent liver since the activity can normally only be demonstrated in Golgi apparatus using CMP as substrate and may be the manifestation of an enzyme different from acid phosphatase (Eppler and Morré, 1982).

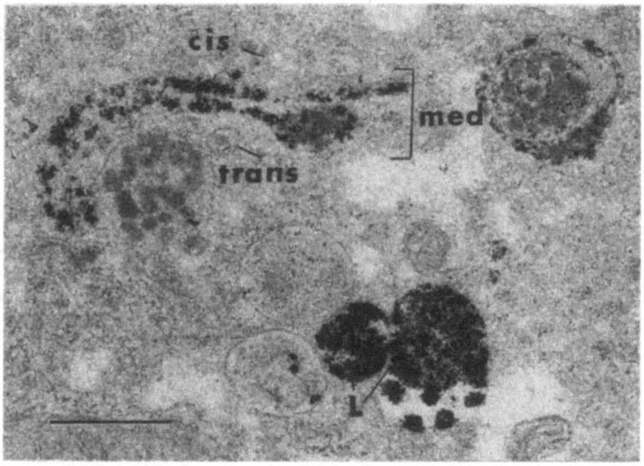

FIGURE 2. Cytochemical localization of NADPase in rat liver Golgi apparatus by the Gomori method involving precipitation of insoluble lead phosphate. Activity is most concentrated in lysosomes (L) and in median (med) Golgi apparatus cisternae in this and other (Smith 1980) tissues. From Navas et al., (1986). Scale = 0.5 μm.

SUBCELLULAR DISTRIBUTION OF THE CATION-INDEPENDENT PHOSPHOMANNOSYL RECEPTOR (PMR^{CI})

The PMR^{CI} has been localized both by immunocytochemistry and by cell fractionation where it is found primarily to be present in endoplasmic reticulum and the Golgi apparatus (Fisher *et al.*, 1980). It is excluded from lysosomes and digestive vacuoles.

Within the Golgi apparatus PMR^{CI} was first shown by immunoperoxidase immunocytochemistry to have a primary location at the cis Golgi apparatus with some activity being present, based on published micrographs, in trans compartments (Brown *et al.*, 1986). Geuze *et al.*, (1984), using colloidal gold labeling, reported heavy immunolocalization over trans Golgi apparatus elements with lesser accumulations ascribed to cis elements. Comparing free-flow electrophoresis subfractions, Minnifield *et al.*, (1986) also located phosphomannosyl receptor activity in both cis and trans cisternal fractions. The evidence would suggest that the PMR^{CI} never becomes concentrated in the medial Golgi apparatus compartments.

Pulse-chase and immunoprecipitation studies show synthesis of PMR^{CI} in endoplasmic reticulum and incorporation into endoplasmic reticulum membranes (Minnifield *et al.*, 1986). The radioactivity of the immunoprecipitated 215 kD PMR^{CI} chase quickly from the endoplasmic reticulum where a portion appears in the cis Golgi apparatus derived fractions (Minnifield *et al.*, 1986, unpublished).

SUBCELLULAR DISTRIBUTION OF LYSOSOMAL MEMBRANE PROTEINS

Although proteins of lysosomal membranes clearly are translated on polyribosomes associated with the rough endoplasmic reticulum, there is only preliminary evidence that they pass through the Golgi apparatus. Virtually nothing is known about the route followed to the lysosome or to digestive vacuoles.

SORTING OF SECRETORY PROTEINS

The sorting signal for lysosomal enzymes entering the Golgi apparatus is apparently the mannose 6-phosphate recognition marker. Lysosomal enzymes may still reach lysosomes even in the absence of this recognition signal by some other mechanism (von Figura and Hasilik, 1986).

Discharge of secretory proteins (enzymes) along the export route is apparently the default pathway. Additional specificity is achieved at the endoplasmic reticulum via a set of signal sequences different from those for lysosomal enzymes.

Analyses of three resident endoplasmic reticulum luminal protein sequences show that these three proteins share the carboxy-terminal sequence of lys-asp-glu-leu (KDEL) (Munro and Pelham, 1987). By deletion or extension of the tail for one protein, grp78, they demonstrated that the mutant protein, when expressed in COS cells, was secreted into the medium while a derivative of chicken lysozyme containing the last six amino acids of grp78 failed to be secreted and instead accumulated in the endoplasmic reticulum lumen. They, therefore, concluded that the KDEL sequence marks the proteins that are to be retained in the endoplasmic reticulum lumen. Similarly, in an attempt to identify the region specifying the intracellular localization of the E19 protein (a viral-coded transmembrane endoplasmic reticulum resident) of adenovirus, Pääbo *et al.*, (1987) showed that shortening of the 15-amino acid member tail (the entire cytoplasmic domain) by 8 amino acids (FIDEKKMP) promoted intracellular transport of the protein to the plasma membrane. By transfer of the cytoplasmic domain of the E19 protein to interleukin-2 receptor (a plasma membrane protein), they showed that the receptor protein derivative was specifically confined to the endoplasmic reticulum when expressed in host cells. Additionally, it appeared that the cytoplasmic domain (or a subset of the cytoplasmic domain) was sufficient for the localization of integral membrane proteins to the endoplasmic reticulum. Thus, for retention of proteins in the endoplasmic reticulum and Golgi apparatus as permanent residents, specific retention signals appear to be employed. In contrast, no apparent signal is needed for rapid and efficient transport from endoplasmic reticulum to the Golgi apparatus to the cell surface.

An alternative explanation to retention of resident endoplasmic reticulum proteins would be their transfer to the cis Golgi apparatus compartment followed by a return mechanism also involving specific signal recognition sequences. For example, many authors favour such an explanation where some or all membrane proteins delivered to the cis Golgi apparatus in the transfer of secretory products are transferred back or recycled to the endoplasmic reticulum by a mechanism involving the return of empty vesicles (see Farquhar, 1978 and 1981 for recent reviews). Since it is known that cells tend to conserve their membrane constituents, it is possible that there is recycling of membrane proteins of the vesicles involved in the transport between the endoplasmic reticulum to the Golgi complex as well as the normal secretory and lysosomal transport. It has become clear that cells repeatedly reutilize or recycle the vesicular membranes involved in carrying out other transport operations (Brown *et al.*, 1983) including transport via coated vesicles (Orci *et al.*, 1986). Either mechanism would yield the same end result and there has been no direct evidence on this point since it is difficult to label and track such constantly mobile membrane elements. However, the advent of new techniques of molecular biology coupled with those of cell biology now may make it possible to test critically and quantitate recycling of membrane proteins and/or secretory protein between the endoplasmic reticulum and the Golgi apparatus.

OTHER SORTING SIGNALS

The primary targeting of nascent proteins to the endoplasmic reticulum, of course, involves a sorting signal, the signal sequence (Walter *et al.*, 1984). Other signals (hydrophobic domains, stop transfer sequences, etc.) help to determine specific orientations of proteins destined to become integral membrane proteins. Import of proteins into mitochondria, plastids, microbodies (peroxisomes), and the nucleus all involve targeting signals rapidly being elucidated (Verner and Schatz, 1988).

SORTING SIGNALS IN CELL-FREE SYSTEMS

A major current research emphasis in cell biology is on the use of cell free systems to dissect the secretory and the endocytic trafficking pathways into discrete steps and reproduce these events in appropriate cell free systems. Included are endosomal fusions (Davey *et al.*, 1985), intercompartment transfer within the Golgi apparatus (Dunphy and Rothman, 1985) and the step between endoplasmic reticulum and Golgi apparatus. The latter has been studied, for example, both in perforated cells to which both large and small molecular weight constituents may be added (Beckers *et al.*, 1987; Simons and Virta, 1987) and in a system reconstituted from isolated transitional donor elements of the endoplasmic reticulum and cis

Golgi apparatus acceptor membranes (Nowack *et al.*, 1987). The total transfer step in the reconstituted system requires ATP and is facilitated by cytosol. Plant systems competent for transfer in a cell-free environment have recently been obtained from etiolated seedlings of soybean and maize (Morré *et al.*, manuscript in preparation).

The transition vesicles formed in the cell free system have been isolated (Paulik *et al.*, 1988). In liver, the isolated vesicles contain a high concentration of phosphomannosyl receptor, in keeping with these vesicles as the major trafficking vehicle between the endoplasmic reticulum and the cis Golgi apparatus compartment (Minnifield *et al.*, 1988).

Palade (1983) was among the first to suggest that the transfer of membrane and other materials between the endoplasmic reticulum and the Golgi apparatus would involve specific signals to correctly target membranes. While no such signaling (receptor) molecules have yet been characterized using the cell-free systems, vesicular traffic from transitional endoplasmic reticulum to the cis Golgi apparatus in the cell-free system does exhibit a degree of acceptor specificity and dependency consistent with involvement of some type of specific molecular targeting mechanism.

SYNTHESIS, SECRETION AND TARGETING OF CELL SEPARATION ENZYMES IN PLANTS

Plant enzymes involved in cell separation are by definition targeted to the cell wall which must be dissolved or softened for separation to occur. Involved are glycosidases for major cell wall structural (cellulase) or cementing (pectinase) materials or (in fungi) esterases involved in the destruction of cuticular materials to aid host penetration. Their study has been made difficult by their synthesis and secretion only in certain cell types and developmental stages. For example, in leaf abscission, break strength changes involve secretion of pectinase by a layer of only a few cells to allow separation of bean leaves at a clearly defined point of petiolar-pulvinar junction (Morré, 1968).

Since the tissue regions involved are so small, special sensitive assays had to be developed even to measure the secreted polygalacturonases (Mussel and Morré, 1969). Therefore, most investigators have turned to the cell wall dissolution processes associated with the ripening of fruits as models for cell separation amenable to hydrolase isolation and characterization where the ultrastructural features of the cytoplasm is similar to that seen in abscission zones (Fig. 3).

Dallman *et al.*, have used protein A gold and a rabbit anticellulase antibody (provided by Dr. Alan Bennett, Department of Vegetable Crops, University of California, Davis) to study cellulase localization in the mesocarp of ripening avocado fruits (in press). Bennett and Christofferson (1986) had previously treated *in vitro* translated cellulase with a signal peptidase to remove any signal sequence present and also treated membrane-bound cellulase with endo-H. They identified four molecular weight forms of cellulase and suggested that the processing and transport of cellulase includes co-translational removal of the signal sequence and addition of oligosaccharide chains in the endoplasmic reticulum. A role for Golgi apparatus was less clear but a role in subsequent carbohydrate trimming was hypothesized. Dallman *et al.*, (in press) found a transition in labeling from the endoplasmic reticulum to the plasmodesmata and cell wall during the period between the late portion of the climacteric rise and the day after the climacteric peak. Very few dictyosomes were either present or gold labeled regardless of the stage of fruit ripening. However, coincident with breakdown of cell wall, Platt-Aloia and Thomson (1980, 1981) found an ultrastructural transformation in the morphology of the endoplasmic reticulum from a sheet-like form to a branching tubular form. Similar structures occur associated with cells considered to be involved in petinase secretion during late stages of leaf abscission (unpublished observations). In the avocado fruit, Dallman *et al.*, (in press) found cellulase in such membrane transitions and adjacent to and entering into branching forms of plasmodesmata. Their evidence would be consistent with cellulase movement to the wall or from cell to cell via a plasmodesmatal route as postulated by Juniper (1977).

Polygalacturonase (PG) or pectinase is one plant cell wall hydrolase amenable to electron microscope cytochemical localization. The method is based on the deposition of insoluble copper salts coming from the reaction of Benedict's solution at 100° with galacturonic acid released from polygalacturonic acid by the action of PG (Nessler and Allen, 1987).

Shown in Fig. 4 are cells of tomato fruits incubated 30 min in a

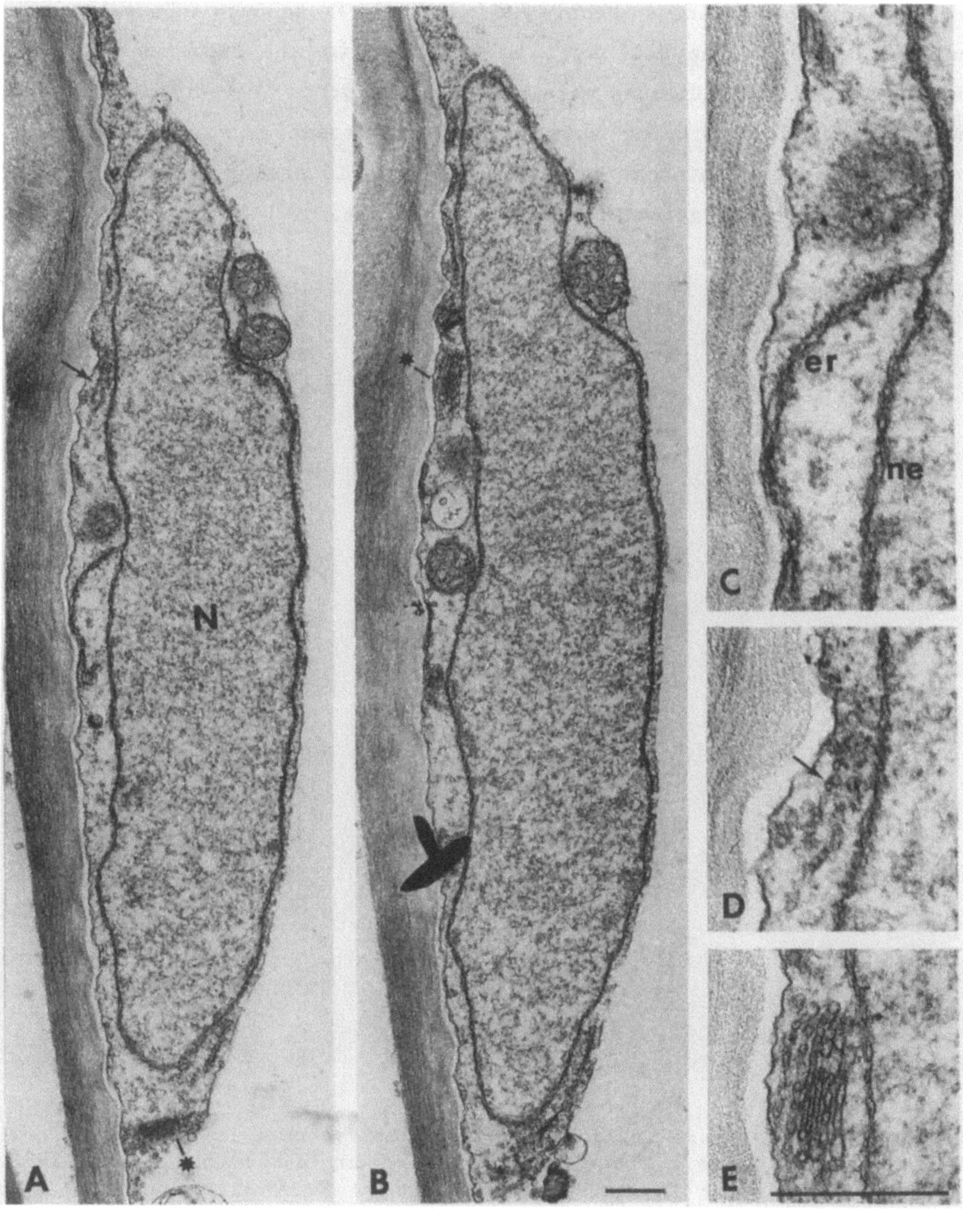

FIGURE 3. Skipped serial sections through the abscission zone of garden bean (Phaseolus vulgaris var. Bountiful). Most of the cell volume is occupied by a large central vacuole. Illustrated are endomembrane components typical of a hydrolase secreting plant cell and include nuclear envelope (NE) which may be continuous with endoplasmic reticulum (ER) (A and C) structures presumed to be annulate lamellae (nuclear envelope extensions containing pore complexes) (arrow A, D), dictyosomes of the Golgi apparatus (asterisks in A and B, E) as well as various small vesicles. Golgi apparatus were most often adjacent to the nucleus (N) and were encountered only rarely elsewhere in the cytoplasm. Scale bar = 0.5 μm.

solution of 0.5% citrus pectin (Sigma) as exogenous substrate. The location of PG in the cell wall is indicated by the deposits of reduced copper complex. The tissues were not fixed prior to incubation with pectin so it is difficult, using this method, to detect internal sites of PG localization, i.e. internal cell compartments, if sealed, would not permit the entry of the polygalacturonic acid substrate.

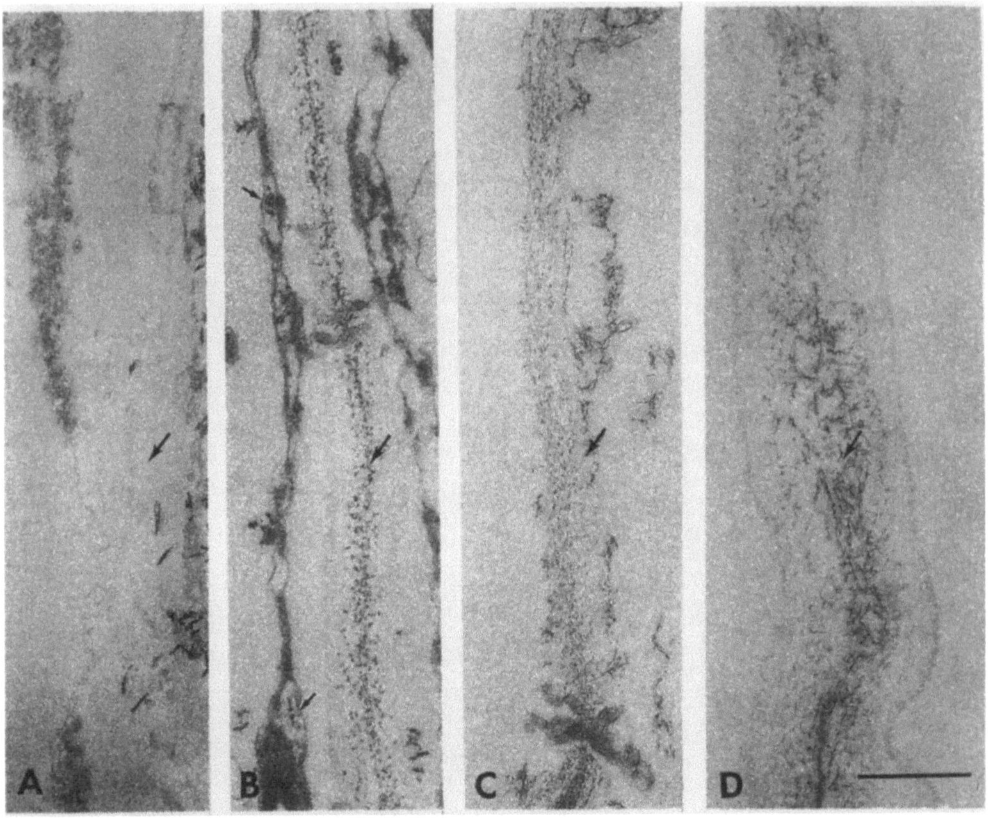

FIGURE 4. Cytochemical localization of pectinase activity in tomato fruit pericarp according to a procedure modified from Nessler and Allen (1987). Unfixed tissues were incubated for 30 min in 0.5% citrus pectin as an exogenous substrate. The sections were then boiled in Benedict's solution. Sites where the pectin was hydrolyzed to release galacturonic acid were marked by electron-dense deposits of a reduced copper complex with sodium-potassium tartrate. Walls of immature (green) fruits (A) were free of complex. With ripening (pink) fruits (B, C) walls were reactive. Deposits also were seen in swollen lumina of the endoplasmic reticulum (small arrows, B). With ripe (red) fruits (D), deposits indicative of pectinase activity also were found in the cell wall (large arrows). Bar = 0.5 µm.

To further localize the subcellular sites of PG localization in the tomato fruits, we used immunogold localization at the electron microscope level using an antibody to tomato fruit PG provided by Drs. Dean DellaPenna and Alan Bennett, Department of Vegetable Crops, University of California, Davis. PG activity increased markedly during the ripening of tomato fruits as a result of de novo enzyme synthesis (Brady *et al.*, 1982; DellaPenna *et al.*, 1986) and PG has been implicated as a major enzyme mediating wall softening during tomato fruit ripening. Based on studies with DNA clones, the steady state level of PG m-RNA increases over 2000- fold (DellaPenna *et al.*, 1986). The ripe fruit contains three isozymic PG forms. All three are glycoproteins and cross react with the same antibody. Results from *in vitro* synthesis and processing studies suggest that proteolytic processing of PG occurs in two steps. In the first, the 24 amino acid signal sequence would be removed co-translationally. This would then be followed by a presumed post-translational removal of a highly charged 47 amino acid prosequence (DellaPenna and Bennett, 1988) in the endoplasmic reticulum, Golgi apparatus or even after secretion into the cell wall. As suggested by DellaPenna and Bennett, one function of the prosequence might be to direct the protein to the cell wall. Information of the targeting of carboxypeptidase Y-invertase fusion proteins to the yeast vacuole is guided by information in a discrete amino acid sequence near the N-terminus of the protein (Johnson *et al.*, 1987). A lack of endoglycosidase H sensitivity of the mature PG isozyme oligosaccharide side chains indicates that the oligosaccharides are modified during the secretory process (DellaPenna and Bennett, 1988). This would, of course, be consistent with passage of PG through the Golgi apparatus during its secretion into the cell wall.

The endoplasmic reticulum of ripening tomato fruits is well developed. In contrast, Golgi apparatus are scarce. Immunogold localization of PG antibody showed PG throughout the endoplasmic reticulum of ripening tomato fruits (Fig. 5). In mature and ripening fruits, the enzyme was present in the cell wall as well but in green fruits there was little or no cell wall labeling (Fig. 5).

The range of possibilities for secretion and transport of cell wall hydrolases in plants is illustrated in Fig. 6. For tomato PG, molecular evidence favours a classic route whereby the 54 kD polypeptide translated on endoplasmic reticulum-bound polyribosomes would be co-translationally glycosylated with removal of the signal sequence and sequestered into the

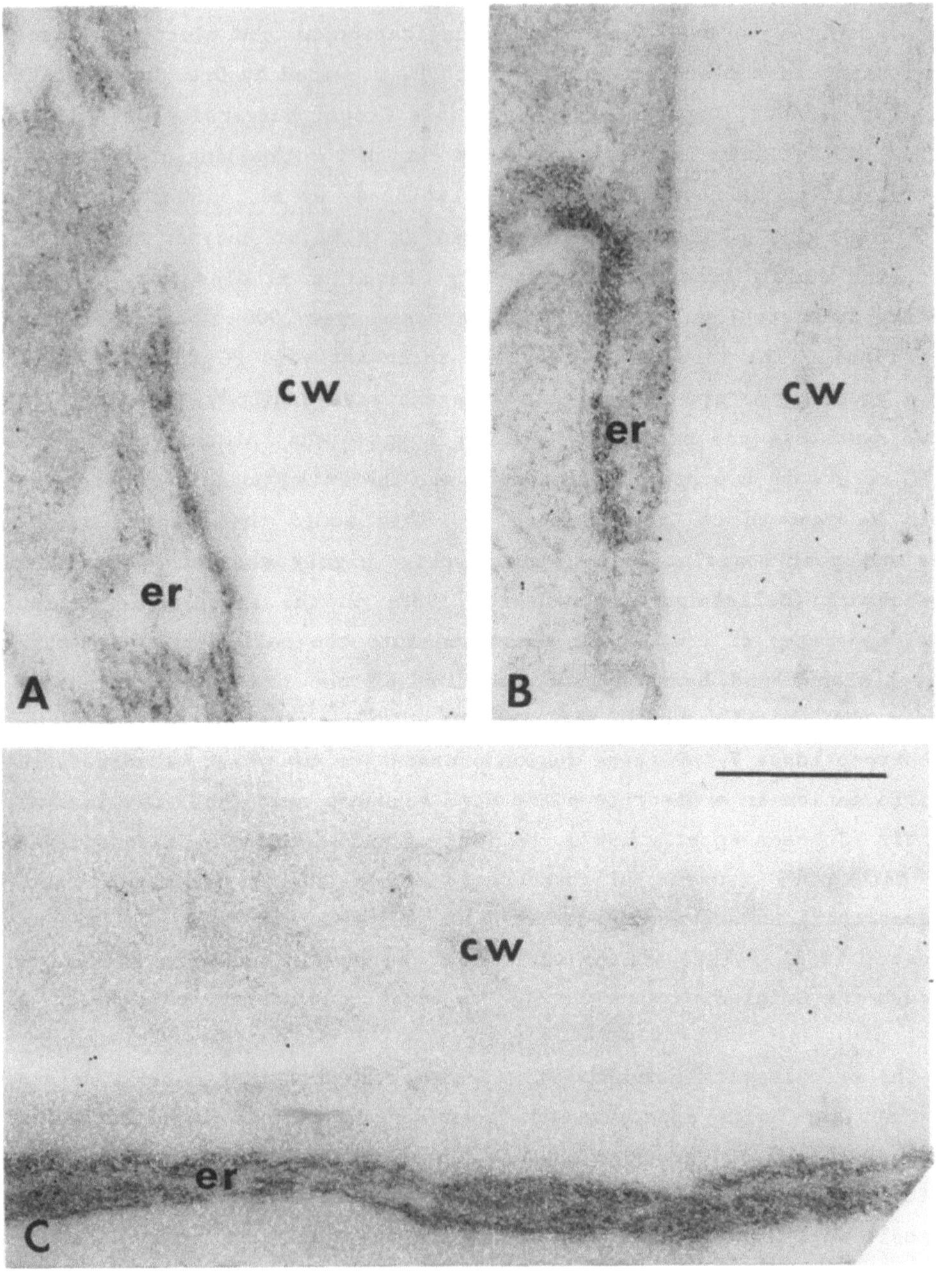

FIGURE 5. Immunogold localization of polygalacturonase in pericarp of garden tomato (var. Better Boy). In the immature (green) fruit (A), 6 nm gold coupled to antirabbit IgG was scarce or absent over the cell wall (CW) but was observed in the cytoplasm mostly associated with vesicles or cisternae of endoplasmic reticulum (ER). In ripening (pink) fruit (B) or in mature (red) fruit (C), gold grains were increased both in the CW and in the ER. Golgi apparatus were scarce or absent. In B, gold is seen also in the space between ER and the CW possibly in small vesicles. Bar = 0.5 μm.

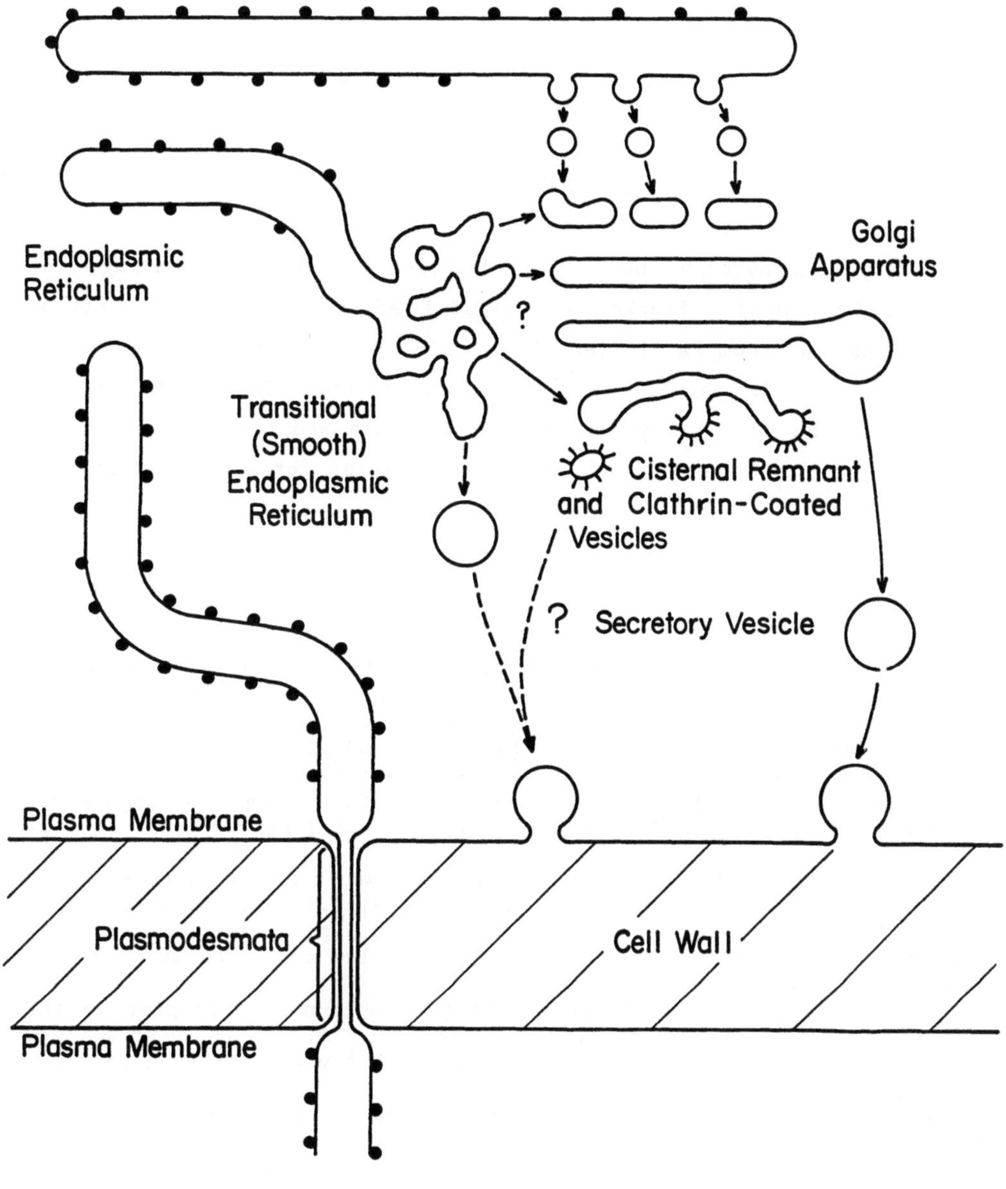

FIGURE 6. Potential subcellular routes for synthesis, processing and transport of hydrolases of cell separation in higher plants. The enzymes appear to be sequestered in the lumens of the endoplasmic reticulum and from there reach the cell wall. The pathway taken has not been shown to obligatorily involve the Golgi apparatus in plants and, if not, a peripheral route would be favoured from available morphological evidence. Endoplasmic reticulum lumens are continuous from cell to cell via plasmodesmata (Juniper, 1977). The plasmodesmatal route may be followed, for example, in the movement of hydrolases between adjacent cells.

lumen of the endoplasmic reticulum. Glycosylation would continue (at the Golgi apparatus) with delivery of the mature form to the cell wall before or after final removal of the 47 amino acid prosequence.

ACKNOWLEDGMENTS

I thank Karin Thornbrough, Dorothy Werderitsh, and Richard Bentlage for assistance, Dr. Charles Bracker for helpful advice and use of electron microscope and darkroom facilities, Drs. T. F. Dallman, W. W. Thomson, E. L. Eaks and E. A. Nothnagel, University of California, Riverside and Nita Minnifield, Purdue University for unpublished data and Drs. D. DellaPenna and A. B. Bennett, University of California, Davis, for polygalacturonase antibody.

LITERATURE CITED

Beckers CJM, Keller DS, Balch WE (1987) Semi-intact cells permeable to macromolecules: Use in Reconstitution of protein transport from the endoplasmic reticulum to the Golgi complex. Cell 50:523-534

Bennett AB, Christofferson RE (1986) Synthesis and processing of cellulase from ripening avocado fruit. Plant Physiol 81:830-835

Brady CJ, MacAlpine G, McGlasson WB, Ueda Y (1982) Polygalacturonase in tomato fruits and the induction of ripening. Aus J Plant Physiol 9:171-178

Brown MS, Anderson RGW, Goldstein JL (1983) Recycling receptors: the round-trip itinerary of migrant membrane proteins. Cell 32:663-667

Brown WJ, Goodhouse J, Farquhar M (1986) Mannose 6-phosphate receptors for lysosomal enzymes cycle between the Golgi complex and endosomes. J Cell Biol 103:1235-1247

Creek KE, Sly WS (1984) The role of the phosphomannosyl receptor in the transport of acid hydrolases to lysosomes. In: Dingle JT, Dean RT, Sly W (eds) Lysosomes in biology and pathology. Elsevier/North-Holland, New York, pp 62-82

Dallman TF, Thomson WF, Eaks IL, Nothnagel EA (1989) Expression and transport of cellulase in avocado mesocarp during ripening. Protoplasma (in press)

Davey J, Hurtley SM, Warren G (1985) Reconstitution of an endocytic fusion event in a cell-free system. Cell 43:643-652

DellaPenna D, Alexander DG, Bennett AB (1986) Molecular cloning of tomato fruit polygalacturonase: analysis of polygalacturonase mRNA levels during ripening. Proc Natl Acad Sci USA 86:6420-6424

DellaPenna D, Bennett AB (1988) In vitro synthesis and processing of tomato fruit polygalacturonase. Plant Physiol 86:1057-1063

Deutscher SL, Creek KE, Merion M, Hirschberg CB (1983) Subfractionation of rat liver Golgi apparatus: Separation of enzyme activities involved in the biosynthesis of the phosphomannosyl recognition marker in lysosomal enzymes. Proc Natl Acad Sci USA 80:3938-3942

Dunphy WG, Rothman JE (1985) Compartmental organization of the Golgi stack. Cell 42:13-21

Eppler CM, Morré DJ (1982) Flow kinetics of a nucleoside phosphatase common to endoplasmic reticulum, Golgi apparatus and plasma membrane of rat liver. Eur J Cell Biol 29:13-23

Farquhar MG (1978) Recovery of surface membrane in anterior pituitary cells. Variations in traffic detected with anionic and cationic ferritin. J Cell Biol 77:R35-R42

Farquhar MG (1981) Membrane recycling in secretory cells: Implications for traffic of products and specialized membranes within the Golgi complex. Methods Cell Biol 23:399-427

Fischer, HD, Gonzalez-Noriega S, Sly WS, Morré DJ (1980) Phosphomannosyl-enzyme receptors in rat liver. Subcellular distribution and role in intracellular transport of lysosomal enzymes. J Biol Chem 255:9608-9615

Gabel CA, Goldberg DE, Kornfeld S (1982) Lysosomal enzyme oligosaccharide phosphorylation in mouse lymphoma cells: Specificity and kinetics of binding to the mannose 6-phosphate receptor *in vivo*. J Cell Biol 95:536-542

Geisow M (1982) Lysosome proton pump identified. Nature 298:515-516

Geuze HJ, Slot JW, Strous GJAM, Hasilik A, von Figura K (1984) Ultrastructural localization of the mannose 6-phosphate receptor in rat liver. J Cell Biol 98:2047-2054

Geuze HJ, Slot JW, Strous GJAM, Hasilik A, von Figura K (1985) Possible pathways for lysosomal enzyme delivery. J Cell Biol 101:2253-2262

Goldberg DE, Kornfeld S (1983) Evidence for extensive subcellular organization of asparagine-linked oligosaccharide processing and lysosomal enzyme phosphorylation. J Biol Chem 258:3159-3165

Griffiths G, Simons K (1986) The trans Golgi network: sorting at the exit site of the Golgi complex. Science 234:438-443

Hasilik A, Waheed A, von Figura K (1981) Enzymatic phosphorylation of lysosomal enzymes in the presence of UDP-N acetylglucosamine. Absence of the activity in I-cell fibroblasts. Biochem Biophys Res Commun 98:761-767

Hoflack B, Fujimoto K, Kornfeld S (1987) The interaction of phosphorylated oligosaccharides and lysosomal enzymes with bovine liver cation-dependent mannose 6-phosphate receptor. J Biol Chem 262:123-129

Johnson LM, Bankaitis VA, Emr SD (1987) Distinct sequence determinants direct intracellular sorting and modification of yeast vacuolar protease. Cell 48:875-885

Juniper BE (1977) Some speculations on the possible roles of the plasmodesmata in the control of differentiation. J Theor Biol 66:583-592

Lang L, Reitman M, Tang J, Roberts RM, Kornfeld S (1984) Lysosomal enzyme phosphorylation. Recognition of a protein-dependent determinant allows specific phosphorylation of oligosaccharides present on lysosomal enzymes. J Biol Chem 259:14663-14671

Minnifield N, Creek KE, Navas P, Morré DJ (1986) Involvement of cis and trans Golgi apparatus elements in the intracellular sorting and targeting of acid hydrolases to lysosomes. Eur J Cell Biol 42:92-100

Minnifield N, Safranski K, Morré DJ (1988) Phosphomannosyl receptor binding is localized in vesicles separated by free-flow electrophoresis from rough endoplasmic reticulum preparations of rat liver. In: Morré DJ, Howell KE, Cook GMW, Evans WH (eds) Cell-free analysis of membrane traffic, Alan R. Liss, New York, pp 417-419

Mollenhauer HH (1971) Fragmentation of mature dictyosome cisternae. J Cell Biol 49: 212-214

Morré DJ (1968) Cell wall dissolution and enzyme secretion during leaf abcission. Plant Physiol 43:1545-1559

Morré DJ, Creek KE, Matyas GR, Minnifield N, Sun I, Baudoin P, Morré DM, Crane FL (1984) Free-flow electrophoresis for subfractionation of rat liver Golgi apparatus. BioTechniques 2:224-233
Morré DJ, Morré DM, Heidrich HG (1983) Subfractionation of rat liver Golgi apparatus by free-flow electrophoresis. Eur J Cell Biol 31:263-274
Munro S, Pelham HRB (1987) A C-terminal signal prevents secretion of luminal ER proteins. Cell 48:899-907
Mussell HW, Morré DJ (1969) A quantitative bioassay specific for polygalacturonases. Anal Biochem 28:353-360
Navas P, Minnifield N, Sun I, Morré DJ (1986) NADP phosphatase: A marker in free-flow electrophoretic separations for cisternae of the Golgi apparatus midregion. Biochim Biophys Acta 881:1-9
Nessler CL, Allen RD (1987) Pectinase In: Vaughn KC (ed) CRC handbook of plant cytochemistry. CRC Press, Baca Raton, FL Vol 1 pp 149-157
Neufeld EF, Sando GN, Garvin AJ, Rome LH (1977) The transport of lysosomal enzymes. J Supramol Struct 6:95-101
Novikoff AB (1976) The endoplasmic reticulum: a cytochemist's view (a review). Proc Natl Acad Sci USA 73:2781-2787
Nowack DD, Morré DM, Paulik M. Keenan TW, Morré DJ (1987) Intracellular membrane flow: Reconstitution of transition vesicle formation and function in a cell-free system. Proc Natl Acad Sci USA 84:6098-6102
Orci L, Glick BS, Rothman JE (1986) A new type of coated vesicular carrier that appears not to contain clathrin: Its possible role in protein transport within the Golgi stack. Cell 46:171-184
Pääbo S, Bhat BM, Wold WSM, Peterson PA (1987) A short sequence in the COOH-terminus makes an adenovirus membrane glycoprotein a resident of the endoplasmic reticulum. Cell 50:311-317
Palade GE (1983) Membrane biogenesis: An overview. Meth Enzymol 69:xxiv-lv
Paulik M, Nowack DD, Morré DJ (1988) Isolation of a vesicular intermediate in the cell-free transfer of membrane from transitional elements of the endoplasmic reticulum to Golgi apparatus cisternae of rat liver. J Biol Chem 263:17738-17748
Pavelka M (1987) Functional morphology of the Golgi apparatus. In: Beck F, Hild W, Kriz W, Ortmann R, Pauly JE, Schiebler TH (eds) Advances in anatomy, embryology and cell biology. Springer-Verlag, Heidelberg 106:1-94
Pesacreta TC, Lucas WJ (1984) The plasma membrane coat and a coated vesicle-associated reticulum of membranes: their structure and possible interrelationship in Chara coralling. J Cell Biol 98:1537-1545
Platt-Aloia KA, Thomson WW (1980) Aspects of the three-dimensional organization of avocado mesocarp cells as revealed by scanning electron microscopy. Protoplasma 104:157-165
Platt-Aloia KA, Thomson WW (1981) Ultrastructure of the mesocarp of mature avocado fruit and changes associated with ripening. Ann Bot 48:451-465
Pohlmann R, Waheed A, Hasilik A, von Figura K (1982) Synthesis of phosphorylated recognition marker in lysosomal enzymes is located in the cis part of Golgi apparatus. J Biol Chem 257:5323-5325
Pohlmann R, Nagel G, Schmidt B, Stein M, Lorkowski G, Krentler C, Cully J, Meyer HE, Grzechik KH, Mersmann G, Hasilik A, von Figura K (1987) Cloning of a cDNA encoding the human cation dependent mannose 6-phosphate-specific receptor. Proc Natl Acad Sci USA 84:5575-79
Reitman ML, Kornfeld S (1981a) UDP-N-acetylglucosamine:glycoprotein N-acetylglycosamine-1-phosphotransferase. Proposed enzyme for the phosphorylation of the high mannose oligosaccharide units of lysosomal enzymes. J Biol Chem 256:4275-4281
Reitman ML, Kornfeld S (1981b) Lysosomal enzyme targeting. N-acetylglucosaminylphosphotransferase selectively phosphorylate native

lysosomal enzymes. J Biol Chem 256:11977-11980
Rosenfeld MG, Kreibich G, Popov D, Kato K, Sabatini DD (1982) Biosynthesis of lysosomal hydrolases: their synthesis in bound polysomes and the role of co- and post-translational processing in determining their subcellular distribution. J Cell Biol 93:135-143
Simons K, Virta H (1987) Perforated MDCK cells support intracellular transport. EMBO J 6:2241-2247
Sly WS, Fischer HD (1982) The phosphomannosyl recognition system for intracellular and intercellular transport of lysosomal enzymes. J Cell Biochem 18:67-85
Sly WS, Stahl P (1978) Receptor mediated uptake of lysosomal enzymes. In: Silverstein SC (ed) Transport of macromolecules in cellular systems. Life Sciences Research Report II Dahlem Konferenzen, Berlin, pp 229-244
Smith CE (1980) Ultrastructural localization of nicotinamide adenine dinucleotide phosphatase (NADPase) activity to the intermediate saccules of the Golgi apparatus in rat incisor ameloblasts. J Histochem Cytochem 28:16-26
Tong PY, Tollefsen SE, Kornfeld S (1988) The cation-independent mannose 6-phosphate receptor binds insulin-like growth factor II. J Biol Chem 6:2585-2588
Varki A, Kornfeld S (1980) Identification of a rat liver α-N-acetylglucosaminyl phosphodiesterase capable of removing "blocking" α-N-acetylglucosamine residues from phosphorylated high mannose oligosaccharides of lysosomal enzymes. J Biol Chem 255:8398-8401
Varki A, Kornfeld S (1981) Purification and characterization of rat liver α-N-acetylglucosaminyl phosphodiesterase. J Biol Chem 256:9937-9943
Verner K, Schatz G (1988) Protein translocation across membranes. Science 241:1307-1313
von Figura K, Hasilik A (1986) Lysosomal enzymes and their receptors. Ann Rev Biochem 55:167-193
Waheed A, Hasilik AG, von Figura K (1981a) Processing of the phosphorylated recognition marker in lysosomal enzymes. Characterization and partial purification of a microsomal α-N-acetylglucosaminyl phosphodiesterase. J Biol Chem 256:5717-5721
Waheed A, Pohlman R, von Figura K (1981b) Subcellular location of two enzymes involved in the synthesis of phosphorylated recognition markers in lysosomal enzymes. J Biol Chem 256:4150-4152
Waheed A, Hasilik A, von Figura K (1982) UDP-N-acetylglucosamine: lysosomal enzyme precursor N-acetylglucosamine-1-phosphotransferase. Partial purification and characterization of the rat liver Golgi enzyme. J Biol Chem 257:12322-12331
Walter P, Gilmore R, Blobel G (1984) Protein translocation across the endoplasmic reticulum. Cell 38:5-8

EXTENSIN PEROXIDASE TIES THE KNOTS IN THE EXTENSIN NETWORK

Derek T.A. Lamport
MSU-DOE Plant Research Laboratory
Michigan State University
East Lansing, MI 48824-1312, USA

INTRODUCTION

I have been interested in cell separation and primary cell walls since 1958 when I cut down a sycamore-maple tree in Madingley Wood, Cambridge U.K. and induced the cambial tissues to proliferate, first as a friable callus, and soon after as the first pipettable cell suspension culture to be used by a biochemist. Incidentally this was the first tree to be grown in log phase! (Lamport 1960; 1964) Of course this merely demonstrates their algal ancestry, so instead of asking why or how cells of higher plants separate, it may be just as instructive to examine the opposite side of the same coin and ask how they stick together. In other words, what is it about cell walls that ties cells together? If we know that perhaps we shall know what sorts of questions to be asking about cell separation.

The primary cell wall is an exquisitely designed micro-composite of interpenetrating polymeric networks (IPNs) whose overall properties help predestine cell shape and size. Therefore, we must describe intermolecular reactions between these IPNs to understand how the cell wall works, like other cell organelles [Lamport 1964]. This view is quite contrary to archaic descriptions of cellulose microfibrils embedded in an "amorphous matrix," a contradiction in terms, which ignores insights gleaned from modern materials science.

In general terms, IPN interactions involve crosslinks both covalent and non-covalent. Thus for the network to grow and accommodate changes in cell shape we have to answer three general questions:

What are the crosslinks?

How are they made?

How is the network loosened?

That is a tall order, so here I shall cut the problem down to manageable proportions by focusing on one structural component in particular, a protein which hypothetically mechanically couples the load-bearing polymers in the dicot primary cell wall. I shall

NATO ASI Series, Vol. H 35
Cell Separation in Plants
Edited by D. J. Osborne and M. B. Jackson

emphasize possible intermolecular protein-protein crosslinks although no doubt carboxylate-amino salt bridges play an important role in pectin-extensin association.

In outline this presentation begins by defining the major subject: what is extensin? Then, I shall describe the origin of the warp-weft model and summarize evidence for intra and intermolecular crosslinks of extensin, followed by recent data showing *in vitro* crosslinkage of extensin, which give some clues for chemical identification of the intermolecular crosslinks. Then follow some speculations about molecular mechanisms for assembly of the extensin network *in muro*, which may involve reptating rods of high tacticity, and physiological control mechanisms, including network loosening.

Finally, I shall suggest possible multifunctional aspects of extensin at three levels of organization, including the possibility that some plants may have found alternatives to extensin.

EXTENSIN DEFINED

Extensin is a small family of basic hydroxyproline-rich glycoproteins (HRGP) also rich in serine, lysine/histidine and tyrosine (Smith *et al.,* 1984). Generally, most of the hydroxyproline residues are O-glycosylated by short arabino-oligosaccharides (Lamport 1967) and some of the serine residues are O-galactosylated (Lamport *et al.,* 1973) by a single sugar residue. Thus as predicted by circular dichroism spectroscopy (Lamport 1977, Van Holst and Varner 1984) and confirmed by electron microscopy (Van Holst and Varner, 1984; Heckman *et al.,* 1988) extensin monomers exist as flexuous (80 nm) rods, with sugar accounting for nearly two-thirds of the mass. Although the bulk of extensin is generally insoluble, even in anhydrous hydrogen fluoride (Mort and Lamport 1977), the growing wall has a small pool of soluble extensin monomers which are precursors to the insoluble network. Salt solutions rapidly elute this pool from the walls of intact cells; the elution occurs even at 4°C so monomer release is an ionic desorption from a pectic polyanion rather than secretion (Smith *et al.,* 1984).

Purification of these monomers from tomato cell suspension cultures clearly showed for the first time that there was more than one extensin. Peptide maps and sequences of the major peptides display a marked periodicity (Smith *et al.,* 1986); high yields of the major tryptic peptides show that each peptide occurs several times in the molecule. Each of the major peptides also displays a striking pentameric motif: Ser-Hyp-Hyp-Hyp-Hyp seen as Ser-Pro-Pro-Pro-Pro in genomic sequences (Showalter and Varner, 1988). This motif is currently considered as a diagnostic ***sine qua non*** of extensin, but this "rule" is based on a very limited experience of three dicotyledonous families. Our new data show that the rule has to be bent from "all or nothing" to "sometimes," as we shall now see:

Because of our interest in the phyllogeny of extensin we are exploring extensins of the monocots (Kieliszewski and Lamport, 1987; Hood and Varner, 1988) and also primitive dicots represented by families such as the Chenopodiaceae. Very recently we isolated a putative extensin monomer from intact sugar beet cell suspension cultures (Li *et al.*, 1988). Compositionally, sugar beet extensin appeared typical but to our great suprise none of the major tryptic peptides contained the Ser-$(Hyp)_4$ motif. On the other hand, close homology with extensin P1 of tomato is apparent from the identical seven residue sequence: Hyp-Hyp-Thr-Hyp-Val-Tyr-Lys common to the tryptic peptides of both tomato H_{20} and sugar beet H_{14}.

Why doesn't sugar beet extensin have Ser-$(Hyp)_4$? Because it is cryptic! Here is the explanation. Consider the two major repetitive peptides of tomato P1 extensin:

Ser-Hyp-Hyp-Hyp-Hyp-Thr-Hyp-Val-Tyr-Lys [H_5] and

Ser-Hyp-Hyp-Hyp-Hyp-**Val-Lys-Pro-Tyr-His-Pro**-Thr-Hyp-Val-Tyr-Lys [H_{20}] which differs from [H_5] only by a hexapeptide insertion after the Ser-$(Hyp)_4$. Now closely examine the major sugar beet peptide [H_{14}]:

Ser-Hyp-Hyp-Val-His-Glu-Tyr-Pro-Hyp-Hyp-Thr-Hyp-Val-Tyr-Lys and you see that removal of the **Val-His-Glu-Tyr-Pro** "insertion" restores the familiar sequence of tomato peptide H_5 and the Ser-$(Hyp)_4$ pentamer. Thus, in chenopod extensin, the insertion sequence **splits** the rigid tetrahydroxyproline block while in higher dicots the insertion sequence comes immediately after the rigid block. These data also illustrate remarkable evolutionary conservation of the major P1 motif represented by tomato H_5.

Well what does all this arcane lore signify? It may help us answer three general questions:

1. What defines the extensin family and subfamilies P1, P2, P3, ...Pn?
2. Do insertion sequences represent crosslink domains?
3. What determines molecular flexibility?

First, sugar beet extensin monomers show that we cannot dismiss an HRGP and deny its family membership just because it happens to lack tetrahydroxyproline blocks or a preceeding serine residue (A single base change relates Ala Thr Ser & Pro). For example, the soybean genomic repeat Pro-Pro-Val-Tyr-Lys (Hong *et al.*, 1987; Averyhart-Fullard

et al., 1988) is clearly highly homologous with the repeating octapeptide:
Ser-Hyp-Hyp-Hyp-Hyp-Val-Tyr-Lys of tomato extensin P2, while the recently reported nodulin N75 (Franssen *et al.,* 1987) repeating octapeptide:
Pro-Pro-**[His-Glu-Lys-Pro]**-Pro-Pro is comparable to sugarbeet extensin:
Hyp-Hyp-[Val-His-Glu-Tyr-Pro]-Hyp-Hyp. Similarly one can also relate the wound-induced protein P33 (Chen and Varner, 1985; Tierney *et al.,* 1988) which has numerous repeat units:
Pro-Pro-**[X-Y-Z]**-Pro-Pro where the X-Y-Z tripeptide is often: Val-Tyr-Thr (3 repeats), **Val-His-Lys** (5 repeats), and **Ile-His-Lys** (3 repeats).

Secondly, we emphasize direct peptide sequencing rather than genomic sequencing because post-translational modification of proline to hydroxyproline defines the tetrahydroxyproline blocks, while a single *unmodified* proline residue helps define the insertion sequence. Because this insertion sequence is non-glycosylated and generally contains tyrosine, it is a candidate crosslink domain.

A third reason to focus on the insertion sequences may be because they represent a mechanism for fine-tuning molecular flexibility. This can be measured approximately by measuring contour length and persistence length of molecules electron microscopically (Heckman *et al.,* 1988). The results show relatively stiff molecules designed to diffuse not by random tumbling but by more orderly reptation *i.e.* by endwise migration of linear macromolecules. Why should that be of any interest?

THE WARP-WEFT MODEL

A few years ago we discovered that cell suspension cultures contained a small pool of extensin monomers *in muro* which could be salt-eluted from the intact cells. This elution raises a problem. How does a macromolecule 80 nm in length move through cell wall pores of around 5 - 6 nm diameter? Transformation of protoplasts after electroporation is a similar problem. In both instances, linear macromolecules wiggle through small pores, i.e. they **reptate.** And because elution of extensin monomers occurs so rapidly I suggested that extensin was initially inserted into the wall as **transmural protein rods** orthogonal to the cellulose microfibrils (Lamport and Epstein, 1983; Lamport 1986). Crosslinked extensin network would almost inescapably have to crosslink around the microfibrils, so the limiting thickness of the primary cell wall and the 80 nm length of the monomer may reflect design rather than coincidence (Heckman *et al.,* 1988). Naturally, attention must now focus on the intermolecular crosslinks because they define the porosity of the network and may be involved in regulating primary wall growth.

EXTENSIN CROSSLINKS

Our two approaches to identification of crosslinks involve: first, an attempt to identify them in various wall fractions, and second, attempts to identify them after their enzymic formation *in vitro*.

So far the first approach has yielded only an *intra*molecular linkage (Epstein and Lamport, 1984)], namely isodityrosine (IDT) which occurs in extensins P2 and P3 represented by tryptic peptides:

Ser-Hyp-Hyp-Hyp-Hyp-Val-$^{1}/_{2}$IDT-Lys-$^{1}/_{2}$IDT-Lys (P2), and:

Ser-Hyp-Hyp-Hyp-Hyp-Ser-Hyp-Ser-Hyp-Hyp-Hyp-Hyp-$^{1}/_{2}$IDT-Tyr-$^{1}/_{2}$IDT-Lys (P3). In other words the direct analytical approach has not yielded the elusive *inter*molecular linkage (crosslinking extensin), commonly assumed to be IDT, but now somewhat questionable.

We have also been pursuing a parallel but more indirect approach by attempting to generate crosslinks enzymically *in vitro*, reasoning that if we can elute monomeric extensin precursors from intact cells, then the crosslinking reaction occurs *in muro*, and we should therefore also be able to elute the crosslinking enzymes from intact cells.

CROSSLINKAGE *in vitro*

First, we devised an assay to monitor increased size of the crosslinked products via fast protein liquid chromatography on Superose-6. (Superose-6 is a 4% crosslinked agarose column, with an exclusion limit up to 40 megadaltons, far beyond the size limits imposed by SDS-PAGE.) Then we began to assay salt eluates of intact cells. As a working hypothesis first raised in 1980 (Lamport 1980), we suggested a peroxidatic crosslinking reaction to account for the (then) unknown tyrosine derivative, later identified as IDT (Fry 1982). Thus the same concentration of salt which releases extensin monomers should also release the enzyme. From a rather slow beginning when we first observed *in vitro* extensin multimer formation we have progressed to a stage (Everdeen *et al.*, 1988) where we know the optimum pH, (6 - 6.5). The optimum peroxide concentration (30-60 microMolar) is much lower than that of an artificial substrate such as ABTS, 2,2'-azino-bis-(3-ethylbenzothiazoline-6-sulphonic acid). The macromolecular substrate concentration saturates the enzyme only at relatively high levels. Most recently we have purified the enzyme to homogeneity (Dey *et al.*, 1988). It is basic, with a pI = 9. Quite surprisingly it is essentially the only peroxidase eluted from intact cells by 50 mM $CaCl_2$. However, much higher (plasmolysing) salt concentrations elute two additional peroxidases (pI = 8.5 and 9.6) but we have not yet been able to test those for crosslinking activity. The cells also secrete a soluble acidic peroxidase (pI = 4.3) into the medium.

At this stage two questions beg to be answered: Is the enzyme, tentatively identified as extensin peroxidase, specific for extensin, and what is the chemical identity of the crosslink formed? More boldly we should ask: are we studying an artifact? Certainly, peroxidases at high, probably stoichiometric levels, can crosslink proteins, but our enzyme is highly specific for extensin. It does not crosslink other proteins such as bovine serum albumin, apomyoglobin, trypsin inhibitor, ribonuclease, thryoglobulin or aldolase (Everdeen *et al.*, 1988). Nor does it crosslink deglycosylated extensin. Of course with an enzyme like peroxidase you can never do enough controls. Naturally we have examined the commercially available Sigma Horseradish peroxidase isozymes. Interestingly type VI (basic; pI = 9) at a comparable level, does show some activity (ca. 50%) towards P1 extensin monomers from tomato, while two acidic isozymes vary from 12 - 75% activity. Unfortunately we have not been able to prepare horseradish extensin.

Well, what is the chemical identity of the crosslink formed?

CHEMISTRY OF EXTENSIN INTERMOLECULAR CROSSLINKS

Although there seems little doubt of their existence, currently we cannot chemically identify the intermolecular crosslinks of extensin. However we do have some clues: First, we cannot detect IDT as the *in vitro* product of P1 crosslinkage. Second, the crosslinked product shows anomalous behaviour in 2% sodium dodecylsulphate...it tends to get lost on the column! Third, cyanoborohydride reduction "stabilizes" the crosslinked product. This means either we have an interesting aggregation artifact, or that an oxidation step creates the crosslink and then a subsequent reduction stabilizes it; oxidation of a lysine epsilon amino to an aldehyde would allow formation of the (unstable) Schiff base stabilized by reduction; for example two crosslinked lysine residues would yield lysinonorleucine. Having stabilized the enzymically generated crosslink we can proceed with further enzymic digestion and isolation of crosslinked peptides. This strategy depends on the stability of the stabilized crosslink to the anhydrous HF required for deglycosylation which greatly enhances trypsinolysis.

We may be able to corroborate those results by peptide mapping the low levels of small-oligomers (< 10%) present in monomer preparations, provided we can generate sufficient material. Presumably the small oligomers elute from the wall because they are small enough to escape entrapment. (Eluates of intact cells never contain large oligomers and this suggests entrapment of oligomers above a critical size.) There is an implicit assumption here that addition of extensin (P1 or P2) to the wall occurs by insertion of the rodlike monomers; *i.e.* intussusception rather than apposition. As extensin P3 is non-elutable it may be built into the wall by apposition (Smith *et al.*, 1984; 1986).

EXTENSIN NETWORK ASSEMBLY *in muro*

If there is *transmural* insertion of extensin monomers, what are the driving forces? The general question of macromolecule migration through a porous matrix excites much current interest (Maddox 1987; Rubinstein 1987) not least because it is the key to the separation of very high molecular weight DNA. And the key concept is *reptation*, *i.e.* endwise migration of linear molecules. For example, we interpret the anomalously low retardation of extensin monomers on Superose-6 gel filtration, in terms of endwise insertion into the gel matrix rather than due to a large Stokes radius (Heckman *et al.*, 1988). Because electrical fields assist reptation and because the elution evidence for ionic desorption of polycationic extensin from pectic polyanions is clear, then wall-bound pectin methylesterase may create a gradient of de-esterified pectin, hence an electrostatic gradient, increased even further by proton secretion during "acid growth"; the outermost wall layer being oldest will be the most negatively charged. But there may be even more to it than that, because general mechanical principles also apply at the molecular level [Jarosch 1987]. Until recently we and others assumed that hydroxyproline glycosylation served to stabilize the polyproline II backbone conformation of extensin, but the molecular rod clearly remains intact after HF-deglycosylation (Heckman *et al.*, 1988). Perhaps we should consider additional roles for the hydroxyproline arabinosides. Molecular modelling suggests that they H-bond with and "nest" along the polypeptide backbone (Lamport 1980). So perhaps one should view extensin monomers as molecular screws whose helical thread, enhanced by glycosylation, assists reptation into the wall matrix!

The peptide periodicity of these molecular screws creates still another remarkable characteristic referred to by polymer chemists as *tacticity* - the orderly non-random arrangement in space, of polymer side groups (Baer 1986). Synthetic polymers of high tacticity readily crystallize, so tacticity in a biological polymer is evidence of a molecule designed for highly ordered interactions necessary for self-assembly. Of those, the covalent crosslinks are probably the easiest to deal with at the moment; they certainly have important physiological implications which we shall now discuss.

CROSSLINK PHYSIOLOGY

The possibility that crosslink formation involves a two step oxidation/reduction fits quite well with known wall enzymology. It seems almost certain that the wall can generate H_2O_2 via peroxidase NADH oxidase activity:

$2NADH + O_2 \dashrightarrow 2NAD^+ + H_2O_2$

Another possibility might involve a cell surface NAD(P)H oxidase:

$NADH + O_2 \dashrightarrow NAD^+ + O_2^{-\cdot}$

and superoxide dismutase:

$2O_2^{-\cdot} + 2H^+ \dashrightarrow H_2O_2 + O_2$

Other plasma membrane redox reactions also occur; for example intact cell suspensions reduce ferricyanide probably involving an NADH-ferricyanide reductase, although Pantoja and Willmer (1988) suggest an alternative explanation: peroxidase-generated superoxide reduction of ferricyanide. And there is the recent exciting report of an auxin-stimulated NADH oxidase (Brightman *et al.*, 1988) which utilizes oxygen but not ferricyanide and is perhaps coupled to proton extrusion, hence wall acidification. Regeneration of NADH probably occurs by a specific wall bound malic dehydrogenase:

$malate + NAD^+ \dashrightarrow NADH + oxaloacetate$

In addition a whole host of effectors may act as fine controls regulating the activity of extensin peroxidase. These include manganese ions, phenolics and especially ascorbate, which at low levels inhibits extensin peroxidase *in vitro*. Hence regulation of crosslinking might depend on the *in muro* ascorbate level, which in turn, would depend on the rate of ascorbate secretion, the activity of *in muro* ascorbic acid oxidase, and possibly also on the activity of the plasma membrane ferricyanide reduction system recently demonstrated in cell suspension cultures.

Of course we should not ignore the obvious control mechanism: direct regulation of extensin peroxidase levels *in muro*. There is good evidence that this happens in whole plants. Indeed, Ridge and Osborne (1971) first pointed out that increased wall hydroxyproline which occurs during cell maturation, or after treatment with ethylene, is invariably accompanied by an increase in the activity of wall-bound peroxidase activity. And quite recently Lagrimini and Rothstein (1987) wounded tobacco (leaves and stem) and noted induction of a previously absent basic isoperoxidase; its pI 8.9 corresponds very closely to extensin peroxidase (pI 9.0) from tomato. However the 8.9 isoenzyme is constitutive in young tobacco roots; that seems significant because known high levels of root extensin correlate well with its probable defence role (Esquerre-Tugaye *et al.* 1979). Very recent histochemical detection (tissue blots) of ethylene-induced increases in extensin and peroxidase of pea epicotyl epidermis and cortex (Cassab and Varner 1988) nicely corroborates the Ridge and Osborne story. But we are still unable to test the attractive hypothesis that: (Ridge and Osborne 1970) "Local increases in ethylene production by (pea epicotyl) hook cells could so enhance the hydroxyproline levels in the walls of these cells as to predetermine their subsequent expansion growth."

EXTENSIN MULTIFUNCTIONAL ROLE

We are now at the stage where the idea of crosslinked extensin networks is acceptable, while their functions remain loosely defined and hypothetical. This is because the networks themselves are not defined, yet we need this structural information to visualize interactions with other wall components. Hence the importance of extensin peroxidase which by crosslinking extensin will help us define the crosslink domains and the network itself. But how does elucidation of network structure begin to define function?

The primary wall involves extracellular assembly, and it serves many functions; it must therefore be highly organized. What are the principles? The approach taken here views the wall as a composite whose properties have been optimized by the molecular design of polymers for specific functions. For example side group regularity or tacticity, enhances orderly polymer interactions (Baer 1986).

Periodicity and tacticity are related. Thus extensins and their relatives which show high fidelity periodicity such as the tomato P2 repeat:
Ser-Hyp-Hyp-Hyp-Hyp-Val-Tyr-Lys-Tyr-Lys, the soybean: **Pro-Pro-Val-Tyr-Lys**, and the nodulin: **Pro-Pro-His-Glu-Lys-Pro-Pro-Pro** must exhibit a high tacticity. Tacticity determines molecular packing (Baer 1986). And *inhibition* of extensin hydroxylation or glycosylation inhibits cell wall regeneration in protoplasts (Cooper 1984). Thus at one level extensin networks of high tacticity might act as *organizers* of primary cell walls by providing a highly

ordered skeletal framework which guides extracellular assembly of the polysaccharide matrix.

This suggestion is at least consistent with the highly ordered glycoprotein cell wall of *Chlamydomonas*, a primitive protist whose self-assembly paradigm (Roberts *et al.* 1985) depends entirely on interactions between rodlike hydroxyproline-rich glycoproteins, forming a porous network (Goodenough and Heuser, 1985). This probably set the evolutionary stage for the development of more complex and stronger cell walls by the interpenetration of microfibrillar polysaccharides; hence the "invention" of turgor pressure (Raven 1982) allowing the transition from aquatic to terrestrial habitat.

The further evolution of tissue differentiation required different walls optimized for each tissue. That suggests the need for multiple organizers of molecular assembly and packing. Thus in general terms we can account for multiple extensins, and the variation in their relative amounts, at three functional levels: the lowest amounts, exemplified by monocot cell walls, involved purely as an **organizing** framework; intermediate amounts involved in coupling load-bearing polymers and **growth regulation**; and the highest amounts involved in **disease resistance** by creating a tightly knit meshwork to prevent ingress of pathogens. There is increasing evidence of function-specific extensins resulting from differential gene expression. Thus, seed coat formation (Cassab *et al.* 1985), root nodule formation (Franssen *et al.* 1987), wounding (Chen and Varner, 1985) and ethylene effects all probably turn on a specific extensin or set of extensins.

NETWORK LOOSENING

So far we have discussed the evidence for network formation. Is it a one-way process? Does the network ever break down? Certainly wall protein content increases as epicotyl growth decreases, which is evidence for a role of extensin in the cessation of elongation. On the other hand rapidly growing dicot suspension cultures often have very high levels of wall-bound extensin. As extensin does not inflict *rigor mortis* on the wall there are probably mechanisms for relaxing the network. Proteases are unlikely, although specific peptide bond cleavage involving the highly acid-labile **Asp-Pro** linkage of P3 is an intriguing possibility. We inferred the existence of the non-elutable "brace-protein" P3 from wall tryptic peptides. Genomic sequences confirm the existence of P3 and also show a central Asp-Pro, supporting the idea of a brace protein cleavable under acid-growth conditions. Other possibilities exist for loosening the network without breaking covalent bonds. For example salt-bridge coupling between extensin and pectin would decrease at low pH, possibly releasing steric restraints to allow slippage of cellulose microfibrils through extensin network pores (Lamport 1986).

ALTERNATIVES TO EXTENSIN

Virtually all the work on extensin is concentrated in a relatively few advanced families of dicots. Indeed only in the past year have we obtained convincing evidence for extensin *per se* in graminaceous monocots (Kieliszewski and Lamport 1987, Hood and Varner 1988). However extensin levels in monocots are generally at least an order of magnitude lower than in dicots. Why? One can speculate that low level monocot extensins may act as cell wall organizers, but what replaces the role played by the high levels of dicot extensins? Does the construction of the pectin-poor monocot cell wall allow it to largely dispense with extensin? Is there a monocot substitute for the crosslinking role of extensin? Perhaps phenolic crosslinking plays a larger role in the monocots, but there is also increasing evidence of other structural proteins, possibly glycine-rich (Condit and Meagher 1986). Future work must also integrate these into a wall model. A wrong model is better than no model at all (Goethe in "Analyse und Synthese").

ACKNOWLEDGMENTS

It is a pleasure to acknowledge the following colleagues for their valuable contributions to various aspects of the work: Hasan Alizadeh, Prakash Dey, Daniel Everdeen, Stephanie Kiefer, Marcia Kieliszewski, Xiong-Biao Li, Patrick Muldoon, and David Van Sandt.
This work was supported by DOE Grant # DE-AC02-76ERO-1338, NSF Grant # DCB-8801713 and USDA Competitive Research Grant # 88-37261-3682.

REFERENCES

Averyhart-Fullard VK, Datta, Marcus A (1988) A hydroxyproline-rich protein in the soybean cell wall. Proc Natl Acad Sci USA 85:1082-1085
Baer E, (1986) Advanced polymers. Scientific American 255:179-190
Brightman AO, Barr R, Crane FL, Morre JD (1988) Auxin-stimulated NADH oxidase purified from plasma membrane of soybean. Plant Physiol 86:1264-1269
Cassab GI, Nieto-Sotelo N, Cooper JB, van Holst GJ, Varner JE (1985) A developmentally regulated hydroxyproline-rich glycoprotein from the cell walls of soybean seed coats. Plant Physiol 77:532-535
Cassab GI, Lin J-J, Varner JE (1988) Ethylene effect on extensin and peroxidase distribution in the subapical region of pea epicotyls. Plant Physiol. 88: in press.

Chen J, Varner JE (1985) Isolation and characterization of cDNA clones for carrot extensin and a proline-rich 33 kD protein. Proc Natl Acad Sci USA 82:4399-4403
Condit CM, Meagher RB (1986) A gene encoding a novel glycine-rich structural protein of petunia. Nature 323:178-181
Cooper JB (1984) Hydroxyproline synthesis is required for cell wall regeneration. In: Dugger WM, Bartnicki-Garcia S (eds) Structure, function and biosynthesis of plant cell walls. University of California, Riverside, p 397
Dey PM, Muldoon EP, Lamport DTA (1988) unpublished data
Epstein L, Lamport DTA (1984) An intramolecular linkage involving isodityrosine in extensin. Phytochemistry 23:1241-1246
Esquerre-Tugaye M-T, Lamport DTA (1979) Cell surfaces in plant-microorganism interactions. Plant Physiol 64:314-319
Everdeen DS, Kiefer S, Willard JJ, Muldoon EP, Dey PM, Li X, Lamport DTA (1988) Enzymic cross-linking of monomeric extensin precursors in vitro. Plant Physiol 87:616-621
Franssen HJ, Nap JP, Gloudemans T, Stiekema W, van Dam H, Govers F, Louwerse J, van Kammen A, Bisseling T (1987) Characterization of cDNA for nodulin-75 of soybean: A gene product involved in early stage of root nodule development. Proc Natl Acad Sci USA 84:4495-4499
Fry SC (1982) Isodityrosine, a new cross-linking amino acid from plant cell-wall glycoprotein. Biochem J 204:449-455
Goodenough UW, Heuser JE (1985) The Chlamydomonas cell wall and its constituent glycoproteins analyzed by the quick-freeze, deep-etch technique. J Cell Biology 101:1550-1568
Heckman Jr JW, Terhune BT, Lamport DTA (1988) Characterization of native and modified extensin monomers and oligomers by electronmicroscopy and gel filtration. Plant Physiol 86:848-856
Hong JC, Nagao RT, Key JL (1987) Characterization and sequence analysis of a developmentally regulated putative cell wall protein gene isolated from soybean. J Biological Chemistry 262:8367-8376
Hood EE, Shen QX, Varner JE (1988) A developmentally regulated hydroxyproline-rich glycoprotein in maize pericarp cell walls. Plant Physiol. 87:138-142
Jarosch R (1987) Screw-mechanical models for the action of actin and tubulin-containing filaments.
Kieliszewski M, Lamport DTA (1987) Purification and partial characterization of a hydroxyproline-rich glycoprotein in a Graminaceous monocot, Zea mays. Plant Physiol. 85:823-827
Lagrimini ML, Rothstein S (1987) Tissue specificity of tobacco peroxidase isozymes and their induction by wounding and tobacco mosaic virus infection. Plant Physiol. 84:438-442
Lamport DTA, Northcote DH (1960) The use of tissue cultures for the study of plant cell walls. Biochem J 76:52P
Lamport DTA (1964) Cell suspension cultures of higher plants, isolation and growth energetics. Exp Cell Res 33:195-206
Lamport DTA (1967) Hydroxyproline-O-glycosidic linkage of the plant cell wall glycoprotein extensin. Nature 216:1322-1324
Lamport DTA (1977) Structure, biosynthesis and significance of cell wall glycoproteins. In: Loewus FA, Runeckles VC (eds) Recent advances in Phytochemistry, vol 11. Plenum publishing corporation, pp 79-115

Lamport DTA (1980) Structure and function of plant glycoproteins. In Preiss J (ed) The biochemistry of plants, vol 3. Academic Press, pp 501-504
Lamport DTA (1986) The primary cell wall: A new model. In: Young RA, Rowell RM (eds) Cellulose: Structure, modification, and hydrolysis. Young RA, Rowell RM John Wiley and Sons, Inc, pp 77-90
Li X-B, Kieliszewski M, Lamport DTA (1988) in preparation.
Maddox J (1987) New ways with reptating polymers. Nature 330:11
Mort AJ, Lamport DTA (1977) Anhydrous hydrogen fluoride deglycosylates glycoproteins. Analytical Biochemistry 82:289-309
Pantoja O, Willmer CM (1988) Redox activity and peroxidase activity associated with the plasma membrane of guard-cell protoplasts. Planta 174:44-50
Raven JA (1982) The energetics of freshwater algae: Energy requirements for biosynthesis and volume regulation. New Phytol 92:1-20
Ridge I, Osborne DJ (1970) Hydroxyproline and peroxidase in cell walls of Pisum sativum: regulation by ethylene. J Experimental Botany 21:843-856
Ridge I, Osborne DJ (1971) Role of peroxidase when hydroxyproline-rich protein in plant cell wall is increased by ethylene. Nature New Biology 229:205-208
Roberts KC, Grief GJ Hills, Shaw PJ (1985) Cell wall glycoproteins: Structure and function. J Cell Sci Suppl 2:105-127
Rubinstein M (1987) Discretized model of entangled-polymer dynamics. Physical Review Letters 59:1946-1949
Showalter AM, Varner JE (1988) Biology and molecular biology of plant hydroxyproline-rich glycoproteins. In: Marcus A (ed) The biochemistry of plants: A comprehensive treatise, vol 15. in press
Smith JJ, Muldoon EP, Lamport DTA (1984) Isolation of extensin precursors by direct elution of intact tomato cell suspension cultures. Phytochemistry 23:1233-1239
Smith JJ, Muldoon EP, Willard JJ, Lamport DTA (1986) Tomato extensin precursors P1 and P2 are highly periodic structures. Phytochemistry 25:1021-1030
Tierney ML, Wiechert J, Pluymers D (1988) Analysis of the expression of extensin and p33-related cell wall proteins in carrot and soybean. Mol Gen Genet 211:393-399
Van Holst GJ, Varner JE (1984) Reinforced polyproline II conformation in a hydroxyproline-rich cell wall glycoprotein from carrot root. Plant Physiol 74:247-251

CELL WALL-DERIVED ELICITORS - ARE THEY NATURAL (ENDOGENOUS) REGULATORS?

Alan Campbell and John Labavitch
Department of Pòmology
University of California
Davis, California 95616
U.S.A.

INTRODUCTION

Development of the idea that carbohydrates digested from cell walls could carry the information necessary to "activate" host defense responses to pathogens began in the 1970s. Early work (Ayers *et al.*, 1976) demonstrated that a glucan derived from cell walls of the soybean pathogen *Phytophthora megasperma* f. sp. *glycinea* could elicit the synthesis of phytoalexins by soybean tissues. Subsequent studies have identified a heptasaccharide subfragment of the hyphal wall glucan, presumably generated by the action of soybean-produced glucan hydrolases, as the most active component (Sharp *et al.*, 1984). Studies in recent years have identified similar responses in several cell culture and tissue systems. The production of phytoalexins and other presumed "defense" compounds in response to elicitors derived from host, as well as pathogen, cell walls has been described (Darvill and Albersheim, 1984; Lamb *et al.*, 1986). Most recently these studies have utilized the techniques of molecular genetics to follow responses at the level of gene transcription (e.g., Bell *et al.*, 1986). Genome activation with a lag time as short as 5 to 10 minutes has been shown (Lawton and Lamb, 1987). Questions about the means by which cells perceive the presence of elicitors (Schmidt and Ebel, 1987) and then mobilize their responses remain. However, it is clear that the biochemical confrontation at the infection site often includes many enzymes that can generate elicitor-active wall fragments.

Cell wall-derived elicitors may also play a role in regulating plant developmental events in the absence of pathogens (and their wall-digesting enzymes). The results of York *et al.* (1984) suggest a role for an oligosaccharide digested from cell wall xyloglucan in regulating cell growth. Xyloglucan turnover, perhaps brought about by the action of an endogenous endo-glucanase (Hayashi and Maclechlan, 1984), seems to be an important, early event in the stimulation of growth by auxin (Labavitch

NATO ASI Series, Vol. H 35
Cell Separation in Plants
Edited by D. J. Osborne and M. B. Jackson

and Ray, 1974). Cell wall digestion is an integral part of developmental events that involve cell separation and it is therefore reasonable to assume that wall breakdown products could play an important role in the regulation of these processes. Wall breakdown in abscission (Rasmussen and Bukovac, 1969), duct formation (Morrison and Polito, 1985), and dehiscence (Weis *et al.*, 1988) is limited to a few cell layers and, hence, its characterization is potentially obscured by the large amount of "non-reacting" cell wall. However, wall metabolism in ripening fruits is pronounced and eventually involves the entire organ. In addition, the pectins, which are the source of all of the host-derived elicitors of defense responses described thus far (Darvill and Abersheim, 1984), are the wall components most clearly altered during ripening. As a consequence a number of laboratories have recently begun studies to determine whether endogenous elicitors serve as regulators of the ripening syndrome.

Our laboratory's interest in this area evolved from an examination of the development of xylem vessel occlusions which followed infections by vascular wilt pathogens. We (VanderMolen *et al.*, 1983) had shown that vessels of castor bean leaves became plugged in response to infection, introduction of pathogen-produced wall degrading enzymes, or exposure to ethylene. Introduction of enzymes to the leaves stimulated a transient increase in ethylene production; plugging of vessels in response to enzymes did not occur if ethylene synthesis had been inhibited. Because ethylene synthesis and wall metabolism are closely associated during fruit ripening and invasion of fruits by rot pathogens, we wondered if introduction of wall-digesting enzymes to mature, non-ripening fruits would stimulate their ethylene synthesis. Injections of mixtures of wall hydrolases from fungal cultures into horticulturally-mature tomatoes, kiwifruits, pears and oranges caused an increase in ethylene production that generally accelerated the onset of ripening (Labavitch, 1983). Unfortunately the injection technique we used often triggered wound ethylene production by the fruits. As a result we turned to the use of suspension-cultured pear cells as a test system. Tong *et al.* (1986) showed that both wall degrading enzymes and materials digested from pear cell walls by those enzymes could elicit ethylene production from the cultures. This observation has led us to the current work in which we describe the effects of separate elicitor-containing preparations (produced *in vitro* or *in vivo*) on ethylene production and other aspects of ripening-related cell metabolism.

SOURCES OF ELICITORS

The pectic oligomers used in this study were homoligomers of α-1.4-D-galacturonic acid prepared by Mike Saxton of the UC Davis Plant Growth Laboratory. They were derived from citrus pectin by acid hydrolysis and separated into mixtures of smaller (GUA-S) and larger (GUA-L) oligomers by gel filtration chromatography. Anion-exchange chromatography on QAE-Sephadex, with an imidazole gradient, revealed that the two mixtures contain some oligomers in common but are clearly different in average oligomer size (Fig. 1).

The endogenous elicitors used were in crude mixtures derived from ripening pears and tomatoes. Fruit at different stages of ripeness were juiced into hot ethanol. Elicitor mixtures were isolated by solubility in 60% ethanol, insolubility in 80% ethanol with 0.2 M NaCl, and by passage through a 10 kD ultrafiltration membrane. (These procedures would collect more than 90% of the GUA-S and -L materials if they were present in extracts. While the purified extracts contain primarily galacturonic acid, some protein and neutral sugar was also collected. The distribution of size classes represented by the extracts has not been tested. As a consequence these preparations cannot yet be described as "*pectin* oligomers".)

Pectic oligomers elicited responses from both pear cells in suspension and tomato cells in tissue disks. Extracts from pear elicited ethylene from pear cells in suspension, but were not tested on tomato disks. Extracts from tomato elicited responses from tomato cells in tissue disks, but not from pear cells in suspension.

PEAR CELL SUSPENSION STUDIES

<u>Materials and Methods</u>. Pear cells from 'Passe Crassane' (*Pyrus communis* L.) fruit were maintained by weekly subculturing in suspension culture with modified Murashige and Skoog nutrients and 1 mg/ml 2.4-D (Puschmann and Romani, 1983). Cells for experiments were subcultured to the same medium without 2,4-D, grown for 7 to 9 days, then washed with fresh medium

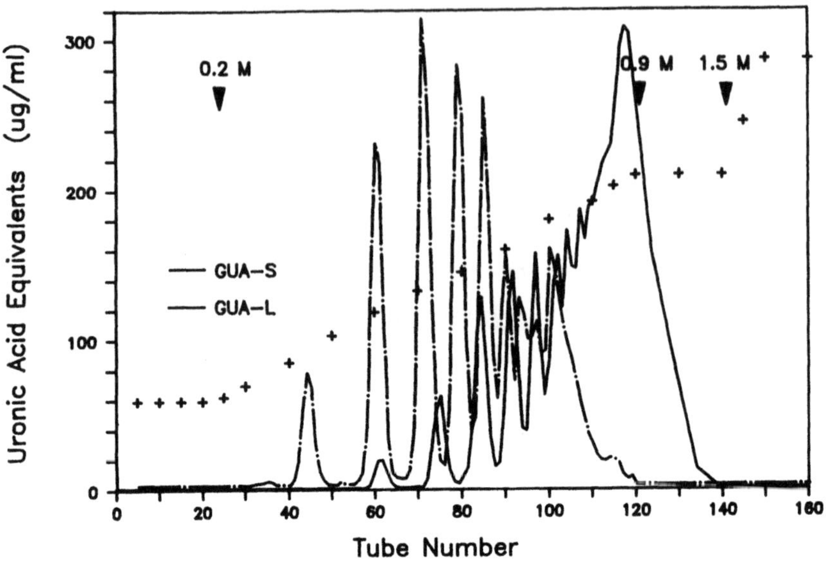

FIGURE. 1. QAE-Sephadex ion-exchange chromatography of GUA-L and GUA-S in an imidazole-HCl (pH 7.0) gradient. The fractions were assayed for uronic acids (Blumenkrantz and Asboe-Hansen, 1973) and for conductivity (+) in order to assess the shape of the gradient.

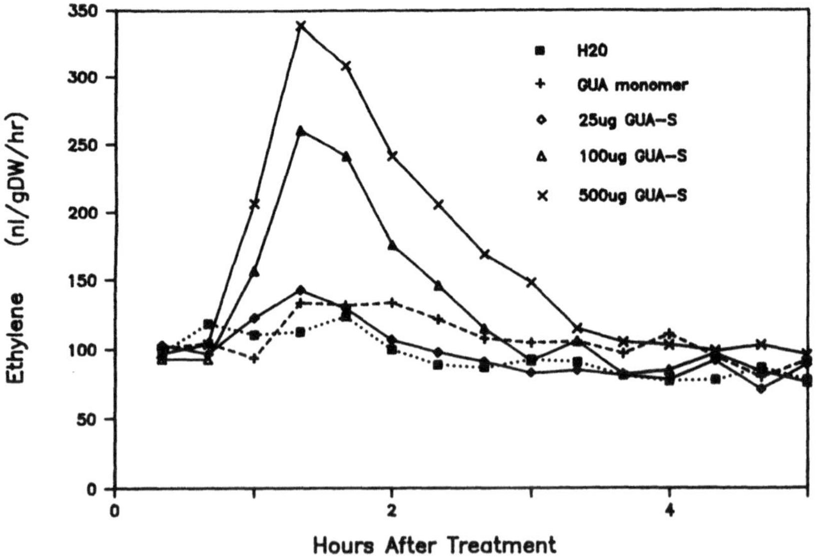

FIGURE 2. Ethylene production by pear cell cultures in response to smaller pectic oligomers. Oligomers were added at time zero and ethylene production was measured at 20 minute intervals.

and diluted to a concentration of 5 mg dry weight/ml. 5 ml aliquots were transferred to 25 ml flasks and treatments added in 0.5 ml doses. For measurement of ethylene synthesis, the flasks were capped for 15 minute intervals and gas samples were analyzed by gas chromatography.

Results. Pectic oligomers GUA-S and GUA-L induce a rapid, but transient, increase in ethylene biosynthesis in pear suspension cultures. This reaches a maximum between 1 and 2 hours after addition (Fig. 2). The maximum rate is dose-sensitive; it increases with oligomer concentration up to 500 μg per flask. The timing of the transient response is not dose-sensitive: the pattern of increase, maximum, and decline is similar for all concentrations. The response to GUA-L parallels the response to GUA-S, but with a reduced maximum. Galacturonic acid monomer (GUA-1) elicits no change in ethylene production. The ethylene response is not sensitive to Ca^{++} concentration in the medium; additions of Ca^{++} up to fifty times the concentration normally found in the medium do not alter the ethylene response.

The increase in ethylene biosynthesis is accompanied by an insensitivity to further changes in oligomer concentration; the degree of induced insensitivity increases with initial oligomer concentration (Fig. 3).

The ability of pear cells to respond to added pectic oligomers changes over the growth cycle in suspension culture. Pear cells normally produce ethylene during their growth phase after subculturing. Pectic oligomers induce the largest significant increase in ethylene synthesis over this changing basal rate in the 1 or 2 day interval at the end of the growth phase, as the basal rate of ethylene synthesis declines.

TOMATO DISK STUDIES

Materials and Methods. Large, unblemished fruit from field-grown 'Castlemart' (*Lycopersicon esculentum* L.) plants were washed, equilibrated at 21°C, and monitored for rate of ethylene synthesis to estimate state of maturity.

Disk Preparation. Whole fruit were surface sterilized in 1% sodium hypochlorite and rinsed with sterile water, and exact state of maturity determined (Kader *et al.*, 1977). One centimeter cores were cut from the pericarp surface, usually from a medial band, and sliced by hand into disks approximately 3 mm thick and 250 mg in weight, with epidermis

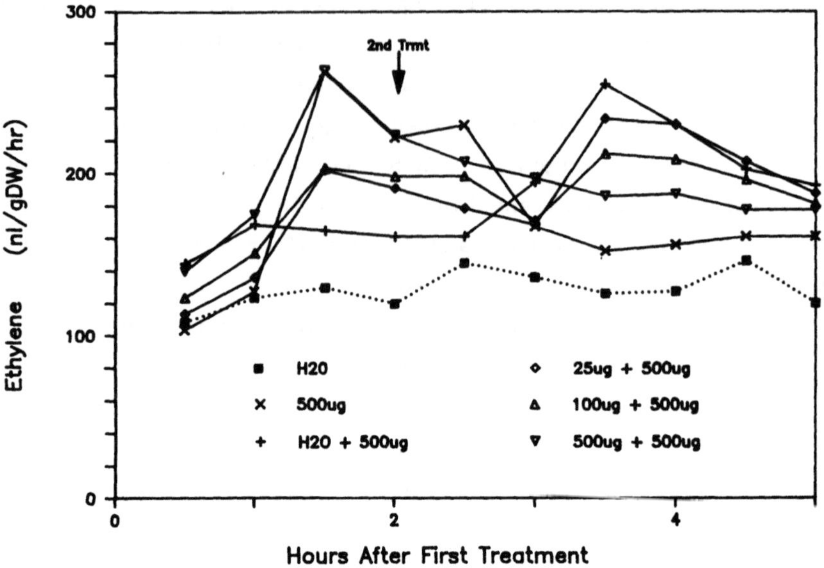

FIGURE 3. Effect of GUA-S treatment of pear cell cultures at time zero on pear cell ethylene production in response to a second oligomer treatment. Treatments with H_2O (+) and 25 (◇), 100 (Δ), and 500 µg (∇) GUA-S at time zero were followed by an optimal (500 µg) dose of GUA-S at 2 hours.

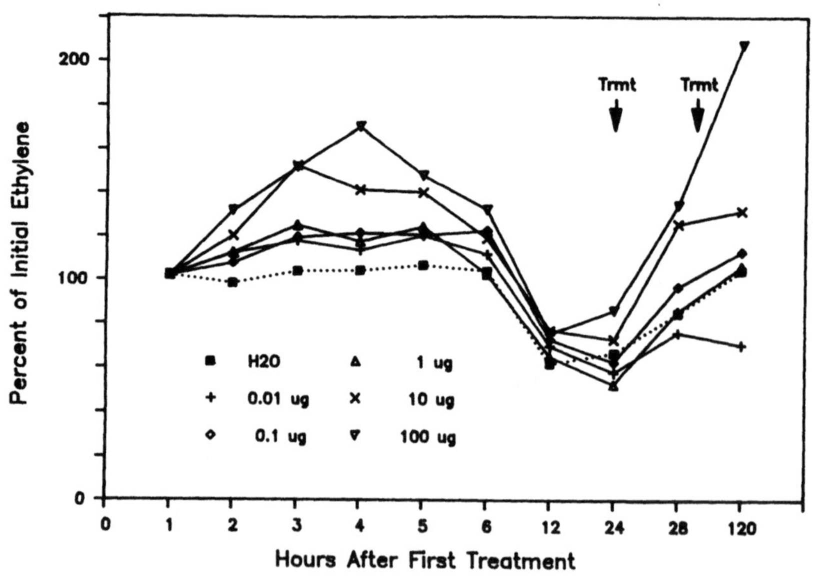

FIGURE 4. Ethylene biosynthesis by tomato pericarp disks after treatment with larger pectic oligomers (GUA-L). Treatments were dissolved in 10 µl of water, and applied three times at 24 hour intervals to disks from mature-green three stage fruit. Initial rates of ethylene synthesis were measured 2 hours before first treatment. Note that the time axis is not linear.

included. The disks were rinsed in sterile water, drained, and blotted dry. For most experiments, disks were then transferred, epidermis down, into individual wells of sterile 24-well, clear plastic, microtitre plates, in which they remained through all subsequent storage, treatments, and ethylene measurements. In a few experiments, disks were placed in groups in 5 cm petri dishes. The plates or dishes of disks were held at 21°C under a slow flow of ethylene-free, water-saturated air.

Treatment. Treatments were applied to the disk surface as sterile, unbuffered, aqueous solutions in 10 μl aliquots, beginning 48 hours after disk excision. Treatment compounds included GUA-L, GUA-S, GUA-1, and the extract from turning stage tomato fruit. Treatments of no solution (air), water, and 5 mM 1-aminocyclopropane-1-carboxylic acid (ACC) were also applied as controls.

Ethylene Measurement. The ethylene production of individual disks was measured periodically. The separate wells of a plate were sealed with a sterile rubber gasket. After a brief accumulation period, a 1-ml sample of gases was drawn from each well and analyzed by gas chromatography.

Colour Measurement. The colour of disk tissue was measured for lightness (L) and red to green tint (+a to -a) using a reflectance colourimeter. Intermediate estimates of browning were made by eye using a 6 point subjective scale from green to dark brown (1-5) or wet (6).

Results. Pectic oligomers induced both a transient rise in ethylene biosynthesis and long-term changes in biosynthesis of phenolic compounds, ethylene, and lycopene in tomato pericarp disks.

Wound Ethylene. Ethylene synthesis increases sharply after disk excision, rising several hundred-fold to a maximum 2 hours after the first cuts (from below 0.1 nl/gFW/hr to over 10 nl/gFW/hr in a typical fruit). Ethylene synthesis declines slowly to a new plateau by 48 hours, when initial treatments were applied.

Ethylene Transient. Pectic oligomers GUA-L and GUA-S, as well as GUA-1, induce a rapid, transient rise in ethylene biosynthesis in tomato disks, reaching a maximum about 4 hours after addition (Fig. 4). The pattern of transient increase is similar to that in pear suspension cultures, but the time to maximum is doubled. The maximum rate is dose-sensitive; it increases with concentration up to 100 μg per disk. The tomato disks are sensitive to smaller amounts of pectic oligomers than the suspension culture system. The timing of the transient is not dose-sensitive; the patterns of increase, maximum, and decline are similar for all concentrations, as in the suspension culture system. Insensitivity to further changes in oligomer concentration during the transient has not been tested in the tissue system but is suggested by the dose-insensitive pattern. Disks do respond to a second or third treatment with oligomer applied at 24 or 48 hours. Treatment of disks with 5 mM ACC produces a persistent 6- to 8-fold increase in ethylene synthesis.

Induction of Delayed Ethylene Biosynthesis. GUA-L and GUA-S, and to a lesser extent GUA-1, induce a delayed increase in ethylene synthesis by tomato disks (Fig. 4). This secondary rise begins at least two days after the initial treatment. Treatment with ACC does not induce this secondary rise.

Induction of Phenolic Biosynthesis. GUA-L and GUA-S induce a significant accumulation of dark pigments in tomato disks over a period of days, probably through induction of phenolic compound biosynthesis. The accumulation of pigments is progressive (Fig. 5). GUA-1 and ACC both induce somewhat lower levels of browning.

Induction of Lycopene Biosynthesis. Pectic oligomers GUA-L and GUA-S can induce a significant increase in the accumulation of red pigments in tomato disks, probably through induction of lycopene biosynthesis (Fig. 6). This accumulation is dose-sensitive, and is also dependent on the conditions under which the disks are maintained. Treatment with ACC does not induce this increase in lycopene production.

REPONSES TO FRUIT EXTRACTS

Elicitors prepared from turning stage tomatoes stimulate both ethylene synthesis and phenolic pigment accumulation of disks (measured at 6 days). Effects on short term (transient) ethylene synthesis have not been tested. Concentrations of extracts which elicit responses in disks closely approximate concentrations in intact fruit (fresh weight basis).

Extracts prepared from firm pears which had just begun to soften stimulated ethylene production of pear cells. In this case, no measurement of "physiologically-relevant" concentration can be made.

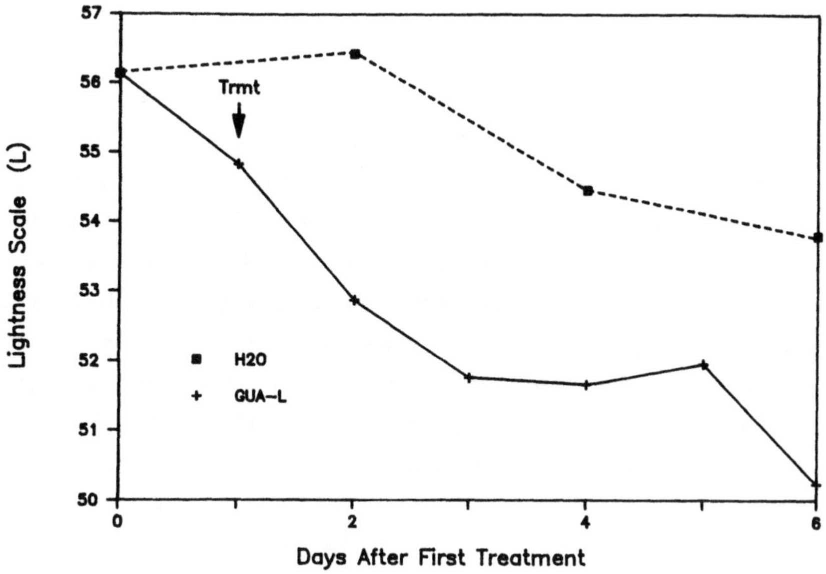

FIGURE 5. *Change in lightness (L) of tomato pericarp disks after treatment with larger pectic oligomers. Treatments of 100 µg were applied 24 hours apart to disks from mature-green two stage fruit. Color of disk tissue was measured on a L-a-b scale with a reflectance colorimeter. Change in lightness appears due to accumulation of phenolic materials.*

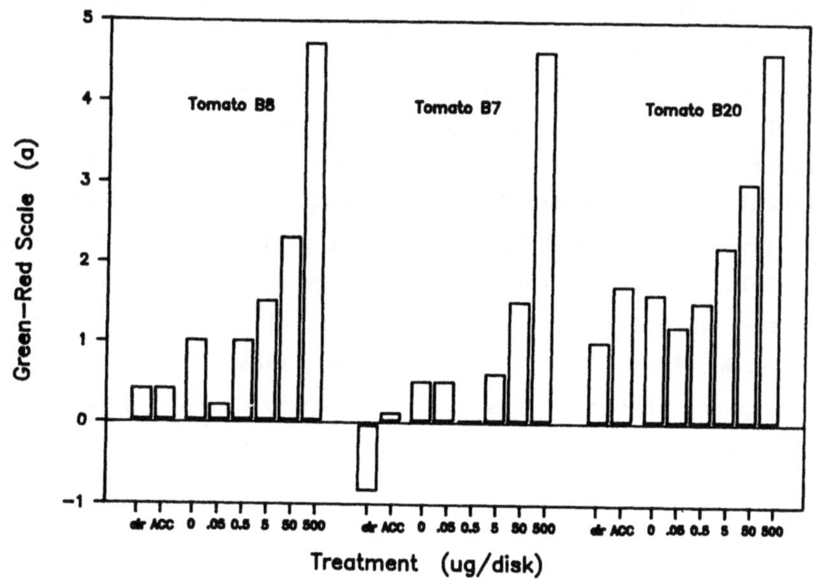

FIGURE 6. *Redness of tomato pericarp disks 6 days after first treatment with smaller pectic oligomers. Treatments were applied three times at 24 hour intervals to disks from mature-green second stage fruit. Color of disk tissue was measured on an L-a-b scale with a reflectance colorimeter.*

DISCUSSION

Our work with the more chemically-defined pectin oligomers (GUA-S and -L) shows clearly that fruit cells and tissues can respond to these elicitors in ways that are characteristic of ripening fruits. Pericarp disks are consistently more "sensitive" and slow-to-respond than the cultured cells. Whether the differences in the minimum dose required for stimulation of ethylene synthesis truly indicates that the more organized disk system is more responsive is unclear. It is possible that the cells became less "sensitive" because they are constantly bathed in a pear wall pectin-containing culture medium. This point is currently under investigation. It is interesting that the later responses of disks (ethylene synthesis and accumulations of pigments and phenolics) are not strictly linked to their early ethylene production. Disks treated with ACC produced several times as much ethylene as elicitor-treated disks and yet their later responses are much delayed and reduced. This suggests that the oligomer signal is perceived in several ways.

Work with the galacturonide oligomer-containing extracts of ripening pears and tomatoes supports the idea that naturally-produced (presumed) wall fragments could be regulators of aspects of fruit ripening ("stage 2" ethylene synthesis, pigment production). This idea is also supported by work from other laboratories. Brecht and Huber (1988) have reported that pectin-containing materials released during autolysis of tomato cell walls will stimulate ripening when infiltrated into green fruit. Tong and Gross (unpublished) report that pectin-containing fractions extracted from walls of tomato fruits in the MG2 to breaker stages will also promote ripening. It is possible that all of these elicitors are produced (at least in part) by the action of polygalacturonase, which itself stimulates tomato fruit ethylene production (Baldwin and Pressey, 1988). One of the most important aspects of our work is the fact that the tomato disks respond to concentrations of the "endogenous" elicitors that are similar to those found in intact fruits.

The most immediate goal of our work is to establish whether the elicitor activity of pectin oligomers and fruit extracts can be tied to specific components of the preparations. Such a finding would make simpler the definition of a biologically-active dose. This in turn would smooth the analyses of successively-riper fruits in an effort to learn when (if at all) potentially-active oligomers accumulate during ripening.

REFERENCES

Ayers A, Valent B, Ebel J, Albersheim P (1976) Host-pathogeninteractions XI. Composition and strcuture of wall-released elicitor fractions. Plant Physiol 57:766-774

Baldwin E, Pressey R (1988) Tomato polygalacturonase elicits ethylene production in tomato fruit. J Am Soc Hort Sci 113:92-95

Bell J, Ryder T, Wingate V, Bailey J, Lamb C (1986) Differential accumulation of plant defense gene transcripts in a compatible and incompatible plant-pathogen interaction. Mol Cell Biol 6:1615-1623

Blumenkrantz N, Asboe-Hansen G (1973) A new method for quantitative determination of uronic acids. Anal Biochem 54:484-489

Brecht J, Huber D (1988) Products released from enzymically active cell wall stimulate ethylene production and ripening in preclimacteric tomato (Lycopersicon esculentum Mill.) fruit. Plant Physiol 88:1037-1041

Darvill, A, Albersheim P (1984) Phytoalexins and their elicitors - A defense against microbial infection in plants. Ann Rev Plant Physiol 35:243-275

Hayashi T, Wong Y, Maclachlan G (1984) Pear xyloglucan and cellulose. II. Hydrolysis by pear endo-1,4-β-glucanases. Plant Physiol 75:605-610

Kader A, Stevens MA, Albright-Holton M (1977) Effect of fruit ripeness when picked on flavor and composition in fresh market tomatoes. J Am SocHort Sci 102:724-731

Labavitch J (1983) Stimulation of fruit ethylene production. Plant Physiol 72 (Suppl):158

Labavitch J, Ray P (1974) Turnover of polysaccharies in elongating pea stem segments. Plant Physiol 53:669-673

Lamb C, Corbin D, Lawton M, Sauer N, Wingate V (1986) Recognition and response in plant:pathogen interactions. In: Lugtenberg B (ed) Recognition in Microbe-Plant Symbiotic and Pathogenic Interactions, Springer-Verlag, Berlin Heidelberg pp 333-344

Lawton M, Lamb C (1987) Transcriptional activation of plant defense genes by fungal elicitor, wounding and infection. Mol Cell Biol 7:335-341.

Morrison J, Polito V (1985) Gum duct formation in almond fruit, Prunus dulcis (Mill.) D.W. Webb. Bot Gaz 146:15-25

Puschmann R, Romani R (1983) Ethylene production by auxin-deprived, suspension-cultured pear fruit cells in response to auxins, stress or precursor. Plant Physiol 73:1013-1019

Rasmussen H, Buckovac M (1969) A histochemical study of abscission layer formation in the bean. Am J Bot 56:69-82

Schmidt W, Ebel J (1987) Specific binding of a fungal glucan phytoalexin elicitor to membrane fractions from soybean Glycine max. Proc Natl Acad Sci USA 84:4117-4121

Sharp J, McNeil M, Albersheim P (1984) The primary structure of one elicitor-active and seven elicitor-inactive hexa (β-D-glucopyranosyl)-D-glucitols isolated from the mycelial wall of Phytophthora megasperma f. sp. glycinea. J Biol Chem 259:11321-11326

Tong C, Labavitch J, Yang S (1986) The induction of ethylene production from pear cell culture by cell wall fragments. Plant Physiol 81:929-930

VanderMolen G, Labavitch J, Strand L, DeVay J (1983) Pathogen-induced vascular gels. Ethylene as a host intermediate. Physiol Plant 59:573-580

Weis K, Polito V, Labavitch J (1988) Microfluorometry of pectic materials in the dehiscense zone of almond (Prunus dulcis [Mill.] DA Webb) fruits. J Histochem Cytochem 36:1037-1041
York W, Darvill A, Albersheim P (1984) Inhibition of 2,4-dichlorophenoxyacetic acid-stimulated elongation of pea stem segments by a xyloglucan oligosaccharide. Plant Physiol 75:295-297

POLYGALACTURONASE ACTIVITY IN THE RIPENING TOMATO FRUIT

Johan Bruinsma, Erik Knegt and Evert Vermeer
Department of Plant Physiology
Agricultural University
Arboretumlaan 4
NL-6703 BD Wageningen
The Netherlands

INTRODUCTION

A ripening tomato fruit softens because of cell separation in the pericarp tissue. This cell separation is brought about by solubilization of the middle lamellae. The middle lamellae consist largely of pectin that is degraded by the combined actions of pectinesterase (E.C.3.1.1.11) and polygalacturonase (E.C.3.2.1.15). The former enzyme removes the methyl groups from the polyuronide chain, whereupon the latter enzyme depolymerizes the polygalacturonic acid by cutting it into oligomeric parts. It thus acts as an endo-enzyme.

Pectinesterase is present already in the unripe fruit and does not show spectacular changes in activity during ripening. Also an exo-polygalacturonase, that cleaves monomer units from the non-reducing end of polygalacturonic acid, occurs at a very low level throughout the development of the fruit (Pressey, 1987). Also such glycosidases as α-D-mannosidase (E.C.3.2.1.24), and α-D-(E.C.3.2.22) and β-D-galactosidase (E.C.3.2.1.23) are present in the unripe fruit already. Their activities may increase during ripening, but seem not to be related with the softening of the fruit. For instance, mannosidase may play a role in the turnover of glycoproteins and glycolipids of membranes (Watkins *et al.*, 1988).

On the contrary, endo-polygalacturonase (PG) activity cannot be found prior to ripening. It is induced when the autocatalytic ethylene evolution surpasses a certain threshold value (Bruinsma, 1983, Su *et al* ., 1984). This is why the enzyme is so appropriate for molecular-biological studies on gene expression. However, the enzymology of the PG is also of interest in order to understand the molecular basis of fruit softening and, thereby , of this type of cell separation.

NATO ASI Series, Vol. H 35
Cell Separation in Plants
Edited by D. J. Osborne and M. B. Jackson

It is the aim of the present paper to discuss the problems connected with the study of this enzyme *in vitro*: its isolation, concentration, and determination, particularly in connection with its existence in two forms, PG1 and PG2, and the interconversions between the two iso-enzymes. We will also evaluate the relevance of these *in vitro* biochemical studies for understanding the mechanism of cell separation *in vivo* .

PROPERTIES OF ENDO-POLYGALACTURONASE IN VITRO

Determination of enzyme activity. The first problem in the study of PG *in vitro* is to determine its activity. The natural substrate is the tomato pericarp cell wall, particularly the middle lamella part of it. The precipitate from a tissue homogenate in acetone gives a cell wall preparation that shows autolysis because of endogenous pectolytic activity. This activity can be removed by stirring the acetone-insoluble powder in a mixture of phenol:acetic acid:water = 2:1:1 (w/v/v) (Seymour *et al*., 1987). However, the resulting substrate, although enzymatically inactive, consists of more or less fine particles of middle-lamella-containing cell wall fragments, the pectin fraction of which is, moreover, highly methylated. Because the PG-enzyme is most active on pectins with a low degree of esterification, the substrate has to be pre-treated with pectinesterase. Alternatively, the natural substrate can be abandoned altogether by replacing it by commercially available polygalacturonic acid. We used pectic acid with an average molecular mass of 10 KDa from K&K laboratories (ICN Inc., New York, U.S.A.) The product sold by Sigma may interfere with the determination method.

The pectic acid was stirred in water under slow addition of a 0.1 M NaOH solution to maintain the pH at 5.0 for ready dissolution. After clearing by centrifugation and filtration through a 450 nm membrane filter, the pH was brought to 3.6 or 4.4 with 0.1 M HCl. The stock solution, kept at 2 °C, contained 0.1% polygalacturonic acid; in the reaction medium the final concentration was 0.05% (Knegt *et al*., 1988).

In principle, two parameters are available for the determination of PG-activity: a decrease in viscosity of a solution of polygalacturonic acid caused by the reduction of the chain length of the polymer; and accumulation of the reducing groups that appear upon hydrolysis of the 1-4-α-bond between the galacturonic acid units. Viscometry seems preferable to measure endo-enzyme activity. At low activities, however, enzyme dilution led to a more than proportional decrease in the viscometrically determined activity. This deviation could only partially be corrected by the addition of salt (0.1 M NaCl). It was therefore decided to abandon this time-consuming and laborious method and instead to determine reducing groups.

Methods for determining reducing saccharides have usually been developed for clinical purposes. Their conversion to a microscale often generates problems similar to those mentioned above for the viscometric determination. This concerns both the commonly used Nelson-Somogyi copper reduction and the dinitrosalicylic acid complex formation, which in our hands failed to give reproducible results at low uronide concentrations. In contrast the method using 2-cyanoacetamide, introduced by Gross (1982), turned out to be simple, reliable, and sensitive. The complex formed has an absorption peak at 276 nm.

Sodium azide and merthiolate, added to media to prevent microbial contamination that might exert spurious enzymatic activities, were omitted in our experiments because they were found to interfere with the PG-activity.

Enzyme activity was compared with the absorbancy of 100 nmol galacturonic acid added to the same medium, and expressed as the number of moles reducing groups per sec per litre (kat l^{-1}).

<u>Extraction and concentration</u>. The extraction of PG-activity from homogenized pericarp tissue depends on salt concentration and pH. When a homogenate in water is brought to pH 3 with 0.1 M HCl and centrifuged, the supernatant shows only 2% of the maximum obtainable activity (Pressey, 1986). At lower, but particularly at higher pH values, more PG-activity can be extracted. At neutral pH, little PG-activity only is obtained in water. Moreover, concentration of the extract may greatly reduce the activity and also change the ratio of the two iso-enzymes, PG I and PG II. This ratio can be determined after their separation by gel filtration on a Sephadex G-100 column eluted with 100 mM sodium acetate + 200 mM sodium chloride at pH 4.4.

Extraction upon homogenization of green tomato pericarp tissue in an equal amount of water and centrifugation gave no PG-activity in the supernatant. In the course of ripening activity could be detected, but only of PG

TABLE 1. Contents of PG I and PG II, in nkat g^{-1} fresh weight, of tomato pericarp, subsequently extracted with increaing NaCl concentrations. From Knegt et al. 1988.

Stage of ripeness	Extraction with H_2O		0.5 M NaCl		1.25 M NaCl	
	PG I	PG II	PG I	PG II	PG I	PG II
Green	-	-	0.0	0.1	0.0	0.0
Breaker	-	-	0.2	0.2	0.2	0.0
Turning	0.0	0.1	0.0	1.5	2.0	0.2
Pink	0.0	0.3	0.7	4.2	2.0	0.3
Red	0.0	0.7	0.3	11.7	3.5	1.2
Dark red	0.0	0.7	0.3	12.1	3.8	1.8

II (Table 1). When the pellets were resuspended in a mild salt solution of 0.5 M NaCl, the suspension stirred, and centrifuged again, then the supernatant contained considerably more activity, again particularly of PG II. In the course of ripening PG II can be detected before PG I at the mature green state, just as the fruit is about to change colour. When the pellets from the 0.5 M NaCl solutions are resuspended in a stronger salt solution, of 1.25 M NaCl, a little more PG II activity is extracted, but now the apparently stronger bound PG I is set free. Further extractions are no longer effective. If no fractionation is required, extraction with 1.25 M salt solution suffices. This extraction was often performed on the water-insoluble pellet to remove large amounts of water-soluble, low-molecular substances, as well as many proteins.

The development of the activities of PG I and II during ripening is presented in Fig. 1; they amount to 17.4 and 18.7 nkat g^{-1} at the red and dark-red stages, respectively. These are much higher values than encountered

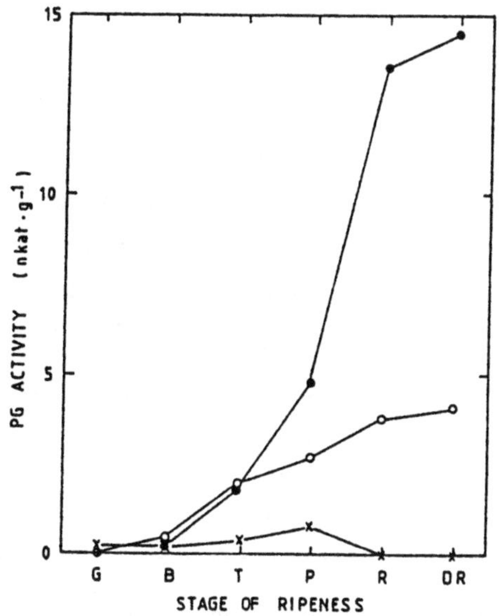

FIGURE 1. Activities of PG I (open dots), PG II (closed dots) and free convertor (x) in tomato fruit pericarp during ripening. For abbreviated stages of ripeness, see Table 1.

in other studies. When converted to the same system of units, Pressey (1986) found a maximum activity of 12.3 nkat g^{-1}, Tucker *et al.* (1980) arrived at values of 1.9 and 4.1 nkat g^{-1} for the cultivars 'Potentate' and 'Ailsa Craig', respectively. It may be that varietal differences account for these differences in enzyme activity. Alternatively, however, the differences may at least partly be due to differences in concentration methods.

Both Pressey (1986) and ourselves (Knegt *et al.*, 1988) have found that precipitation at 80% saturation with ammonium sulphate gives considerable losses in enzyme activity. Up to one third of PG I activity was lost and up to half of the activity of PG II had also vanished. Lyophilization gave much better results, but the large amount of salt renders it highly impractical. Good results were also obtained, by Pressey and by ourselves with ultra flow dialysis through 10 KDa cutoff membranes (Amicon Corporation, Danvers, MA, USA). Therefore this is the preferred method.

The two iso-enzymes and the convertor. The first clear demonstration of the occurrence of two iso-enzymes of PG in ripening tomato fruits was presented by Pressey and Avants (1973). They named the iso-enzymes I and II, determined their molecular weights by gel filtration to be 84 and 44 KDa, respec-

TABLE 2. Properties of PG isoenzymes. n.d. = not determined. After Knegt et al. 1988.

PG enzyme	pH optimum	[NaCl] optimum (mM)	Temp. half-inact-ivated (°C)	Temp. totally inact-ivated (°C)	MW gel filtr-ation (kDa)	IEP chromato-focusing
PG2	4.4	200	57	63	41	8.0
PG1	3.6	300	79	90	106	7.0
PGx	4.4	n.d.	83	91	74	7.2

tively, and evaluated their stabilities at different temperatures and pH-values. Most of their data could be confirmed by us except that the optimum pH-values were found to be distinctly different, *viz.* 3.6 and 4.4 instead of 4.3 and 4.8 for PG I and PG II, respectively. This may be explained by differences in concentrations of substrate, salt etc, as discussed above. Some properties are recorded in Table 2, which shows that the smaller PG II is more heat-labile than PG I; after 5 minutes heating at 65 °C only the activity of the larger PG I remains. Measurement of PG-activity before and after this temperature treatment offers the possibility of determining the amounts of both isoenzymes without actually separating them on a column.

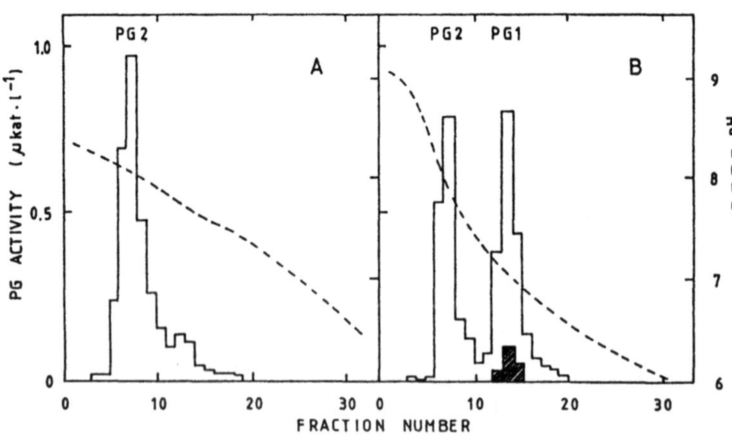

FIGURE 2. Re-chromatography of A: PG II fractions and, of B: PG I fractions on a chromatofocusing column. The hatched area is convertor activity.

However, although PG I is less sensitive to higher temperatures than PG II, it shows another lability not found with PG II: *in vitro* it gradually decomposes into PG II. This can be seen after chromatofocusing of chromatographically purified PG I, when part of the PG-activity reappears in the PG II fractions (Fig. 2). Destruction of all PG-activity in the PG I fractions by heating for 7 to 10 minutes at 100 °C produces a medium in which PG II is altered to another PG-isoenzyme. The substance, of about the same molecular size as PG I, that is able to convert PG II was discovered by Tucker *et al.* (1981) and called the PG-convertor by Pressey (1984), who tried further to purify and characterize the substance. It turned out to be a glycoprotein, labile at alkaline conditions and located in the cell walls and in other parts of the tomato plant. Its amount can be determined upon the supply of an overdose of PG II, removal of the excess PG II at 65 °C, and measurement of the remaining PG activity.

It then turns out, that the convertor is not completely stable at 100 °C. It rapidly accumulates during the first minutes when PG I is inactivated, but then also gradually disappears. Extrapolation to zero time shows that per unit PG I-activity one unit convertor appears and that after about 40 minutes half of this amount has vanished (Knegt *et al.*, 1988).

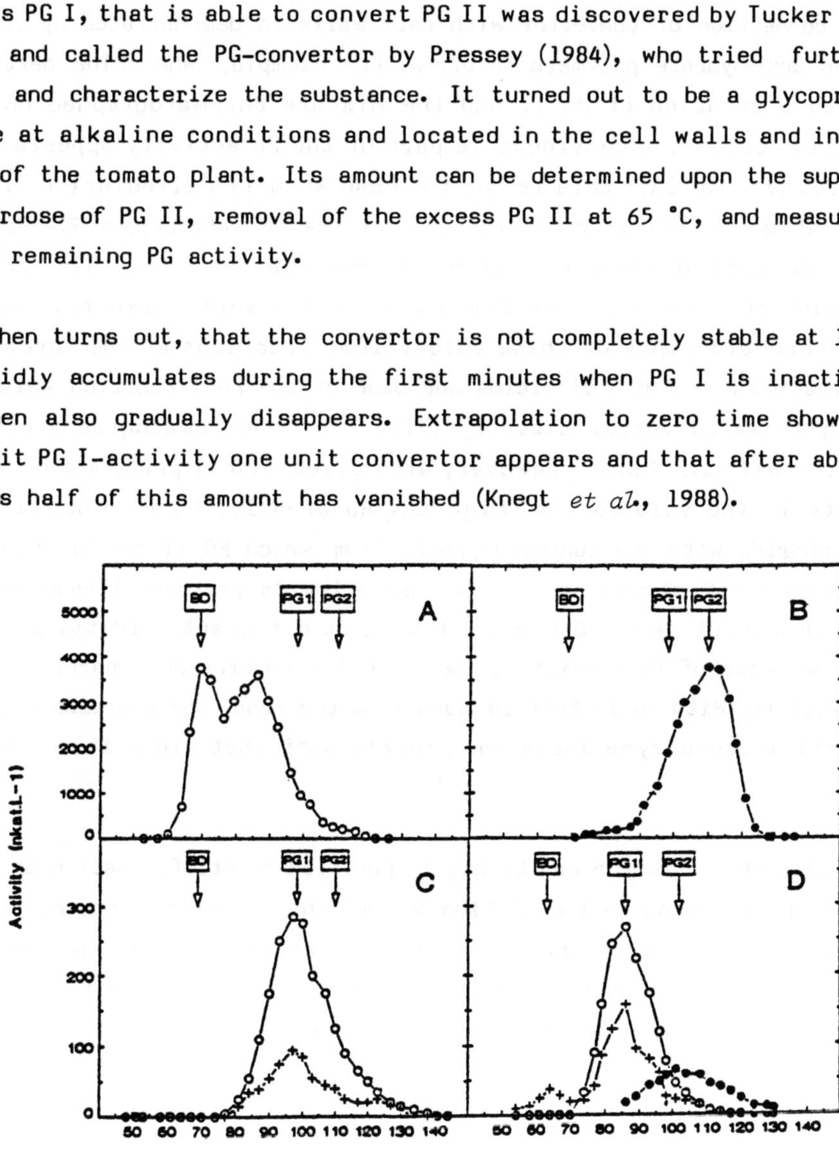

FIGURE 3. Chromatography and re-chromatography of mixtures of PG I and PG II with blue dextran. Explanation in text.

The earlier mentioned authors believed that upon supply of PG II to the convertor, PG I is formed again. However, although the newly formed isoenzyme has the temperature stability of PG I (Table 2), the curve of its activity at different pH-values differs distinctly from that of PG I and matches closely that of PG II. We therefore distinguish this in vitro formed isoenzyme as PGx from PG I (Knegt *et al.*, 1988).

The connection of convertor with cell walls is demonstrated by its binding to carbohydrate polymers *in vitro*. For example, when blue dextran is added to a solution of PG I, and the mixture chromatographed on a gel filtration column, a considerable part of the PG activity appears in the void volume (Fig. 3A); this is not the case when a PG II-solution is used instead (Fig. 3B). At re-chromatography of the PG I-fractions 91-98, these fractions elute at the same elution volume. However, when they are again mixed with blue dextran, they then appear in the void-volume fractions 65-71. Re-chromatography of these void-volume fractions at the normal salt concentration of 0.2 M NaCl yields one peak at the PG I position which upon heating shows convertor activity (Fig. 3C). Re-chromatography of these fractions with 1.5 M NaCl, however, also gives a PG II peak, and convertor activity in the void volume (Fig. 3D). Apparently, PG I is bound to the polysaccharide with its convertor part, from which PG II can be split off. Also, when purified cell walls from green tomato pericarp tissue are suspended in a solution of PG I in 0.2 M NaCl at different acidity, a considerable decrease of PG activity remains in the supernatant, particularly at low pH-values. With PG II this is much less the case, again probably because this latter iso-enzyme lacks the protein part that binds it to its substrate.

Finally, the convertor can be highly purified by gel filtration on Sephacryl S-300. According to the elution volume the molecular mass corresponds with that of a globular protein of 81 KDa, which is much lower than the value determined by Pressey (1984) of 102 KDa. However, since the convertor was produced by denaturation of PG I, it is possibly a more oblong molecule of a much lower molecular weight.

THE PHYSIOLOGICAL ACTION OF ENDO-POLYGALACTURONASE

What do these biochemical studies reveal of the participation of endopolygalacturonase in the processes of tomato fruit ripening and, particularly, of softening by cell separation?

It has been surmised by Tigchelaar and McGlasson (1977) that the enzyme plays a primary role in the ripening process, even prior to the effect of ethylene. The main rise in PG activity occurs at high levels of internal ethylene, above 10 ppm (Fig. 4A). However, water-soluble pectins appear already at very low levels of PG-activity (Fig. 4B) and, in fact, Fig. 4A shows a small increase of PG-activity already at much lower ethylene concentrations. Also, Table 1 shows that PG II occurs already at the onset of ripening, perhaps such low activities suffice already to induce an early effect. However, on the one hand, Su *et al.* (1984) demonstrated that an increase in ACC-synthase activity in the early mature green stage distinctly preceded the occurrence of PG-activity in the breaker stage. On the other hand, induction of ethylene biosynthesis by cell wall-derived elicitors, as found in many pathogenic systems and recently indicated to be evoked also in tomato fruits (Baldwin and Pressey, 1988), is probably not involved in tomato fruit ripening. The increase in autocatalytic ethylene biosynthesis at the onset of ripening is probably triggered by the decrease of a ripening inhibitor below a threshold value at which the sensitivity of the fruit tissues to ethylene increases (Sawamura *et al.*, 1978). This increased sensitivity then leads to the induction of ACC-synthase activity.

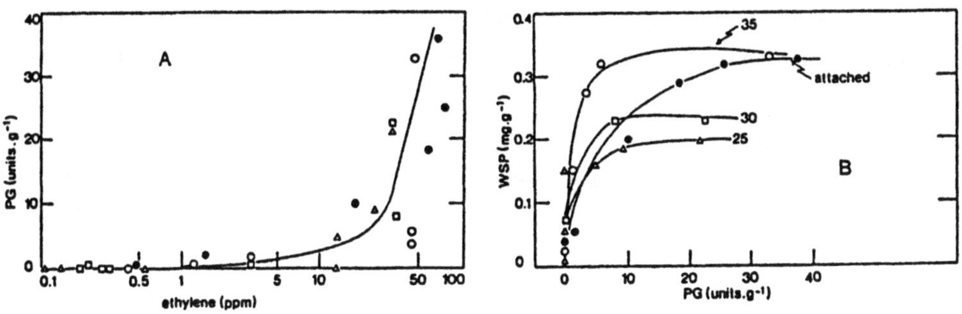

FIGURE 4. Relationship between A: ethylene content and PG activity, and B: PG activity and content of water-soluble pectin (WSP) in tomato pericarp during ripening
After Sawamura et al., 1978.

When we confine our attention to the role of PG in cell wall solubilization, the question arises as to how far the data obtained *in vitro* can be applied to the situation *in vivo* Seymour *et al.* (1987) found that the combination of PE and PG could mimic cell wall autolysis *in vitro* but that *in vivo* the combined action of these two enzymes must be considerably restricted. This is little wonder when we consider that the acidity at the pericarp cell walls is about 6, at which pH the two PG isoenzymes are actually inactive *in vitro*. Rushing and Huber (1987) found optimum autolysis of tomato cell walls in vitro at pH-values of 2.5 and 4.5. At the latter acidity calcium ions were found to inhibit autolysis. Burns and Pressey (1987) determined calcium in the cell walls and middle lamellae of tomato pericarp and found it to increase considerably during ripening. The role of this factor and of the other conditions in the cell walls on the activity of PG is far from clear. Possibly the PE-activity liberating carboxyl groups of the polygalacturonic-acid chain locally decreases the pH at its site of action sufficiently to enable PG to exert its depolymerizing action.

The next and here final question then is: of the two known PG isoenzymes, which one is responsible for cell separation? Subcellular localization of the two iso-enzymes might answer this question, but until now this has not been possible. Immunologically, the two isoenzymes cannot be distinguished. We tried to isolate cell wall-located enzymes by vacuum-infiltration of the tissue with salt solutions and recovery of the intercellular liquid, but the determination of PG-activity in this liquid gave very irreproducible results. Alternatively, we tried to obtain the intracellular enzymes by the isolation of protoplasts. However, even protoplasts from green pericarp tissues showed high PG-activity, either from uptake from the maceration solution or induced by the stress during the over-night incubation in this solution.

Where a direct determination has therefore been unsuccessful up to now, data from the *in vitro* experiments might give indications. The recent view of Pressey (1988) is that because PG II can be isolated first and convertor is always present in the cell walls of the tomato fruits and other organs, PG I is formed as an artifact *in vitro* when PG II and the convertor are solubilized simultaneously, so that PG II is the only endogenous PG.

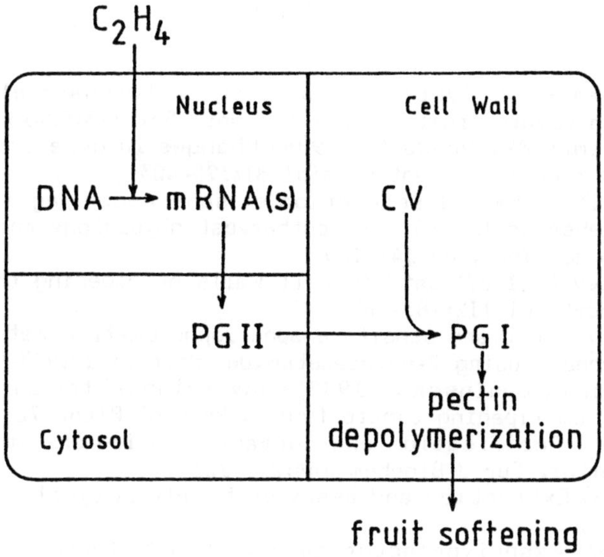

FIGURE 5. Hypothetical scheme of the role of the ethylene-induced PG isoenzymes in cell separation of ripening tomato pericarp tissue.

This view neither takes into account that PG I is labile *in vitro* and desintegrates into PG II and the convertor, nor that when PG II and convertor are combined *in vitro* another isoenzyme, PGx, appears. From a physiological point of view, it is very likely that an enzyme active in the cell wall, is kept there by binding to a cell wall-adhesive component such as the convertor. In our view (Fig. 5), therefore, the smaller isoenzyme PG II is produced from its messenger RNA in the cytosol (Biggs *et al.*, 1986), moves through the plasmamembrane into the cell wall, where it is fixed to its substrate by being bound to the convertor. According to this view, isoenzyme PG I is the truly active form, responsible for the cell separation leading to the softening of the tomato fruit.

REFERENCES

Baldwin EA, Pressey R (1988) Tomato polygalacturonase elicits ethylene production in tomato fruit. J Amer Soc Hort Sci 113:92-95

Biggs MS, Harriman RW, Handa AK (1986) Changes in gene expression during tomato fruit ripening. Plant Physiol 81:395-403

Bruinsma J (1983) Hormonal regulation of senescence, ageing, fading, and ripening. Lieberman M (ed) In: Postharvest physiology and crop preservation. Plenum New York, pp 141-163

Burns JK, Pressey R (1987) Ca^{2+} in cell walls of ripening tomato and peach. J Amer Soc Hort Sci 112:783-787

Gross KC (1982) A rapid and sensitive spectrophotometric method for assaying polygalacturonase using 2-cyanoacetamide. HortSci 17:933-934

Knegt E, Vermeer E, Bruinsma J (1988) Conversion of the polygalacturonase isoenzymes from ripening tomato fruits. Physiol Plant 72:108-114

Pressey R (1984) Purification and characterization of tomato polygalacturonase convertor. Eur J Biochem 144:217-221

Pressey R (1986) Extraction and assay of tomato polygalacturonases. Hort Sci 21:490-492

Pressey R (1987) Exopolygalacturonase in tomato fruit. Phytochem 26:1867-1870

Pressey R (1988) Reevaluation of the changes in polygalacturonases in tomato during ripening. Planta 174:39-43

Pressey R, Avants JK (1973) Two forms of polygalacturonase in tomatoes. Biochim Biophys Acta 309:363-369

Rushing JW, Huber DJ (1987) Effects of NaCl, pH and Ca^{2+} on autolysis of isolated tomato fruit cell walls. Physiol Plant 70:78-84

Sawamura M, Knegt E, Bruinsma J (1978) Levels of endogenous ethylene, carbon dioxide, and soluble pectin, and activities of pectin methylesterase and polygalacturonase of ripening tomato fruits. Plant Cell Physiol 19:1061-1069

Seymour GB, Lasslett Y, Tucker GA (1987) Differential effects of pectolytic enzymes on tomato polyuronides *in vivo* and *in vitro*. Phytochem 26:3137-3139

Su LY, McKeon T, Grierson D, Cantwell M, Yang SF (1984) Development of 1-aminocyclopropane-1-carboxylic acid synthetase and polygalacturonase activities during the maturation and ripening of tomato fruit. HortSci 19:576-578

Tigchelaar EC, McGlasson WB (1977) Tomato ripening mutants; a key role for polygalacturonase in fruit ripening? Plant Physiol 59:supp. 121

Tucker GA, Robertson NG, Grierson D (1980) Changes in polygalacturonase isoenzymes during "ripening" of normal and mutant tomato fruit. Eur J Biochem 112:119-124

Tucker GA, Robertson NG, Grierson D (1981) The conversion of tomato-fruit polygalacturonase isoenzyme 2 into isoenzyme 1 in vitro. Eur J Biochem 115:87-90

Watkins CB, Haki JM, Frenkel C (1988) Activities of polygalacturonase, α-D-mannosidase, and α-D- and β-D-galactosidases in ripening tomato. Hort Sci 23:192-194

HORMONE-REGULATED MODIFICATIONS TO CELL WALL POLYSACCHARIDES: RELEVANCE TO CELL SEPARATION

Y. Masuda, T. Hoson, R. Yamamoto[1] and M. Inouhe[2]
Department of Biology
Faculty of Science,
Osaka City University
Sumiyoshi-ku
Osaka 558
Japan

INTRODUCTION

When plants grow and develop, the cell wall undergoes biochemical modifications. The type of cell wall polysaccharide components subjected to degradation and biosynthesis appears to be dependent upon the stage and the type of growth and development. It also depends upon plant species, due to the difference in matrix composition, and probably upon the hormones involved.

When auxin induces cell extension, it causes cell wall loosening, probably by partially degrading of certain matrix components, and then promoting new cell wall synthesis. In this chapter we describe auxin-regulated modifications of cell wall components in terms of degradation and synthesis.

DEGRADATION OF CELL WALL COMPONENTS

When auxin (IAA, indole-3-acetic acid) induces cell wall loosening, the following results have been reported: In graminaceous coleoptiles, specific degradation of β-glucans, glucuronoarabinoxylans and xyloglucans is caused by auxin (Loescher and Nevins, 1972; Sakurai and Masuda, 1978; Darvill *et al.*, 1978; Inouhe *et al.*, 1984). In stems of dicotyledons, degradation of polyuronides, galactans and xyloglucans is stimulated by auxin (Labavitch and Ray,1974; Terry *et al.*, 1981; Nishitani and Masuda, 1980, 1983).

It is well known that the matrix composition of cell walls is different in graminaceous and in dicotyledonous species (McNeil *et al.*, 1984; Masuda

[1] Food Science Laboratory, Tezukayama College, Gakuen-minami, Nara 631, Japan
[2] Department of Biology, Faculty of Science, Ehime University, Bunkyo-cho, Matsuyama 790, Japan

NATO ASI Series, Vol. H 35
Cell Separation in Plants
Edited by D. J. Osborne and M. B. Jackson

and Yamamoto, 1985). Cell walls of the former contain β-glucans and glucurono- arabinoxylans but not much galactan and only small amounts of xyloglucan and pectic substances. Dicotyledonous cell walls, on the other hand, contain much galactan and xyloglucan but not β-glucan and arabinoxylan. Questions then arise concerning which matrix components are responsible for IAA-induced cell wall loosening in both types of plants. In order to characterize the types of matrix component and oligosaccharides that play crucial roles in wall loosening, we have examined effects, on IAA-induced growth, of lectins and of antibodies raised against sugar residues and oligosaccharide fragments. Lectins and antibodies can specifically bind to certain cell wall components and affect their metabolic turnover as a consequence of their sugar and oligosaccharide specificities.

To secure the penetration of lectins and antibodies into the cell wall of organ segments, we used abraded segments of oat (*Avena sativa* L.) and maize (*Zea mays* L.), and azuki bean (*Vigna angularis* Ohwi and Ohashi). The binding of lectins and antibodies to the cell wall was examined with FITC-lectins and by an indirect immunofluorescence method using FITC-labelled anti-IgG antibodies. The mechanical properties of the cell wall were measured by the stress-relaxation method (Yamamoto *et al.*, 1970).

Of 12 kinds of lectins commercially available, representing six groups of sugar specificity, the ones specific for fucose, i.e. TPA and UEA I, were bound to the cell wall of azuki bean epicotyls but not to the cell wall of oat coleoptiles. It is likely that they react with xyloglucans, since xyloglucans of the cell wall of dicotyledons contain fucose in their side chains while those in graminaceous cell walls do not (Kato and Matsuda, 1976). These lectins were effective in inhibiting IAA-induced elongation of epicotyl segments of azuki bean and cell wall loosening (Hoson and Masuda, 1986, 1987). However, they did not inhibit IAA-induced elongation of oat coleoptile segments. The results suggest that xyloglucans are involved in auxin-induced cell wall loosening and cell elongation in dicotyledons.

Since there are no lectins specific for arabinose available commercially, anti-α-L-arabinofuranose antibodies were raised and purified (Kaku *et al.*, 1986). The antibodies were found to bind to the cell walls of oat coleoptiles and azuki bean epicotyls. They inhibited IAA-induced cell elongation and cell wall loosening of both oat and azuki bean; the effect being stronger in oat coleoptiles than in azuki bean epicotyls (Hoson and Masuda, 1986).

We then raised antibodies against (1→3),(1→4)-β-D-glucans, endo-β-glucanase, and xyloglucans of heptamers, octamers and dimers (iso-primeverose) and tested their effects on auxin-induced growth and cell wall loosening. The antibodies were raised against oat caryopsis β-D-glucans. They bound to the cell walls of graminaceous species and inhibited IAA-induced elongation of maize coleoptile segments (Fig. 1). IAA-induced cell wall loosening (decrease in T_0 value) and decrease in the amount of glucans in the cell wall of maize coleoptiles were also strongly inhibited by the antibodies. The antibodies were not bound to the cell wall of azuki bean epicotyl segments nor did they inhibit IAA-induced growth and cell wall loosening. The antibodies raised against an endoglucanase fraction of maize coleoptile cell walls showed the same effect as the anti-glucan antibodies (Fig. 1). The autolytic activity of maize coleoptile cell wall preparations was also inhibited by those antibodies. These results suggest that β-glucan degradation is involved in auxin-induced cell wall loosening and elongation in graminaceous coleoptiles.

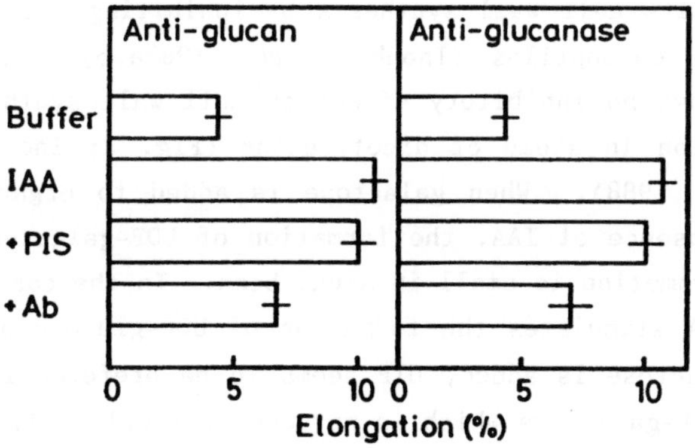

FIGURE 1. Effect of antibodies raised against β-D-glucans and an endo- glucanase on elongation of maize coleoptile segments. Segments 10 mm long were floated on 10 mM MES-KOH buffer, pH 6.0, containing 10 μM IAA with or without 400 μg/ml pre-immune serum (PIS) or antibodies (Ab) for 4 hours. Values are means ± SE (n = 30)

The anti-xyloglucan antibodies inhibited IAA-induced cell elongation and cell wall loosening in azuki bean epicotyls but not in oat coleoptiles.

These antibodies were found to be bound to the cell wall of not only the former but also the latter when examined by fluorescence microscopy. It is not understood why the antibodies were bound to the cell wall of oat coleoptiles, even though they showed no inhibitory effect on growth. At least, it can be said that xyloglucan turnover is involved in auxin-induced elongation and cell wall loosening in the stems of dicotyledons.

The results obtained in the above study support the view that the degradation by wall-associated enzymes of (1→3),(1→4)-β-D-glucans in graminaceous plants (Loescher and Nevins, 1972; Sakurai and Masuda, 1978) and that of xyloglucans in dicotyledons (Labavitch and Ray, 1974; Nishitani and Masuda, 1983) is involved in the cell wall loosening responsible for auxin-induced cell elongation. The lectins and antibodies specific for cell wall components can thus be potent probes for clarifying the mechanism of cell elongation induced by auxin and other hormones.

ROLE OF CELL WALL SYNTHESIS IN AUXIN-INDUCED GROWTH

It has been known that galactose specifically inhibits auxin-induced growth of graminaceous coleoptiles but not of dicotyledonous stems, although barley and pea are exceptions (Yamamoto *et al.*, 1988). It has been reported that galactose inhibits cell wall synthesis by inhibiting the formation of UDP-glucose in oat coleoptiles (Inouhe *et al.*, 1987a,b). Galactose, on the other hand, shows no inhibitory effect on cell wall synthesis, nor on UDP-glucose formation in stems of dicotyledons (Fig. 2; Inouhe *et al.* 1987a,b; Yamamoto *et al.*, 1988). When galactose is added to organ segments in the presence and absence of IAA, the formation of UDP-galactose is promoted in oat but this promotion is small in azuki bean. In the cases of both oat and azuki bean, IAA stimulates the formation of UDP-glucose and UDP-galactose. If in oat, galactose is added, UTP seems to be preferentially used for the formation of UDP-galactose which is not used for cell wall synthesis, and is thus accumulated. This is not the case for azuki bean which may convert UDP-galactose to UDP-glucose by epimerase and possibly utilizes the former for cell wall synthesis. It is considered that the synthesis, recycling and consumption of UTP needed for UDP-glucose formation and thus for cell wall synthesis is involved in the action of IAA and in the differential effect of galactose on auxin-induced growth.

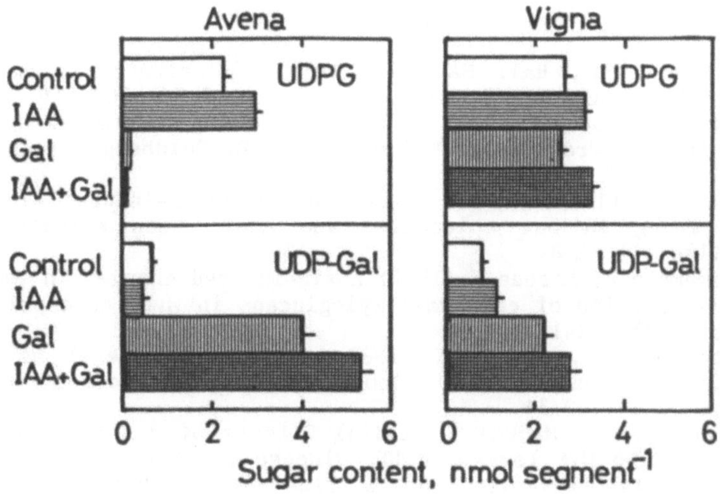

FIGURE 2. Effect of IAA and/or galactose on the levels of UDP-sugars in oat coleoptiles and azuki bean epicotyl segments. Segments 10 mm long were floated on 10 mM potassium phosphate buffer, pH 6.5, containing 10 μM IAA and/or 10 mM galactose for 2 hours. Values are means ± SD (n = 3)

Galactose does not inhibit auxin-induced cell wall loosening, as judged by stress-relaxation studies, even in oat coleoptiles (Yamamoto and Masuda, 1984). Therefore, it can be said that cell wall synthesis may not be involved in cell wall loosening process that are caused by degradation of certain functional polysaccharides in the cell wall.

RELEVANCE TO CELL SEPARATION

Our studies on auxin-regulated cell elongation have been performed using young organs as coleoptiles and stems of usually etiolated seedlings. According to the definition of Osborne (1979), the plant material we use could represent "type 1" target organs. Our technique of using antibodies specific for oligo- and polysaccharides could be used to study cell wall turnover in other "types" of target organs e.g. abscission zones. If cell wall synthesis, in addition to degradation, are possibly involved, changes in the activity of UDP-glucose formation could be a good measure of such cell wall changes.

REFERENCES

Darvill AG, Smith CJ, Hall MA (1978 Cell wall structure and elongation growth in *Zea mays* coleoptile tissue. New Phytol 80: 503-516

Hoson T, Masuda Y (1986) Effects of lectins and sugar-recognizing antibodies on auxin-induced growth. In: Vian B, Reis R, Goldberg R (eds) Cell Wall '86. Paris, p 242

Hoson T, Masuda Y (1987) Effect of lectins on auxin-induced elongation and wall loosening in oat coleoptile and azuki bean epicotyl segments. Physiol Plant 71: 1-8

Inouhe M, Yamamoto R, Masuda Y (1984) Auxin-induced changes in the molecular weight distribution of cell wall xyloglucans in *Avena* coleoptiles. Plant Cell Physiol 25: 1241-1351

Inouhe M, Yamamoto R, Masuda Y (1987a) UDP-glucose level as a limiting factor for IAA-induced cell elongation in *Avena* coleoptile segments. Plant Physiol 69: 49-54

Inouhe M, Yamamoto R, Masuda Y (1987b) Effects of indole-acetic acid and galactose on the UTP level and UDP-glucose formation in *Avena* coleoptile and *Vigna* epicotyl segments. Physiol Plant 69: 579-585

Kaku H, Shibata S, Satsuma Y, Sone Y, Misaki A (1986) Interactions of α-L-arabinofuranose-specific antibody with plant polysaccharides and its histochemical application. Phytochem 25: 2041-2047

Kato Y, Matsuda K (1976) Presence of xyloglucan in the cell wall of *Phaseolus aureus* hypocotyls. Plant Cell Physiol 17: 1185-1198

Labavitch JM, Ray PM (1974) Turnover of cell wall polysaccharides in elongating pea stem segments. Plant Physiol 53: 669-673

Loescher W, Nevins DJ (1972) Auxin-induced changes in *Avena* coleoptile cell wall composition. Plant Physiol 50: 556-563

Masuda Y, Yamamoto R (1985) Cell-wall changes during auxin-induced cell extension. Mechanical properties and constituent polysaccharides of the cell wall. In: Brett CT, Hillman JR (eds) Biochemistry of Plant Cell Walls. Cambridge University Press, Cambridge, p 269

McNeil M, Darvill AG, Fry SC, Albersheim P (1984) Structure and function of the primary cell walls of plants. Ann Rev Biochem 53: 625-663

Nishitani K, Masuda Y (1980) Modifications of cell wall polysaccharides during auxin-induced growth in azuki bean epicotyl segments. Plant Cell Physiol 21: 169-181

Nishitani K, Masuda Y (1983) Auxin-induced changes in the cell wall xyloglucans: effects of auxin on the two different subfractions of xyloglucans in the epicotyl cell wall of *Vigna angularis*. Plant Cell Physiol 24: 345-355

Osborne DJ (1979) Target cells - New concepts for plant regulation in horticulture. Sci Hort 30: 31-43

Sakurai N, Masuda Y (1978) Auxin-induced changes in barley coleoptile cell wall composition. Plant Cell physiol 19: 1217-1223

Terry ME, Jones RL, Bonner BA (1981) Soluble cell wall polysaccharides released from pea stems by centrifugation. I. Effect of auxin. Plant Physiol 68: 531-537

Yamamoto R, Inouhe M, Masuda Y (1988) Galactose inhibition of auxin-induced growth of mono- and dicotyledonous plants. Plant Physiol 86: 1223-1227

Yamamoto R, Masuda Y (1984) Galactose inhibition of auxin-induced elongation in oat coleoptile segments. Physiol Plant 61:321-326

Yamamoto R, Shinozaki K, Masuda Y (1970) Stress-relaxation properties of plant cell walls with special reference to auxin action. Plant Cell Physiol 11: 947-956

ETHYLENE AND POLYGALACTURONASE - WHAT ELSE IS INVOLVED IN CELL SEPARATION IN RIPENING FRUIT?

Michael Knee*
AFRC Institute of Horticultural Research
East Malling
Maidstone
Kent ME19 6BJ
United Kingdom

INTRODUCTION

In the ripening of many fleshy fruits, cells become more easily separable and the tissue softens. Cellular cohesion depends on the properties of the major intercellular polysaccharide, pectin, and its cross-linkage by calcium ions. Conceptually, there are many ways in which cohesion could be lost and a subset of these would explain the increase in concentration of cold-water soluble pectin which is observed in most, if not all softening fruits. Many fruits develop endo-polygalacturonase (PG) activity, and this is often regarded as responsible for both the production of soluble (degraded) pectin and the loss of cell cohesion.

The synthesis of PG protein follows the rise in ethylene production in tomato fruits (Su *et al.* 1984) and ethylene treatment of mature, pre-climacteric tomatoes stimulates enzyme production (Maunders *et al.*, 1987). At first sight we appear to have accounted for cell separation in ripening fruit by providing an agent (PG) and a signal (ethylene) for its induction at the appropriate developmental stage. On further reflection three questions arise: a rise in ethylene itself requires some stimulus or signal, so how is the synthesis of ethylene regulated; is PG directly induced by ethylene, or do other regulatory steps intervene; can the rise in PG activity

*Present address: Department of Horticulture, The Ohio State University, 2001 Fyffe Court, Columbus, OH 43210, USA.

NATO ASI Series, Vol. H 35
Cell Separation in Plants
Edited by D. J. Osborne and M. B. Jackson

account for fruit softening? I will concentrate on the first and third of these questions, as the regulation of PG activity will be dealt with elsewhere in this volume.

ETHYLENE ACTION

Although ethylene is the immediate signal for initiation of ripening in climacteric fruits, an increase in the ability to respond to ethylene precedes the rise in ethylene synthesis. I have discussed elsewhere the control of ethylene action during fruit development and the effect of genetic and chemical factors (Knee, 1989).

OTHER REGULATORS OF FRUIT RIPENING

No plant hormone other than ethylene has been shown to be a natural regulator of fruit ripening. The existence of non-climacteric fruits requires some other factor to explain the co-ordination of their ripening. Recent evidence suggests that a decline in IAA is the signal for ripening, including cell separation in strawberry fruits (Given *et al.*, 1988). Earlier a decline in IAA was said to permit response to endogenous ethylene in pears and other climacteric fruits (Frenkel and Dyck, 1973). In apples, IAA increases prior to, and declines with the rise in ethylene (Mousdale and Knee, 1981), whereas in peach mesocarp tissue, IAA increases continuously before and during the ethylene rise (Miller *et al.*, 1987). A case can be made that endogenous IAA is an inducer of ethylene synthesis in these fruits as it is in vegetative tissue (Yang and Hoffman, 1984).

When fruits are treated with IAA or auxin analogues, ripening processes are inhibited, provided that access of the compound to the tissue is ensured e.g. by vacuum infiltration. However, except in the experiments of Tingwa and Young (1975) on avocado, inhibition has been achieved with auxin at micromolar levels in the tissue and ethylene synthesis was stimulated (Vendrell, 1969; Frenkel and Dyck, 1973). Tingwa and Young

(1975) did not specify the aspects of ripening which were inhibited by IAA, apart from ethylene synthesis and the respiratory rise. It seems that high concentrations of auxin can inhibit various aspects of fruit ripening, but the regulatory role of the endogenous (nanomolar) concentrations remains unclear.

Abscisic acid is the only other plant hormone which has been reliably assayed in fruit tissue. It generally increases during fruit development, but without any consistent relation to the rise in ethylene (McGlasson *et al.*, 1978). Application of ABA to fruits usually hastens ripening, but although the promotive effects of water stress on ripening of avocados are suggestive (Adato and Gazit, 1974a) the natural role of ABA in fruit ripening is far from established.

ENVIRONMENTAL AND OTHER SIGNALS

Recent work with apples suggests that fruits may be pre-disposed to ripen when they attain a certain size, conforming to a general tendency in organisms to undergo developmental changes when a particular size is reached (M. Knee, unpublished). Some, but not all fruits ripen earlier if they are detached from the parent plant than if they are left attached (Adato and Gazit, 1974b). This has led to hypotheses of "tree factors" which inhibit ripening, and decline once fruit is detached. In avocado, the factor seemed to block response to ethylene for some time after harvest (Adato and Gazit, 1974b), but apples on the tree are as responsive to ethylene as detached apples (Knee, 1989). In many pear varieties and at least one apple, ethylene synthesis can be induced by exposure to temperatures below 5℃ (Knee *et al.*, 1983).

Mechanical, water and temperature stress can induce ethylene synthesis in many plant tissues, and it is not surprising that they can induce fruit ripening. These observations fit into a larger pattern of the reproduction of plant hormone effects by many kinds of stress (Trewavas, 1987). In future work it will be important to distinguish merely experimental effects from those with some biological relevance.

ENZYMIC DEGRADATION OF PECTIN

Many fruits develop endo-polygalacturonase activity on ripening, and if attention is confined to these fruits it is tempting to look no further for the agent of fruit softening and pectin solubilization. However, apples (Bartley, 1978) and strawberries (Huber, 1984) lack this activity and their pectin remains of high molecular weight during ripening (Knee, 1978a). Recent careful examination of pectin from tomatoes, which contain endo- polygalacturonase, shows that the action of this enzyme *in vivo* is much more limited than was first thought (Seymour and Harding, 1987).

Apples contain exo-polygalacturonase, and although this enzyme is often ignored, it does release uronic acid from fruit cell walls (Bartley, 1978). Liang *et al.* (1982) report the increase in activity of a polygalacturonase on ripening of apples. Since a reducing sugar assay was used this was presumably the same enzyme reported by Bartley (1978). The enzyme attacks from the non-reducing end of the polymer and should stop when it reaches a rhamnose residue, or a uronic acid residue which is esterified, or carries a neutral sugar side chain. Nevertheless apple exo-polygalacturonase can release more than 15% of the total uronic acid residues of the apple cell wall (Bartley, 1978) and could well account for the loss of total pectin sometimes observed on ripening (Knee, 1978b). Such degradation should affect the physical properties of the cell wall and the tissue, and Bartley (1978) suggested that it could account for the solubilization of pectin *in vivo*. *In vitro* no more polymeric material was released from the wall in the presence of the enzyme than in its absence.

The other glycosidase capable of attacking pectin, which is found in many fruits is β-galactosidase (Bartley, 1978).

Several fruits lose galactose residues from their cell walls during ripening (Gross and Sams, 1984), but the effect of this compositional change is unclear. If the galactose side chains on pectin are attachment sites for other polymers (Keegstra *et al.*, 1973), their removal could weaken wall structure. Turnover of galactose residues seems to be a normal feature of cell expansion in plants (Labavitch, 1981). The decline of wall galactose in ripening could result from the decline of synthesis of a branched polymer against a constant background of β-galactosidase activity (Knee and Bartley, 1982).

PECTIN SYNTHESIS

Discs cut from ripening apples (Knee, 1978b) and pears (Knee, 1982a) incorporate precursors into pectin. In apples, high rates of incorporation from [$^{14}CH_3$] methionine were observed in the mature fruit before ripening; incorporation declined before increasing with the rise in ethylene production by the ripening fruit (Knee, 1978b). The experiments with pears followed storage at -1℃, and there was at least a tenfold increase in incorporation during ripening at 15℃ (Knee, 1982a). Some cell wall deposition, associated with growth might be expected in fruit after harvest, and this would be likely to decline during storage at low temperature. The experiments with pears show that at least the capacity for pectin synthesis develops with ripening.

The results of a pulse-chase experiment with pear tissue were inconclusive because it seemed to be impossible within a reasonable timescale to dilute the precursor pools by supplying excess [$^{12}CH_3$] methionine. The increase in total pectin in the tissue observed in this experiment suggested that pectin synthesis was a feature of wounded tissue. Although pectin synthesis was inhibited under nitrogen (Knee, 1982a), in a subsequent paper other explanations were sought for the inhibition of softening by anoxia (Knee, 1982b). However, incubation under nitrogen does cause a decline in soluble pectin and prevents an increase in total pectin in excised pear tissue (Table 1). The continued decline of the viscosity of the soluble pectin shows that the polygalacturonase activity, which persists under nitrogen, is still functional. The accelerated loss of methyl ester groups suggests that pectinesterase activity was favored by anoxia and the extra carboxyl groups could provide new junction zones to stabilize the middle lamella. This experiment suggests that pectin synthesis is a necessary part of the process of textural change in pears. The softening of apples during storage and the early stages of softening of pears could be reversed by infiltrating tissue with calcium salts (Knee, 1982b). The secretion of newly synthesized pectin to the inner surface of the primary wall could result in a withdrawal of calcium ions from the middle lamella.

TABLE 1. Effect of anoxia on softening and pectin in tissue excised from Conference pears after ripening for 3 days at 18^{0}C.

Incubation	None	6 h air	6 h N_2	SE
Firmness (N)	33.1	26.1	34.2	0.95
Soluble pectin (mg/g)	1.31	1.30	0.90	0.034
η_{sp} of pectin (ml/mg)	1.88	1.29	1.27	0.13
Total pectin (mg/g)	4.69	5.12	4.68	0.13
% methyl ester	63.9	59.4	53.8	0.89

SOLUBLE PECTIN AND SOFTENING

Changes in soluble pectin correlate with some, but not all of the changes in the physical properties of apple tissue. For example storage in low oxygen atmospheres delayed pectin accumulation more consistently than softening in Cox apples (Knee and Sharples, 1981). One consequence of the inhibition of ethylene action by daminozide was an inhibition of both soluble pectin accumulation and softening in stored apples (Knee, 1986). Whereas the build up of pectin was consistent, the extent of softening varied from year to year and this variation was not affected by daminozide. Softening without soluble pectin formation was more clearly seen in storage of Bramley apples (Fig. 1). The extent of this softening was influenced by the ethylene level during storage and, on removal from store, further softening was associated with a rise in both soluble and total pectin. Thus there are recurrent indications of a process involved in softening, other than formation of soluble pectin. the identification of this process will probably require detailed study of the linkages in the wall and their turnover.

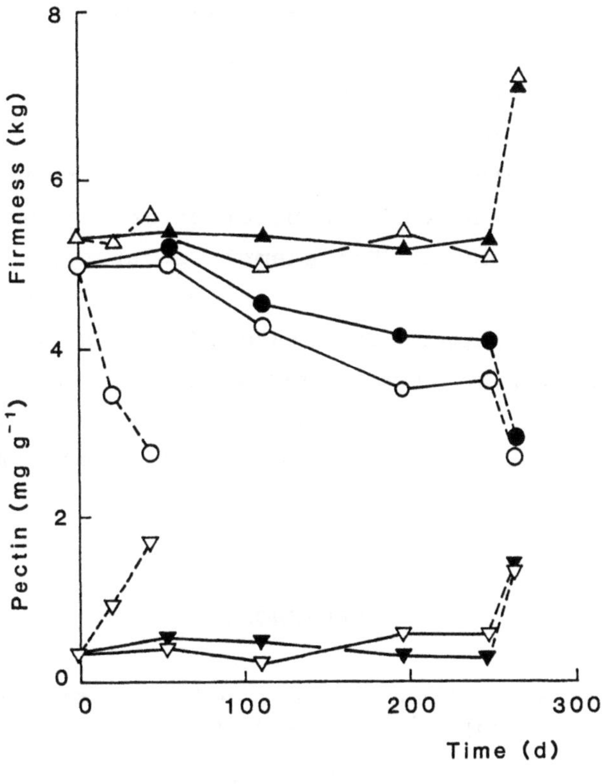

FIGURE 1. Softening and pectin changes in Bramley apples during storage in 9kPaCO_2: 13 kPa O_2 at 3.5^0 (continuous lines) and ripening in air at 15^0 (broken lines). Solid points, ethylene removed by $KMnO_4$; open points, ethylene allowed to accumulate (up to 80 ul 1^{-1}. ▲ △ , total pectin; ● ○ , firmness; ▼ ▽ , soluble pectin.

OTHER CELL SEPARATING SYSTEMS

Synthesis as well as degradation of wall polysaccharides occurs during cell separation elsewhere in plants. The leaf abscission zone of cotton incorporates label from [^{14}C] methionine into pectin more actively than adjacent tissue (Valdovinos and Muir, 1965). The synthesis of specialised acidic polysaccharides by root cap cells is well documented, and its lubricant function is readily understood (Paull and Jones, 1976). The function of polysaccharide synthesis in other cell separating systems is a matter for speculation; we are only beginning to appreciate the complexities of cell wall metabolism, and do not yet have a complete accepted model of cell wall structure. Likewise we have only established a few features of the control system. We are still at the stage of exploratory experiments, and cannot predict the consequences of manipulation of any component of the cell separating process. For example soluble pectin formation and softening can occur in Gloster 69 apples stored for long periods in the preclimacteric state (Knee and Tsantili, 1988); the early indications from experiments with transgenic tomato plants expressing anti-sense RNA to PG messenger are that physical changes in the fruit are unaffected (Smith *et al.*, 1988).

REFERENCES

Adato I, Gazit S (1974a) Water-deficit stress, ethylene production, and ripening in avocado fruits. Plant Physiol 53:45-46.

Adato I, Gazit S (1947b) Postharvest response of avocado fruits of different maturity to delayed ethylene treatments. PlantPhysiol 53:899- 902.

Bartley IM (1974) β-Galactosidase activity in ripening apples. Phytochem 13:2107-2111.

Bartley IM (1978) Exo-polygalacturonase of apple. Phytochem 17:213-216.

Frenkel C, Dyck R (1973) Auxin inhibition of ripening in Bartlett pears. Plant Physiol 51:6-9.

Given NK, Venis MA, Grierson D (1988) Hormonal regulation of ripening in the strawberry, a non-climacteric fruit. Planta 174:402-406.

Gross KC, Sams CE (1984) Changes in cell wall neutral sugar composition during fruit ripening: a species survey. Phytochem 23:2457-2461.
Huber DJ (1984) Strawberry fruit softening: the potential roles of polyuronides and hemicelluloses. J Food Sci 49:1310-1315.
Keegstra K, Talmadge KT, Bauer WD, Albersheim P (1973) The structure of plant cell walls III. A model of the walls of suspension-cultured sycamore cells based on the interconnections of the macromolecular components. Plant Physiol 51:188-196.
Knee M (1978a) Properties of polygalacturonate and cell cohesion in apple fruit cortical tissue. Phytochem 17:1257-1260.
Knee M (1978b) Metabolism of polymethylgalacturonate in apple fruit cortical tissue during ripening. Phytochem 17:1261-1264.
Knee M (1982a) Fruit softening II. Precursor incorporation into pectin by pear tissue slices. J Exp Bot 33:1256-1262.
Knee M (1982b) Fruit softening III. Requirement for oxygen and pH effects. J Exp Bot 3:1263-1269.
Knee M (1986) Opposing effects of daminozide and ethylene on the induction of flesh softening in apple fruits. J Hort Sci 61:15-21.
Knee M (1989) Control of post-harvest action of ethylene on tree fruits. Acta Hort (in press).
Knee M, Bartley IM (1982) Composition and metabolism of cell wall polysaccharides in ripening fruits. In Recent advances in the biochemistry of fruits and vegetables. (Eds J Friend, MJC Rhodes) Academic Press, London. Pp. 133-148.
Knee M, Sharples RO (1981) The influence of controlled atmosphere storage on the ripening of apples in relation to quality. In: Quality in stored and processed vegetables and fruit. (Eds PW Goodenough, RJ Atkin) Academic Press, London. Pp 341-352.
Knee M, Tsantili E (1988) Storage of 'Gloster 69' apples in the preclimacteric state for up to 200 days after harvest. Phys Plant 74:499-503.
Knee M, Looney NE, Hatfield, SGS, Smith SM (1983) Initiation of rapid ethylene synthesis by apple and pear fruits in relation to storage temperature. J Exp Bot 34:1207-1212.
Labavitch JM (1981) Cell wall turnover in plant development. Ann Rev Plant Physiol 32:385-406.
Liang Y, Ao L, Wang M, Yi W (1982) The role of polygalacturonase (PG) in the ripening of apple fruits. Acta Bot Sin 24:143-146.
Maunders MJ, Holdsworth MJ, Slater A, Knapp JE, Bird CR, Schuch W, Grierson D (1987) Ethylene stimulates the accumulation of ripening-related mRNAs in tomatoes. Plant Cell Envir 10:177-184.
McGlasson, WB, Wade NL, Adato I (1978) Phytohormones and fruit ripening. In: Phytohormones and related compounds-A comprehensive treatise Vol II. (Eds DS Letham, PB Goodwin, TJV Higgins) Elsevier, Amsterdam. Pp 447-493.

Miller AN, Walsh CS, Cohen JD (1987) Measurement of indole-3 acetic acid in peach fruits *Prunus persica* L. Batsch cv Redhaven) during development. Plant Physiol 84:491-494.

Mousdale DM, Knee M. (1981) Indolyl-3-acetic acid and ethylene levels in ripening apple fruits. J Exp Bot 32:753-758.

Paull RE, Jones RL (1976) Studies on the secretion of maize root cap slime V. The cell wall as a barrier to secretion. Z Pflanzenphysiol 79:154-164.

Seymour GE, Harding SE (1987) Analysis of the molecular size of tomato *Lycopersicon esculentum* Mill fruit polyuronides by gel filtration and low-speed sedimentation equilibrium. Biochem J 245:463-466.

Smith CJS, Watson CF, Ray J, Bird CR, Morris PC, Schuch W, Grierson D (1988) Antisense RNA inhibition of polygalacturonase gene expression in transgenic tomatoes. Nature 334:724-726.

Su L, McKeon T, Grierson D, Cantwell M, Yang SF (1984) Development of 1-aminocyclopropane-1-carboxylic acid synthase and polygalacturonase activities during the maturation and ripening of tomato fruit. HortSci 19:576-578.

Tingwa PO, Young RE (1975) The effect of indole-3-acetic acid and other growth regulators on the ripening of avocado fruits. Plant Physiol 55:937-940.

Trewavas AJ (1987) Sensitivity and sensory adaptation in growth substance responses. In: Hormone action in plant development - A critical appraisal. (Eds GV Hoad, JR Lenton, MB Jackson, RK Atkin) Butterworths, London. Pp19-38.

Valdovinos JG, Muir RM (1965) Effects of D and L amino acids on foliar abscission. Plant Physiol 40:335-340.

Vendrell M (1969) Reversion of senescence: effects of 2,4-dichloro-phenoxyacetic acid and indole acetic acid on respiration, ethylene production and ripening of banana fruit slices. Aust J Biol Sci 22:601-610.

Yang SF, Hoffman NE (1984) Ethylene biosynthesis and its regulation in higher plants. Ann Rev Plant Physiol 35:155-189.

SECTION III

Pathogenic Cell Separation

SIGNALS AND CELL WALL EVENTS IN PLANT-PATHOGEN INTERACTIONS

M.T. Esquerré-Tugayé, D. Mazau, and D. Rumeau
Centre de Physiologie Végétale
Université Paul Sabatier
118, Route de Narbonne
31062 Toulouse Cédex
France

INTRODUCTION

Cell wall cohesiveness is affected in plants by senescence and pathogen attack. However, the two phenomena are very distinct from each other. Senescence is the result of developmentally regulated processes, whereas pathogen attack is the first of a series of unscheduled events ultimately leading to cell degradation as well as to defense responses. Despite this obvious basic difference, it has been known for a long time that infected tissues share common traits with those undergoing senescence. This includes cell wall hydrolysis, membrane damage, ethylene formation and lipid peroxidation. Cell surfaces play a key role in the two phenomena in that they are involved in, (1) signal generation and recognition, (2) signal transduction, and (3) response to the signal. This paper illustrates the role played by the cell wall and the plasmalemma in plant-microbe interactions.

SIGNAL GENERATION AND RECOGNITION

Plants respond to pathogen attack by developing a wide array of defense reactions (Collinge and Slusarenko, 1987). In the past twenty years, it has been established that cell wall fragments of either pathogen (fungal) or plant origin have the ability to induce the same defense reactions as pathogen attack (Darvill and Albersheim, 1984). Such fragments, also called elicitors, are supposed to be among the earliest signals of the interaction. How these elicitors are released and recognized is the subject of intense research. Figure 1 indicates that elicitor signals might be generated by :

- release from fungal cell wall by the action of host hydrolases (glycosidases) ;
- direct contact between pathogen cell walls and host structures ;

NATO ASI Series, Vol. H 35
Cell Separation in Plants
Edited by D.J. Osborne and M.B. Jackson

- partial degradation of host cell walls by pathogen enzymes such as pectinases.

Several classes of elicitors, glycans, glycopeptides, lipids, have thus been identified. Of particular relevance to cell separation is the effect of endopolygalacturonase on pectic polysaccharides, leading to the release of endogenous elicitors.

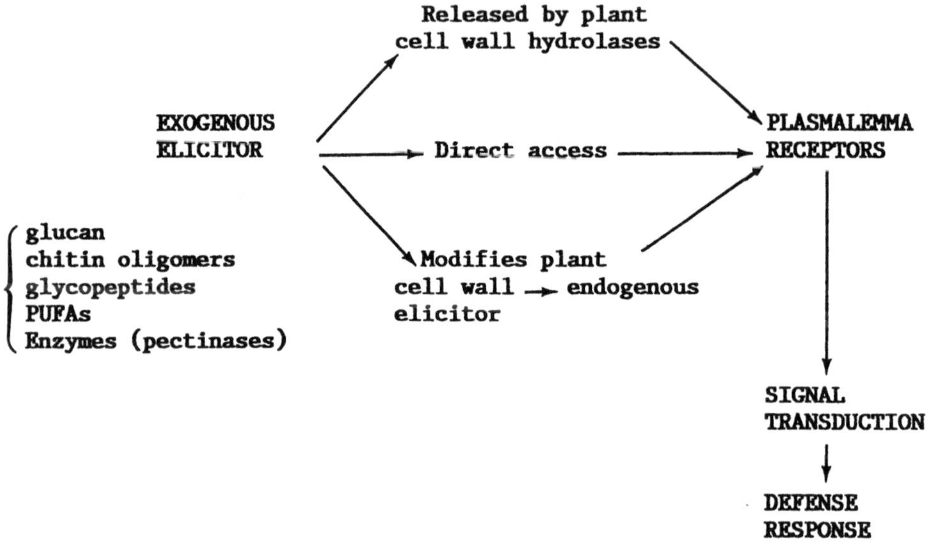

FIGURE 1. Signal generation and recognition

It is likely that elicitors interact with membrane receptors. Recently, Schmidt and Ebel (1987) have characterized membrane proteins with a high affinity to a β-glucan elicitor preparation. It is hypothesized that blocking of the receptor would block signal transduction and the final plant response.

SIGNAL TRANSDUCTION

Signal transduction refers to a series of complex events leading from recognition to physiological responses. Although research in this area is still very limited in plant-micro-organism interactions, a few pathways emerge that may be involved (Fig. 2) :

- changes in membrane potential ;
- phospholipid derived mediators including polyunsaturated fatty acids (PUFAs), their metabolites, and phosphoinositides ;
- ethylene biosynthesis.

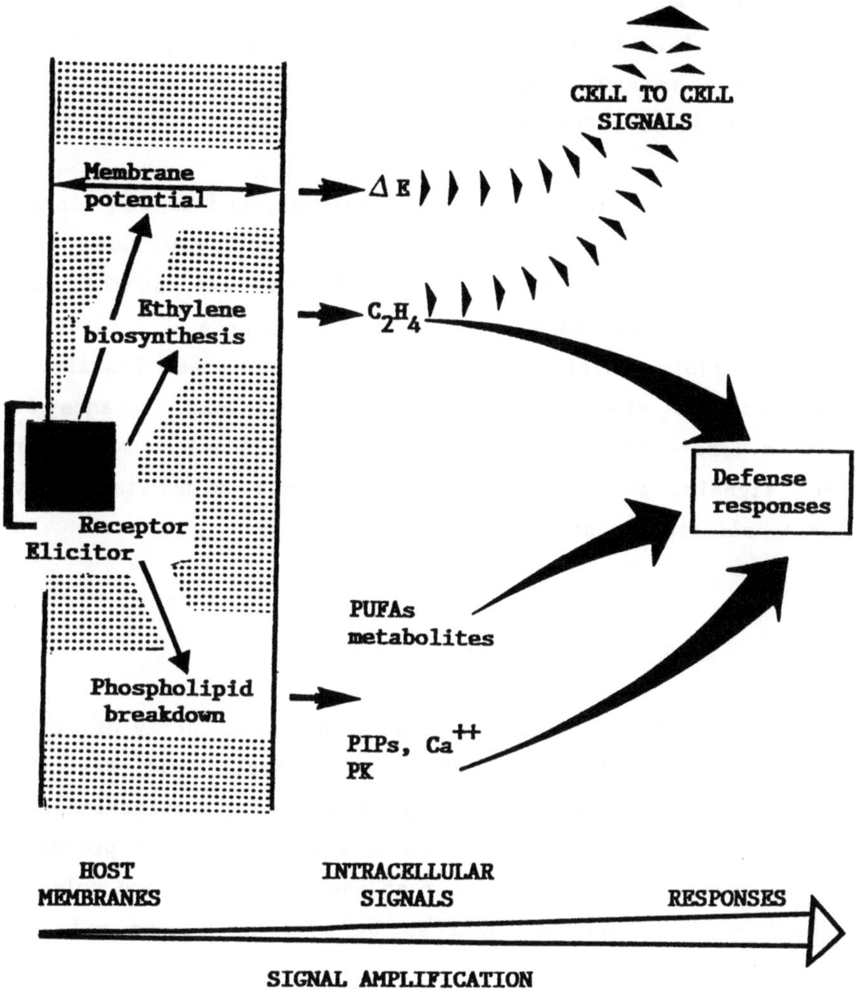

FIGURE 2. Signal transduction in response to elicitors : a proposed scheme. (ΔE = variations in membrane potential : PUFAs = polyunsaturated fatty acids ; PIPs = phosphoinositides ; PK = protein kinases).

We have found that elicitors of Phytophthora parasitica nicotianae and of Colletotrichum lagenarium as well as plant endogenous elicitors, exert a strong depolarizing effect on host plant membranes (Pélissier et al., 1986).

This effect is almost instantaneous, and reversible. It is followed by a rise in ethylene production which occurs after 15 to 30 minutes of incubation in the presence of elicitors (Toppan and Esquerré-Tugayé, 1984). After 3 hours, high lipoxygenase activity is induced (Fournier et al., 1986) and, finally defense responses are detectable after 12 hours or more (Bolwell et al., 1985; Roby et al., 1985,1986; Davis and Hahlbrock, 1987; Rohwer et al., 1987; Roby et al., 1987). Recently, Kurosaki et al. (1987) have reported that phosphoinositide-derived products might act as mediators of the phytoalexin response to endogenous elicitor treatment.

Whether these possible mediators act as a cascade, and if one plays an especially critical role awaits further investigation. Until now, none of the proposed mediators appears to be an absolute requirement i.e. inhibition of ethylene, lipoxygenase, calcium-calmodulin or protein kinase C individually does not result in complete inhibition of the mediated response. Finally, one should also consider that signals might be transported some distance from their origin. It is likely that membrane potentials and ethylene could serve as long distance signals, and account for the systemic responses observed during infection or elicitor treatment.

SIGNAL-INDUCED DEFENSE RESPONSES

Plants have adapted to pathogen attack by developing a wide array of defense responses (Bell, 1981) which involve phytoalexins, cell wall structural components (lignin, suberin, callose, hydroxyproline-rich glycoproteins), and proteins (enzymes, enzyme inhibitors, pathogenesis related or PR proteins). All the defense responses investigated so far at the biochemical and molecular levels are induced by elicitors, as well as by ethylene (Ecker and Davis 1987).

The possible role of ethylene as a mediator of defense responses was discussed in the preceding paragraph. Depending on the plant-pathogen or plant-elicitor system, and on the particular defense reaction, ethylene appears to be more or less involved in mediating the final response (Table 1). In melon for instance, elicitor-induced HRGP (hydroxyproline rich glycoprotein) and ethylene are decreased by AVG (aminoethoxyvinylglycine), an inhibitor of ethylene biosynthesis (Lieberman 1979). This effect can be reversed by adding to the medium ACC (1-aminocyclopropane-1-carboxylic acid), the direct precursor of ethylene (Roby et al., 1985). Elicitation of chitinase is significantly inhibited in melon and bean by AVG but much less so in pea (Mauch et al. 1984; Gaynor and Broglie 1985; Roby et al., 1986). Similarly, inhibition of elicitor-induced PAL (phenylalanine ammonia lyase), a key enzyme in phytoalexin biosynthesis, is considerable in parsley but not in soybean (Paradies et al., 1980; Chappel et al., 1984). Elicitation of proteinase inhibitors is also reduced in melon (Roby et al., 1987).

TABLE 2. Effect of AVG on ethylene biosynthesis and on defense-marker induction in plant-elicitor and plant-fungus interactions

Protein marker of defense	Biological system	Inhibition by AVG	
		Ethylene	Protein marker
HRGP	Melon-Elicitor[a] of Colletotrichum	55 %	45 %
Chitinase	Bean - Colletotrichum[b]	-	57 %
Chitinase	Melon-Elicitor[c] of Colletotrichum	75 %	47 %
Chitinase	Pea-Fusarium[d]	84 %	15 %
Chitinase	Pea-Elicitor[d] chitosan	80 %	15 %
PAL	Soybean-Elicitor[e] of Phytophthora	100 %	10 %
PAL	Parsley-Elicitor[f] of Phytophthora	100 %	55 %
Protease inhibitor	Melon-Elicitor[g] of Colletotrichum	50 %	35 %

Data obtained or recalculated from : a, Roby et al., 1985; b, Gaynor and Broglie, 1985; c, Roby et al., 1986; d, Mauch et al., 1984; e, Paradies et al., 1980; f, Chappell et al., 1984; g, Roby et al., 1987.

It is remarkable that most defense molecules are exported outwards to the cell wall or to the outside the cell ; this is the case for cell wall components and also for phytoalexins and several PR proteins which can be secreted, at least in part, beyond the cell wall. This shows that the cell surface is greatly modified in response to infection related signals.

In most cases these modifications depend on gene expression. Northern blot analysis of HRGP mRNAs, for example, shows that five mRNA species appear in bean and melon in response to infection, elicitors, or ethylene (Showalter et al., 1985; Rumeau et al., 1988). The increased deposition of HRGP molecules has been localised in these host plants by ultrastructural immunocytochemistry. Immunogold labeling techniques using antibodies prepared against melon HRGP (Mazau et al., 1988) have revealed that HRGP accumulates as paramural deposits and within papillae, in close contact to bacterial (Pseudomonas) and fungal (Colletotrichum) pathogens (Brown et al., 1988).

CONCLUSION

Cell wall degradation is one of the earliest events in host plants during plant-pathogen interactions. Among the fragments released by this process, pectic oligosaccharides might play an important role, i.e. the role of host endogenous elicitors. As such, they induce a cascade of endogenous signals, most notably ethylene. In view of the role played by pectinases, pectin degradation and ethylene during cell separation, it would be interesting to compare more thoroughly the signals and the target sites involved in disease and senescence phenomena.

REFERENCES

Bell AA (1981) Biochemical mechanisms of disease resistance. Ann Rev Plant Physiol 32:21-81

Bolwell GP, Robbins MP, Dixon RA (1985) Metabolic changes in elicitor-treated bean cells. Enzymic responses associated with rapid changes in cell wall components. Eur J Biochem 148:571-578

Broglie KE, Gaynor JJ, Broglie RM (1986) Ethylene-regulated gene expression : molecular cloning of the genes encoding an endochitinase from Phaseolus vulgaris. Proc Natl Acad Sci 83:6820-6824

Brown IR, O'Connell RJ, Mansfield JW, Bailey JA, Mazau D, Rumeau D, Esquerré-Tugayé MT (1989) Accumulation of hydroxyproline rich glycoproteins in plants at sites of resistance to pathogens. (Submitted)

Chappell J, Hahlbrock K, Boller T (1984) Rapid induction of ethylene biosynthesis in cultured parsley cells by fungal elicitor and its relationship to the induction of phenylalanine ammonia-lyase. Planta 161:475-480

Collinge DB, Slusarenko AJ (1987) Plant gene expression in response to pathogens. Plant Mol Biol 9:389-410

Davis KR, Hahlbrock K (1987) Induction of defense responses in cultured parsley cells by plant cell wall fragments. Plant Physiol 85:1286-1290

Darvill AG, Albersheim P (1984) Phytoalexins and their elicitors. A defense against microbial infection in plants. Ann Rev Plant Physiol 35:243-275

Ecker J.R., Davis RW (1987) Plant defense genes are regulated by ethylene. Proc Nat Acad Sci 84:5202-5206

Fournier J, Pélissier B, Esquerré-Tugayé MT (1986) Induction d'une activité lipoxygénase dans les cellules de tabac (Nicotiana tabacum) en culture, par des éliciteurs d'éthylène de Phytophthora parasitica var. nicotianae. CR Acad Sc 303:651-654

Gaynor JJ, Broglie R (1985) Defense genes in bean seedlings : induction of chitinase by ethylene. In. Plant Genetics, A Lan R. Liss ed., p 617-627

Kurosaki F, Tsurusawa Y, Nishi A (1987) Breakdown of phosphatidylinositol during the elicitation of phytoalexin production in cultured carrot cells. Plant Physiol 85:601-604

Lieberman M, (1979) Biosynthesis and action of ethylene. Ann Rev Plant Physiol 30:533-591

Mauch F, Hadwiger LA, Boller T (1984) Ethylene : symptom, not signal for the induction of chitinase and β-1,3-glucanase in Pea pods by pathogens and elicitors. Plant Physiol 76:607-611

Mazau D, Rumeau D, Esquerré-Tugayé MT (1988) Two different families of hydroxyproline-rich glycoproteins in melon callus. Biochemical and immunochemical studies. Plant Physiol 86:540-546

Paradies I, Konze JR, Estner E, Paxton J (1980) Ethylene indicator but not inducer of phytoalexin synthesis in soybean

Pélissier B, Thibaud JB, Grignon C, Esquerré-Tugayé MT (1986) Cell surfaces in plant-microorganism interactions. VII. Elicitor preparations from two fungal pathogens depolarize plant membranes. Plant Science 46:103-109

Roby D, Toppan A, Esquerré-Tugayé MT (1985) Cell surfaces in plant-microorganism interactions. V. Elicitors of fungal and of plant origin trigger the synthesis of ethylene and of cell wall hydroxyproline-rich glycoprotein in plants. Plant Physiol 77:700-704

Roby D, Toppan A, Esquerré-Tugayé MT (1986) Cell surfaces in plant-microorganism interactions. VI. Elicitors of ethylene from Colletotrichum lagenarium trigger chitinase activity in melon plants. Plant Physiol 81:228-233

Roby D, Toppan A, Esquerré-Tugayé MT (1987) Cell surfaces in plant micro-organism interactions. VIII. Increased proteinase inhibitor activity in melon plants in response to infection by Colletotrichum lagenarium or to treatment with an elicitor fraction from this fungus. Physiological and Molecular Plant Pathology 30:453-460

Rohwer F, Fritzemeier KH, Scheel D, Hahlbrock K (1987) Biochemical reactions of different tissues of potato (Solanum tuberosum) to zoospores or elicitors from Phytophthora infestans. Accumulation of sesquiterpenoid phytoalexins. Planta 170:556-561

Rumeau D, Mazau D, Panabières F, Delseny M, Esquerré-Tugayé MT (1988) Accumulation of hydroxyproline-rich glycoprotein mRNAs in infected of ethylene treated melon plants. Physiological and Molecular Plant Pathology (in press)

Schmidt WE, Ebel J (1987) Specific binding of a fungal glucan phytoalexin elicitor to membrane fractions from soybean Glycine max. Proc Natl Acad Sci 84:4117-4121

Showalter AM, Bell JN, Cramer CL, Bailey JA, Varner JE, Lamb CJ (1985) Accumulation of hydroxyproline-rich glycoprotein mRNAs in response to fungal elicitor and infection. Proc Natl Acad Sci 82:6551-6555

Toppan A., Esquerré-Tugayé MT (1984) IV. Fungal glycopeptides which elicit the synthesis of ethylene in plants. Plant Physiol 75:1133-1138

HOST CELL WALL LOOSENING AND SEPARATION BY PLANT PATHOGENS

R.M. Cooper
School of Biological Sciences,
University of Bath,
Claverton Down,
Bath
BA2 7AY
U.K.

INTRODUCTION

To achieve infection, many microbial pathogens respond to numerous signals, physical and chemical, from their host plants. Some of these have been identified and are known to activate genes controlling infection; such as certain plant phenolics and expression of virulence genes of *Agrobacterium tumefaciens* and nodulation genes of *Rhizobium* spp., serinol on leaf surfaces of sugar cane which stimulates HS-toxin production by *Helminthosporium sacchari*, and α-tocopherol which results in formation of parasitic mycelium of *Ustilago violacea* (Halverson and Stacey, 1986).

Mono- or dimeric saccharides released from plant cell walls are the signals for induction of enzymes responsible for cell wall loosening or cell separation by some pathogens. Subsequently, certain wall fragments solubilized by these enzymes can function as signals for various responses of the host plant (Cooper, 1983).

Plant cell walls are a potential barrier to microbial pathogens which inevitably encounter them during host colonization, Nevertheless, many parasites penetrate host walls with relative ease and some grow entirely within them, perhaps avoiding resistance reactions triggered by contact with host protoplasts. Most pathogens produce a wide range of cell wall-degrading enzymes (CWDE) corresponding to the diverse polysaccharides which constitute plant cell walls.

Endo-acting polygalacturonan hydrolases and lyases appear to be the only CWDE capable of causing cell separation. Most facultative fungal and bacterial plant parasites produce these enzymes but relatively few cause cell separation during infection. Loosening of cell wall structure is probably a more frequent event, and for fungi with their ability to exert mechanical localized pressure, it is sufficient to facilitate penetration. Other than endo-pectinases, for certain pathogens various evidence

NATO ASI Series, Vol. H 35
Cell Separation in Plants
Edited by D. J. Osborne and M. B. Jackson

implicates neutral polysaccharide-degrading enzymes and non-enzymic mechanisms in wall loosening.

The fundamentally different strategies used by parasites often reflects the nature and extent of host wall breakdown. Because substantial breakdown of primary walls by most necrosis-causing parasites (necrotrophs) usually results in disruption of adjacent protoplasts, it follows that infection by obligate parasites (biotrophs) and mutualistic mycorrhizal fungal involves minimal alterations to host walls (Cooper, 1981; 1983).

This brief account will outline the role of microbial CWDE in infection and cell wall changes and compare certain events with those which occur in healthy plants during cell separations.

SOME PROPERTIES OF MICROBIAL WALL-DEGRADING ENZYMES

The ability to produce enzymes which degrade plant cell wall polymers is not unique to plant pathogens and may reflect saprophytic phases of growth. Most characterized CWDE are small (25-35 Kd), extracellular glycoproteins. Thus the macerating *Erwinia chrysanthemi* secretes over 98% of its main pectinase (Collmer and Keen, 1986). However, increasing evidence is coming to light on involvement of wall-bound forms in organisms with more subtle forms of parasitism as described later.

Single activities usually comprise multiple forms of isoenzymes. Multiplicity can result from post-translational modifications or artefacts but Keen *et al.* (1984) showed that at least two of the five pectate lyases of *E. chrysanthemi* were products of dissimilar genes. Presumably isoenzymes confer flexibility under different conditions of substrate, pH, ions and in the face of host inhibitors. Of various forms of Monilinia, the one with the narrowest specificity for host tissue and region had the least number of isoenzymes (Willetts *et al.*, 1977); however, for the vascular parasites *Verticillium albo-atrum* and *V. dahliae* the relatively few (3-5) polygalacturonase (PG) isoenzymes of the latter do not limit its host range which is similarly wide to that of *V. albo-atrum* which has over 20 apparent forms of PG (Durrands *et al.*, 1988).

Enzyme multiplicity presents considerable problems for detection of mutations in single structural genes because of phenotypic masking (Cooper,

1987). Therefore, there are few genetic analyses of the role of CWDE in pathogenesis, although it is being overcome by *in vitro* mutagenesis of cloned genes of *E. chrysanthemi* (Roeder and Collmer, 1985). Secretory mutants are the most common phenotype obtained, such as for *E. chrysanthemi*, *Xanthomonas campestris* and *V. albo-atrum*. All are non-pathogenic or avirulent (Cooper, 1987; Durrands and Cooper, 1988).

Not all polysaccharidases are effective as CWDE for reasons of size, charge, host inhibitors or substrate inaccessibility. Some CWDE must act in concert or in sequence with others. The interdependence of endo- and exo-cellulases in degradation of crystalline cellulose is well known (Wood and McCrae, 1979); these enzymes and others of related function sometimes exist as complexes such as the cellulase of *Clostridium thermocellum* (Lamed *et al.*, 1983), pectin methylesterase and pectate lyase of *C. multifermentans* (Sheiman *et al.*, 1976) and several wood-degrading enzymes of *Poria placenta* (Highley *et al.*, 1981). Endo-PG activity of *Colletotrichum lindemuthianum* is a prerequisite for degradation of sycamore walls by glycosidases, a cellulase and an exo-PG, and allows markedly enhanced subsequent action by a protease and endo-glucanase (see Cooper *et al.*, 1978). Similarly, arabanases or arabinosidases facilitate arabinoxylan hydrolysis by xylanase by removing the blocking arabinose side chains (Cooper *et al.*, 1988). Coincidentally, fungal CWDE often appear in sequence when pathogens are grown on host cell walls as sole carbon source, in the order pectinase - hemicellulase - cellulase on dicotyledonous walls and arabanase - xylanase - glucanase on cereal walls (Cooper *et al.*, 1988).

Synthesis of most CWDE of plant pathogens is regulated by specific induction by cell wall mono- or disaccharides which are initially released by basal enzyme (Cooper, 1983). Thus pectic enzymes of *V. albo-atrum* and *Fusarium oxysporum* are induced by galacturonic acid, arabanase by arabinose, xylanase by xylose and so on. In view of the abundance of glucose in the free state it is logical that cellulases and amylases are regulated by the corresponding disaccharides, cellobiose and maltose. A dimer, unsaturated digalacturonic acid, is also the most effective inducer of pectate lyases of *Erwinia chrysanthemi* and *E. carotovora* although the final intracellular inducer is deoxyketuronic acid (Collmer *et al.*, 1982). In *E. chrysanthemi*, several endo- and exo-acting, extracellular and intracellular pectic enzymes are involved in substrate degradation and inducer generation; at least two of them, extracellular pectate lyase and exo-PG, are co-ordinately regulated.

Production is also usually subject to various degrees of catabolite repression, caused non-specifically by a wide range of energy sources.

There are many clear cases of looser control of synthesis, such as arabanases of some cereal pathogens which are constitutive and non-repressible (Cooper *et al.*, 1988), which might suggest a key role in parasitic and/or saprophytic stages of the life cycle. The influence on parasitism of these different levels of control is not known but screening media have been developed for selection of various regulatory mutants (Cooper, 1987).

CELL WALL CHANGES DURING PATHOGENESIS

Polygalacturonan degradation. Primary walls are rapidly and extensively degraded in a wide range of diseases involving necrosis, such as lesions on stems and leaves, damping off, soft and dry rots. Most evidence relates to involvement of enzymes which randomly degrade cell wall polygalacturonan. Most pathogens produce polygalacturonan hydrolases and lyases. The key role of this polymer in cell wall structure is evident from the major effects caused by endo-pectinases, which release not only galacturonides but the covalently linked sugars galactose, arabinose and rhamnose (Basham and Bateman, 1976).

Pectinases are the first CWDE produced of a sequence *in vitro* and in infected tissues of several diseases such as *Pyrenochaeta* rot of onions, *Mycosphaerella* leaf spot of pea and *Verticillium* wilt of lucerne (Cooper, 1983). Along with early appearance of these enzymes is the rapid depletion or alteration of galacturonide, such as up to 90% loss from cell walls in developing lesions of *Sclerotium rolfsii* and *Rhizoctonia solani* on bean hypocotyls (Bateman and Basham, 1976), and 93% demethoxylation of wall pectin in sunflower stems by *Sclerotinia sclerotiorum* (Hancock, 1966).

Ultrastructural evidence reveals profound wall changes in epidermal cells during the first few hours of penetration by *Botrytis cinerea* and *B. fabae* (Cooper, 1981) and may relate to the presence of endo-PG in ungerminated spores (Verhoeff and Liem, 1978). Cell wall changes often occur as a selective dissolution of the middle lamella well in advance of pectolytic fungi and bacteria. In contrast to similar images from abscission zones (Sexton and Hall, 1974) and ripening fruit (Ben Arie *et al.*, 1979) this type of wall breakdown always results in disruption of contiguous host cytoplasm (Cooper, 1981). This destructive strategy employed by necrotrophs functions to release water and nutrients and to prevent host defences, such as suppression of *Vicia faba* phytoalexins by *B. fabae* (Mansfield and Richardson, 1981) and of suberin in potato tubers by *E. carotovora* (Zucker and Hankin, 1970).

Studies with purified microbial CWDE show that homogeneous endo-PGs, pectin and pectate lyases have a unique ability to cause cell separation (maceration) and simultaneous damage to protoplasts (Basham and Bateman, 1976). The author is unaware of equivalent investigations with endo-polygalacturonases from plants. Exo-acting pectinases and other enzymes such as arabanases and galactanases are ineffective.

Evidence for a role in maceration also comes from cloning of pectate lyase (PL) genes from *E. chrysanthemi* and *E. carotovora* in *E. coli*. Some PL-overproducing transformants were capable of macerating potato tuber tissue (Collmer and Keen, 1986). Inactivation of pel genes by marker exchange has shown that not all of the 4 or 5 isozymes are required for pathogenesis. Using *Saintpaulia ionantha* inoculated with *E. chrysanthemi* mutants, Boccara *et al.* (1988) found that loss of PL b or PL c had no effect on virulence, whereas PL a^- and PL d^- strains were less virulent and showed reduced systemic colonization and PL d^- was non-invasive.

Cell killing resulting from pectinase activity has long been considered to result from damage to the plasma membrane because the weakened cell wall can no longer contain the turgid protoplast but plasmolysis prevent injury (Basham and Bateman, 1975). However, leakage of ions occurs almost immediately following exposure to endo-pectinases and ultrastructural evidence does not show protoplasts rupturing through damaged walls. Indeed, the overall integrity of cell walls is often evident even when adjacent cellular injury occurs (Hislop *et al.*, 1979; Keon, 1985) and implies that insufficient weakening has occurred at this stage to permit cell rupture. Detailed consideration of alternative mechanisms are beyond the scope of this account but includes the release of toxic oligogalacturonide wall fragments which may act at the plasmalemma - wall interface (Cooper, 1984; Keon *et al.*, 1987; Yamazaki *et al.*, 1983).

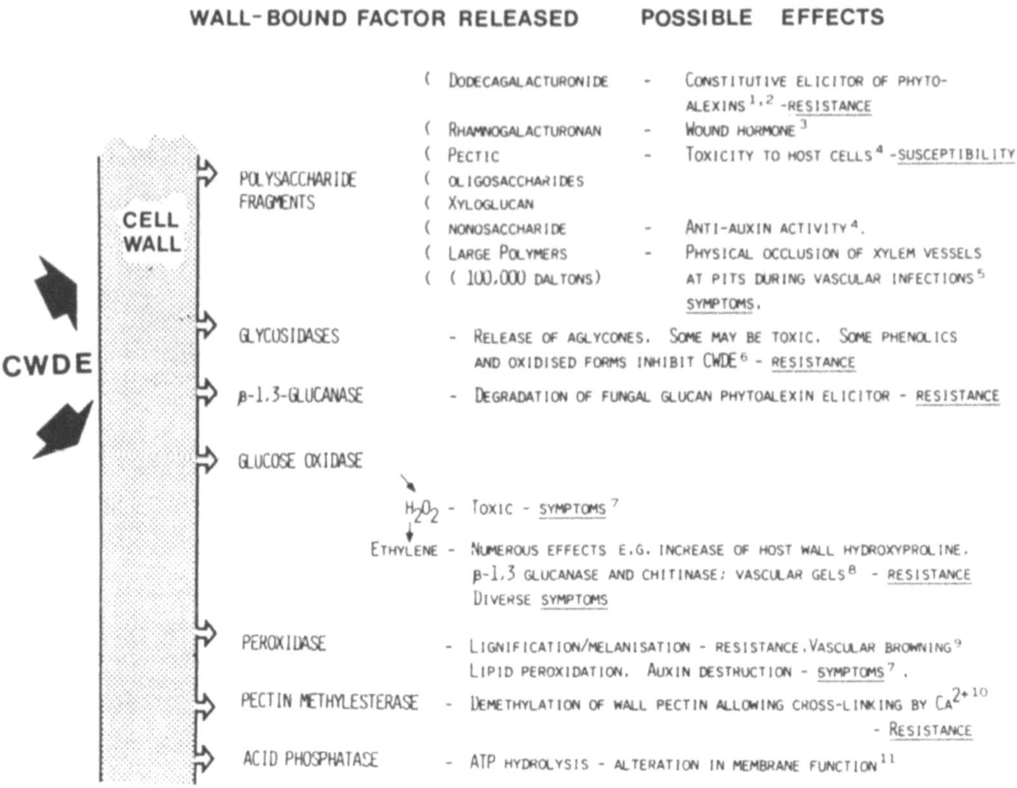

FIGURE 1. Possible diverse effects resulting from enzymic degradation of plant cell walls. Adapted from Cooper (1984). 1, Davis et al. 1984; 2, Nothnagel et al. 1981; 3, Ryan et al. 1981; 4, McNeil et al. 1984; 5, Durrands and Cooper, 1988; 6, Cooper, 1983; 7, Mussell and Strand, 1977, see Cooper, 1984; 8, Lamport, Labavitch this volume; 9, Cooper, 1984; 10, Basham and Bateman, 1976; 11, Hislop et al. 1979.

Many other molecules with potential physiological effects, including oligosaccharides and enzymes, are part of or bound to plant cell walls and may be released or exposed following wall breakdown as summarized in Figure 1. These can give rise to various symptoms or may influence host disease resistance. Thus, pectic fragments are claimed to show regulatory properties by inducing synthesis of a proteinase inhibitor at a distance in tomato plants (Ryan *et al.*, 1981) and others elicit phytoalexin production in several plants such as castor bean and soybean (Davis *et al.*, 1984; Labavitch, this volume). Because endo-pectinases have the potential for elicitation of phytoalexins there must be a fine balance between triggering host resistance in this way and suppression of host defences by cell killing. Successful parasitism is presumably dependent upon either (i) non-production of endo-pectinases as in obligate biotrophs (see below) and some ecto-mycorrhizal fungi, or (ii) strict control of synthesis or activity as in certain facultative biotrophs such as *Colletotrichum lindemuthianum*, *Phytophthora infestans* and certain ecto-mycorrizal fungi, (Keon *et al.*, 1987; Wijesundera *et al.*, 1984) or (iii) rapid and high production of extracellular activities which kill cells in advance of invasion and render wall fragments ineffective as phytoalexin elicitors by degradation to small inactive oligomers (Nothnagel *et al.*, 1983).

Neutral polysaccharides degradation. It is apparent from our recent work with obligate biotrophs that wall penetration can occur without degradation of polygalacturonan. Fungi representing powdery mildews (*Erysiphe graminis*), downy mildews (*Bremia lactucae*), rusts (*Uromyces* spp) and smuts (*Ustilago maydis*) produce a range of glycanases and glycosidases but not endo-PG, pectic lyases or cellulases. In contrast to facultative organisms, activities are generally lower and largely cell-bound rather than extracellular (Cooper, 1983; Keon *et al.*, 1987). This pattern of activity could perhaps have been predicted from the ultrastructural evidence showing highly localized wall erosion during penetration (Cooper, 1981).

CWDE which degrade neutral polymers of the wall matrix may also be used by a wide range of facultative parasites. Enzymes such as galactanases, arabanases and xylanases are readily produced in high levels, and are frequently present in infected tissue corresponding with a depletion in the polysaccharide substrate in host cell walls. Crude mechanical tests suggest that none of these CWDE affect overall wall integrity (e.g. Baker *et al.* 1979; Bauer *et al.*, 1977; Cooper *et al.*, 1978) and consequently they have usually been relegated to a role in providing nutrition for certain parasites. Nevertheless, a pure endo-galactanase of *Sclerotinia sclerotiorum* released *ca.* 56% of the wall matrix from potato tubers (Bauer *et al.*, 1977). β-1,4-galactan constitutes an unusually high proportion of tuber walls and coincidentally endo-galactanases seem to perform a key role in infection by *P. infestans* and *Phoma exigua* (Friend, this volume).

In graminaceous monocots, the predominant matrix polymer is arabinoxylan; non-cellulosic mixed β-1,3- and β-1,4-glucans may also characterize cereal cell walls (McNeil *et al.*, 1984). Fungal pathogens of cereals possess a corresponding spectrum of CWDE. Thus α-arabanase and β-1,4-xylanase, then β-1,3-glucanase are produced in sequence on host cell walls *in vitro* by *Fusarium culmorum*, *Rhizoctonia cerealis* and *Ustilago maydis;* xylanase is the main activity in appressoria of *E. graminis* and xylanase is present in *Rhizoctonia* lesions, maize smut galls and in wheat roots infected with *Gaeumannomyces graminis* (see Cooper *et al.*, 1988). Ishii (1984) claims to have caused cell separation in grass leaves with a pure xylanase.

A constitutive β-1,3-glucanase is produced during infection of cereal florets by *Claviceps purpurea.* In this instance its function may be to degrade wound callose and phloem callose in order to ensure a continued nutrient supply to the developing ergot (Dickerson *et al.*, 1978). *Zea mays* produces an endoglucanase specific for mixed β-glucan in elongating coleoptile tissues (Hatfield and Nevins, 1987). However, the presence of an equivalent enzyme in plant pathogens has not been reported.

α-Arabinosidase has been correlated with pathogenicity of mutants of the brown rot fungus of apple *Monilinia fructigena* by Howell (1957), and the enzyme and its breakdown product arabinose are present in infected fruit.

Cellulose degradation. True cellulase activity is usually absent during early critical stages of infection but may appear in moribund tissue (Cooper, 1983). Exceptions include *R. solani* and *S. sclerotiorum* which degrade wall cellulose within 48 hours (e.g. Lumsden, 1969). Also vascular parasites such as *Verticillium albo-atrum* are in frequent contact with walls of xylem vessels and can erode secondary walls (Bishop and Cooper, 1983).

The general lack of cellulase involvement might be predicted from the refractory nature of cellulose in cell walls. Analogously, cellulases do not seem to be involved in ripening of most fruits in which glucan levels are unaffected (Knee, 1978); however, "cellulases" have been correlated with ripening of avocado (Pesis *et al.*, 1978) and with abscission in some species (Sexton, this volume). It is unfortunate perhaps that most reports from plant physiology have been concerned only with endoglucanase ("Cx" cellulase) which may show no activity against native cellulose. Nevertheless, endoglucanases can reverse the hydrogen bonding of xyloglucan to cellulose fibrils and could affect wall integrity in this way (McNeil *et al.*, 1984). However, the cellulase from abscission zones of *Phaseolus vulgaris* did not degrade tamarind xyloglucan (R. Sexton, pers. comm.).

RESTRICTION OF ENZYME ACTIVITY IN CELL WALLS

During cell separation in plants it is apparent that wall degradation is a controlled, often highly localized process. *Prima facie* maintaining viability of cells adjacent to abscission zones (Sexton, this volume) would ensure a defence capability of the newly exposed surface and would not provide senescing cells from which opportunist pathogens could gain ingress. In addition to obligate parasites described above, some facultative and mycorrhizal fungi do not cause extensive wall changes in spite of high potential to produce endo-pectinases *in vitro.* Indeed, fungi such as *V. albo-atrum* and *C. lindemuthianum* exhibit an initial biotrophic phase (Cooper, 1983). Some possible reasons for restriction of enzyme activity, which might also be applicable to processes in healthy plants, will be considered.

Molecular sieving. Pore diameters in primary walls of plants have been calculated as 35-52 Å and 70 Å; the latter would allow permeation by globular proteins of 60 Kd or less (Cooper, 1983). Low molecular weight pectinases (*ca.* 30 Kd) are well conserved in necrotrophic pathogens, and Knee *et al.* (1975) showed that larger enzymes were less effective in wall degradation; in particular *P. infestans* produces an exceptionally large PG of *ca.* 350 Kd with very low activity on host cell walls, which might explain its initial period of biotrophy.

Ionic binding. At least one of the forms of pectinase secreted by fungal and bacterial pathogens is typically basic with a pI of 8.5 or above (Keon *et al.*, 1987). These CWDE have a positive charge at cell wall pH values and may be effectively immobilised by binding to negatively charged walls. For example *ca.* 98% of a pectin lyase (pI 9) from *F. oxysporum* f.sp. *pisi* binds to pea stem cell walls at pH 4 (Cooper, 1983). Also we have found CWDE ionically bound to walls of some parasites and endo-PG and endo- glucanase remain attached to mycelium of some ectomycorrhizal fungi (Keon *et al.* 1987).

Inhibitors. Wall-bound PG inhibitors have been found in many plants and often implicated in disease resistance by restricting microbial enzymes. It is also feasible that their main purpose is to control wall breakdown by endogenous PG. However, the efficacy and wall-bound nature of these putative inhibitors has been questioned by Cooper in Keon *et al.* (1987).

Mode of enzyme action. As already discussed, the mode of action of pectinases (terminal or random cleavage) determines the extent of wall breakdown and cell damage. "Multiple" attack (combined endo-exo action) enzymes are frequently found but are rarely considered by plant pathologists or physiologists. They have the potential for causing wall loosening without rapid maceration, as shown for a PG of *Aspergillus* by McClendon (1979). The multiple attack PGs of *V. albo-atrum* and *C. linemuthianum* may explain their early biotrophic parasitism (Cooper *et al.* 1978).

NON-ENZYMIC WALL LOOSENING

The integrity of plant cell walls may also be altered during pathogenesis by non-enzymic means.

Chelation of calcium. The decreased ability of many facultative parasites, such as *Pythium* spp. and *R. solani* to invade mature plant tissues is partly due to dramatic changes with age in cell wall polysaccharides, lignification and incorporation of calcium to form PG-resistant cross-linked polygalacturonan (Cooper *et al.*, 1978). *Sclerotium rolfsii* and *S. sclerotiorum* overcome the problem by secreting copious amounts of oxalic acid which both chelates wall calcium and reduces wall pH (from *ca.* 6 to 4) which complements the low pH optima of their CWDE (Bateman and Beer, 1965; Hancock, 1966; Cooper, 1983). At least one bacterial pathogen, *Erwinia amylovora*, produces an extracellular polysaccharide with high affininity for Ca^{2+} ions (A. Smith pers. comm.); the pathogens lacks any CWDE but grows intercellularly and sometimes passes through ruptured (weakened?) host walls (Youle and Cooper, 1987).

Depolymerization by H_2O_2. "Brown rot" fungi of wood use a non-enzymatic cellulose-degrading system mediated by extracellular H_2O_2-Fe^{2+}; lignin breakdown involves H_2O_2-peroxidase. Although not yet implicated in plant pathogenesis recent evidence suggests that the parasite *Botrytis cinerea* uses similar mechanisms in degradation of flax stem fibres (Bratt *et al.*, 1988).

Auxin and ethylene increase. Hyperauxiny characterizes infection by most obligate biotrophs and ethylene is generated in tissues damaged by facultative parasites. Their effects on wall loosening are discussed in other chapters in this volume.

CONCLUSIONS

Although certain aspects of cell separation in healthy and infected plants may differ, plant pathogens of which many have co-evolved with a specific host species, afford an array of diverse, characterized CWDE with which to study the enzymic contribution to cell wall changes. It is likely that in many cases, especially for pathogens which frequently penetrate or grow within host walls, their enzymes are adapted to the wall composition of the host plant; the specialized arabinoxylan-degrading ability of cereal pathogens is one case in point.

Comparison of the two fields is of obvious mutual benefit; knowledge from the better studied, more accessible microbial enzymes suggests that with plants, more attention should be paid to enzymes which might act in sequence or in concert, rather than relating a single activity to a physiological event. Also, appropriate forms of enzymes, active on cell walls and not just on model polymers, should be studied, such as true cellulase complexes rather than only Cx "cellulase" with its convenient soluble substrate. A fascinating enigma is the contrast between killing of plant cells by microbial endo-polygalacturonases and the apparent non-toxicity of endogenous polygalacturonases. This would seem to warrant a comparison of the enzymes' properties in terms of size, charge and mode of action.

REFERENCES

Baker CJ, Whalen CH, Korman RZ and Bateman DF (1979) α-L-Arabinofuranosidase from *Sclerotinia sclerotiorum*: purification, characterization and effects on plant cell walls and tissue. Phytopathology 69: 789-793

Bateman DF and Basham HG (1976) Degradation of plant cell walls and membranes by microbial enzymes. In: Heitefuss R and Williams PH (eds) Encyclopedia of Plant Physiology, vol. IV. Springer-Verlag, New York pp 316-355

Bateman DF and Beer SV (1965) Simultaneous production and synergistic action of oxalic acid and polygalacturonase during pathogenesis by *Sclerotium rolfsii.* Phytopathology 55: 204-211

Bauer WD, Bateman DF and Whalen CH (1977) Purification of an endo-β-1,4-galactanase produced by *Sclerotinia sclerotiorum*: effects on isolated plant cell walls and potato tissue. Phytopathology 67: 862-868

Ben Arie R, Kislev N and Frenkel C (1979) Ultrastructural changes in the cell walls of ripening apple and pear fruit. Plant Physiol 64: 197-202

Bishop CD and Cooper RM (1983) An ultrastructural study of vascular colonization in three vascular diseases. I. Colonization of susceptible varieties. Physiol Plant Pathol 23: 323-343

Boccara M, Diolez A, Rouve M and Kotoujansky A (1988) The role of individual pectate lyases of *Erwinia chrysanthemi* strain 3937 in pathogenicity on Saintpaulia plants. Physiol Molec Plant Pathol 33: 95-104

Bratt RP, Brown AE and Mercer PC (1988) A role for hydrogen peroxide in degradation of flax fibres by *Botrytis cinerea*. Trans Br Mycol Soc 91: 481-488

Collmer A, Berman P and Mount MS (1982) Pectate lyase regulation and bacterial soft-rot pathogenesis. In Mount MS and Lacey GH (eds) Phytopathogenic Prokaryotes, Vol I. Academic Press, London, pp 395-422

Collmer A and Keen NT (1986) Role of pectic enzymes in plant pathogenesis. Ann Rev Phytopath 24: 383-409

Cooper RM (1981) Pathogen-induced changes in host ultrastructure. In: Staples RC and Toenniesen GH (eds) Plant disease control: resistance and susceptibility. Wiley, New York, pp 105-142

Cooper RM (1983) The mechanisms and significance of enzymic degradation of host cell walls by parasites. In: Callow J (ed) Biochemical Plant Pathology. John Wiley & Sons, New York, pp 101-135

Cooper RM (1984) The role of cell wall-degrading enzymes in infection and damage. In Wood RKS and Jellis GJ (eds) Plant diseases: infection damage and loss. Blackwell Scientific Publications, Oxford, pp 13-28

Cooper RM (1987) The use of mutants in exploring depolymerases as determinants of pathogenicity. In: Day PR and Jellis GJ (eds) Genetics and Plant Pathogenesis. Blackwell Scientific Publications, Oxford pp 261-281

Cooper RM, Rankin B and Wood RKS (1978) Cell wall-degrading enzymes of vascular wilt fungi. II. Properties and modes of action of polysaccharidases of *Verticillium albo-atrum* and *Fusarium oxysporum* f.sp. *lycopersici*. Physiol Plant Pathol 13: 101-134

Cooper RM, Longman D, Campbell A, Henry M and Lees PE (1988) Enzymic adaptation of cereal pathogens to the monocotyledonous primary wall. Physiol Molec Plant Pathol 32: 33-47

Davis KR, Lyon GD, Darvill AG and Alberheim P (1984) Host-pathogen interactions XXV. Endopolygalacturonic acid lyase from *Erwinia carotovora* elicits phytoalexin accumulation by releasing plant cell wall fragments. Plant Physiol 74: 52-60

Dickerson AG, Mantle PG, Nisbet LJ and Shaw BI (1978) A role for β-glucanases in the parasitism of cereals by *Claviceps purpurea*. Physiol Plant Pathol 12: 55-62

Durrands PK and Cooper RM (1988) Selection and characterization of pectinase - deficient mutants of the vascular wilt pathogen *Verticillium ablo-atrum*. Physiol Molec Plant Pathol 32: 343-362

Durrands PK, Keene RA, Cooper RM, O'Garro LW and Clarkson JM (1988) Poly-galacturonase isozyme profiles from *Verticillium dahliae* isolates races 1 and 2 from different geographical origins. Trans Br mycol Soc 91: 533-535

Halverson LJ and Stacey G (1986) Signal exchange in plant-microbe inter-actions. Microbial Rev 50: 193-225

Hancock JG (1966) Degradation of pectic substances associated with patho-genesis by *Sclerotinia sclerotiorum* in sunflower and tomato stems. Phytopathology 56: 975-9

Hatfield RD and Nevins DJ (1987) Hydrolytic activity and substrate specificity of an endoglucanase from *Zea mays* seedling cell walls. Plant Physiol 83: 203-207

Highley TL, Wolter KE and Evans FJ (1981) Polysaccharide-degrading complex produced in wood and in liquid media by the brown-rot fungus *Poria placenta.* Wood and Fibre 13: 265-274

Hislop EC, Keon JPR and Fielding AH (1979) Effect of pectin lyase from *Monilinia fructigena* on viability, ultrastructure and localization of acid phosphatase of cultured apple cells. Physiol Plant Pathol 14: 371-81

Howell, HE (1975) Correlation of virulence with secretion *in vitro* of three wall-degrading enzymes in isolates of *Sclerotinia fructigena* obtained after mutagen treatment. J Gen Microbiol 90: 32-40

Ishii S (1984) Cell wall cementing materials of grass leaves. Plant Physiol 76: 959-961

Jarvis MC, Threlfall DR and Friend J (1981) Potato cell wall polysaccharides: degradation with enzymes from *Phytophthora infestans.* J. Exp Bot 32: 1309-1319

Keen NT, Dahlbeck D, Staskawicz B and Belser W (1984) Molecular cloning of pectate lyase genes from *Erwinia chrysanthemi* and their expression in *Escherichia coli.* J Bacteriol 159: 825-831

Keon JPR (1985) Cytological damage and cell wall modification in cultured apple cells following exposure to pectin lyase from *Monilinia fructigena.* Physiol Plant Pathol 26: 11-29

Keon JPR, Byrde RJW and Cooper RM (1987) Some aspects of fungal enzymes that degrade plant cell walls. In: Pegg GF and Ayres PG (eds) Fungal infection in plants. Cambridge University Press, pp 133-157

Knee M (1978) Properties of polygalacturonate and cell cohesion in apple fruit cortical tissue. Phytochem 17: 1257-1260

Knee M, Fielding AH, Archer SA and Laborda F (1975) Enzymic analysis of cell wall structure in apple fruit cortical tissue. Phytochem 14: 2213-2222

Lamed R, Setter E and Bayer EA (1983) Characterization of a cellulose-binding, cellulase-containing complex in *Clostridium thermocellum.* J Bacteriol 156: 828-836

Lumsden RD (1969) *Sclerotinia sclerotiorum* infection of bean and the production of cellulase. Phytopathology 59: 653-657

Mansfield JW and Richardson A (1981) The ultrastructure of interactions between *Botrytis* species and broad bean leaves. Physiol Plant Pathol 19: 41-8

McClendon JH (1979) Subterminal polygalacturonase, a nonmacerating enzyme attacks pectate from the reducing end. Plant Physiol 63: 75-78

McNeil M, Darvill AG, Fry SC and Albersheim P (1984) Structure and function of the primary cell walls of plants. Ann Rev Biochem 53: 625-663

Nothnagel E, McNeil M, Albersheim P and Dell A (1983) Host pathogen interactions. XXII. A galacturonic acid oligosaccharide from plant cell walls elicits phytoalexins. Plant Physiol 71: 916-26

Pesis E, Fuchs Y and Zauberman G (1978) Cellulase activity and fruit softening in avocado. Plant Physiol 61: 416-419

Roeder DL and Collmer A (1985) Marker exchange mutagenesis of a pectate lyase isozyme gene from *Erwinia chrysanthemi.* J Bacteriol 164: 51-56

Ryan CA, Bishop P, Pearce G, Darvill AG, McNeil M and Albersheim P (1981) A sycamore cell wall polysaccharide and a chemically related tomato leaf oligosaccharide possess similar proteinase inhibitor inducing activities. Plant Physiol 68: 616-18

Sexton R and Hall JL (1974) Fine structure and cytochemistry of the abscission zone of *Phaseolus* leaves. I. Ultrastructural changes occurring during abscission. Ann Bot 38: 849-854

Sheiman MI, Macmillan JD, Miller K and Chase T (1976) Coordinated action of pectinesterase and polygalacturonate lyase complex of *Clostridium multifermentans.* Eur J Biochem 64: 565-572

Verhoeff K and Liem JI (1978) Presence of endo-polygalacturonase in conidia of *Botrytis cinerea* before and during germination. Phytopathol Z 91: 110-115

Wijesundera RLC, Bailey JA and Byrde RJW (1984) Production of pectin lyase by *Colletotrichum lindemuthianum* in culture and in infected bean (*Phaseolus vulgaris*) tissue. J Gen Microbiol 130: 285-290

Willetts HJ, Byrde RJW and Fielding AH (1977) The taxonomy of the brown rot fungi (*Monilinia* spp) related to their extracellular wall-degrading enzymes. J Gen Microbiol 103: 77-83

Wood TM and McCrae SI (1979) Synergism between enzymes involved in the solubilization of native cellulose. In: Brown RD and Jurasek L (eds) Advances in Chemistry, vol 181. Hydrolysis of cellulose: mechanisms of enzymatic and acid catalysis. American Chemical Society, pp 181-209

Yamazaki N, Fry SC, Darvill AG and Albersheim P (1983) Host-pathogen interactions. XXIV. Fragments isolated from suspension-cultured sycamore cell walls inhibit the ability of cells to incorporate ^{14}C-leucine into proteins. Plant Physiol 72: 864-9

Youle D and Cooper RM (1987) Possible determinants of pathogenicity of *Erwinia amylovora*: evidence for an induced toxin. Acta Hort 217: 161-166

Zucker M and Hankin L (1970) Regulation of pectate lyase synthesis in *Pseudomonas fluorescens* and *Erwinia carotovora.* J Bacteriol 104: 13-18

THE DEGRADATION OF POTATO CELL WALLS BY PATHOGENS

P.J. Keenan and J. Friend
Department of Applied Biology
The University
Hull, HU6 7RX, UK

INTRODUCTION

In the case of fungal pathogens of potato, the amount of penetration of the tuber differs between necrotrophs, biotrophs and hemibiotrophs. In the case of the first group there is considerable penetration and wall degradation, with the second and third there is initially very little degradation but this is extended in the necrotrophic phase of the third group. In our laboratory we have been investigating the production of cell wall-degrading enzymes by several fungal pathogens and their relationship to the structure of the tuber cell walls. In addition we have looked at the possibility that either the initial wall structure, or changes in structure after infection, may be related to resistance of the tubers to the pathogens.

CELL WALL DEGRADING ENZYMES AND WALL STRUCTURE

The first fungal pathogen which was investigated in Hull was *Phytophthora infestans*, the fungus which causes late blight of potato. It can be seen from both light and electron microscopic examination that there is very little wall degradation in the initial biotrophic phase of the interaction. In order to study the enzymes involved in this degradation it was initially decided to determine the chemical composition of the walls before infection and the changes following infection. The results (Friend and Knee, 1969) showed first of all that the pectic fraction, unlike those of apple and citrus which were the substrates traditionally used by plant pathologists, contained a high proportion of galactose relative to galacturonic acid (Table 1).

Table 1. Monosaccharide composition of potato pectin

Percentage of anhydro sugar by weight of pectin		
Galactose	Galacturonic acid	Arabinose
51.1	13.5	5.0

Adapted from Knee and Friend, 1968

NATO ASI Series, Vol. H 35
Cell Separation in Plants
Edited by D. J. Osborne and M. B. Jackson

Moreover the proportion of galactose in the pectic fraction declined as the disease progressed through the tuber slices which were the tissues used in this investigation. Examination of the enzyme activity in culture filtrates in *P. infestans* showed the presence of an enzyme which caused the release of galactose from potato pectin and also from pectin prepared from white lupin seeds (Knee and Friend, 1968). Since the latter compound was known to be a β-1,4-linked galactan (Hirst, Jones and Walder, 1947) and there was very little hydrolysis of *p*-nitro-phenyl galactoside it was concluded that the enzyme was an endo-galactanase and that the pectic substrate in the potato wall was a β-1,4-galactan.

The properties of the β-1,4-galactanase were investigated in some detail and are described in Knee and Friend (1970). In the final paper of this series Knee (1970) showed that in the early stages of the infection of very thin slices of potato there was a loss of large amounts of galactan from the pectic fraction of the discs. These results thus explained the specific loss of galactose from the tuber cell walls during pathogenesis of discs but did not explain why there was only limited wall degradation during the initial penetration of the discs and also when the hyphae passed from cell to cell. Two possible explanations of this phenomenon are, either the enzyme is firmly attached to the hyphal tips, or the enzyme is so large that it cannot penetrate the host cell walls. It is clear from the paper of Knee *et al.* (1975) that the pectic hydrolase produced by *P. infestans* (Cole, 1970) is too large to diffuse through the walls but it is not clear that the same explanation holds for the galactanase.

Moreover the pectic hydrolase from *P. infestans* gave little wall degradation *in vitro* even when the wall has been previously treated with galactanase or with oxalate-citrate (Jarvis, Threlfall and Friend, 1981); these two treatments were designed to make the galacturonan component of the wall more accessible to the pectic hydrolase. Thus the *P. infestans* enzyme is different from two other pectic hydrolases, one from *Monilinia* which released large quantities of galacturonan and associated polysaccharides (Jarvis *et al.*, 1981) and an endopolygalacturonase from *Aspergillus japonicus* which caused substantial degradation of potato cell walls (Ishii, 1978).

Other pathogens of potato which have been investigated for the cell wall degrading enzymes include *Fusarium solani* var. *coeruleum* and *Phytophthora erythroseptica.* These two fungi are both necrotrophic pathogens, the former causing dry rot and the latter pink rot of potato. In both cases the cell wall degrading enzymes isolated from lesions in infected tuber tissues included galactanases (Sturdy, 1973). The lesions contained macerating activity and although the most effective macerating enzymes isolated were endopolygalacturonate lyases and hydrolases, maceration was also caused by the purified galactanases; the separate galactanases from *F. solani* var. *coeruleum* caused maceration whereas those from *P. erythroseptica* would only cause maceration in combination. It should also be emphasized that maceration caused by the galactanases is far less effective than that caused by the traditional pectic enzyme.

Examination of the reaction of, and the products obtained from, the incubation of cell walls with the pectic hydrolases and the galactanase from the culture filtrate of *P. infestans* has enabled the postulation of a structure for the pectic fraction of potato cell walls (Jarvis, Hall, Threlfall and Friend, 1981; Jarvis, Threlfall and Friend, 1981; Jarvis, 1983). This includes the suggestion that the galacturonan fraction contains large numbers of cross-links through calcium bridges which, unless treated with oxalate-citrate, are unavailable to the pectic hydrolase amd thus only a limited amount of galacturonan-containing material is released. On the other hand, there are many β-1,4-linked galactan side-chains which are accessible to the galactanase; this explains the ease of release of galaactose-containing carbohydrate following enzyme treatment of the cell walls.

The role of cell wall degrading enzymes in the infection of potato tubers by *Phoma exigua* var. *foveata* (potato gangrene) has also been investigated (Keenan, 1984).

Enzymes which degrade galactan, xylan, araban and various pectic substrates (polygalacturonic acid and pectin) were detected when the fungus was cultured in a liquid medium. Low levels of these enzymes were produced in a basal medium; the inclusion of isolated potato tuber cell walls in the medium considerably stimulated fungal growth and enzyme production, in particular galactanase activity. Similar enzymes were extracted from gangrene-infected tuber tissue. A notable difference was that cellulase activity, which was present in infected tissue, was not detected in 10 day old culture filtrates.

It is considered that galactanase activity is important in this disease and in the degradation of the potato tuber cell wall. When the fungus was cultured on liquid medium, galactanase activity was stimulated to the greatest extent by the addition of cell walls, was produced early and was the high enzyme activity detected at all stages of growth. This was an endo-acting enzyme with optimum activity at pH 5 that coincided with the pH of infected tuber tissue. Galactanase was also the highest enzyme activity detected in diseased tissue and production is presumably related to the high galactan content of the tuber cell wall.

Crude enzyme preparations from liquid media and from infected tuber tissue rapidly degraded isolated potato tuber cell walls, releasing carbohydrate which was mainly composed of galactose. The preparations also macerated, and caused ion leakage from, potato tuber tissue.

It was found that enzyme preparations obtained during the early stages of growth on cell walls were very efficient at degrading isolated cell walls *in vitro* but not efficient at degrading standard soluble polysaccharide substrates. Gel filtration separated two groups of enzymes, a high and a low molecular weight fraction each of which contained galactanase, polygalacturonase, xylanase and arabanase; β-galactosidase was confined to the high molecular weight fraction. The low molecular weight fraction was able to degrade isolated potato tuber cell walls but the high molecular weight fraction could not. When enzyme preparations from day 4 to day 10 culture filtrates were examined by gel-filtration, it was found that whereas the day 10 preparation contained both high and low molecular weight peaks of enzyme activity, the day 4 preparations only contained the low molecular weight peak. These results appear to explain the peak of wall degradability in the day 4 culture filtrates. However further work is required to determine the significance of high and low molecular weight enzymes and their role in cell wall degradation during the disease.

It was also found that similar types and levels of cell wall degrading enzymes were produced in infected tissue from both resistant and susceptible tubers, and *in vitro* when the fungus was grown on cell walls isolated from different potato varieties and tissues. The levels of enzymes probably do not contribute to the ultimate size of lesion formed. However, isolated cortical and medullary cell walls from the very susceptible variety (Ulster Sceptre) were more easily degraded than the corresponding walls from the resistant variety, H1/4. This difference in wall degrabability may contribute to the observed difference in varietal resistance which was detected by *in vivo* fungal growth rates and lesion size measurements. As far as we are aware, this is the first recorded instance where the relative resistance or susceptibility of two cultivars of a host plant appears to be related to the ease of degradation of the cell walls of the host cultivar.

THE ROLE OF CELL WALLS IN RESISTANCE

The cell walls are also involved in the active resistance of potato tuber tissue to penetration and degradation by pathogenic fungi and bacteria. The first indication that this might be an important mechanism of resistance was obtained from the experiments of Zucker and Hankin (1970) on the degradation of potato tuber tissue by *Pseudomonas fluorescens* or by the pectic lyases purified from a bacterial culture filtrate. This degradation was inhibited if the tuber discs were incubated in the air for 24 hours before inoculation but not if the discs were incubated in the presence of cycloheximide which inhibited the increased synthesis of phenylalanine ammonia lyase (PAL) normally found in the light. In this case, the surface of the discs remained white unlike those incubated in air which had a tan coloured suberized layer over their cut surface. It was found that if the brown suberized layer of discs incubated in air was removed, the discs were then more susceptible to maceration by pectic lyases. It was therefore concluded that cycloheximide blocks the development of resistance by inhibiting the production of phenolic precursors of suberization, through the inhibition of PAL synthesis.

When the carbohydrate changes in potato tuber discs infected by *Phytophthora infestans* were being examined, it was found that there was an increase in the lignin-like compound (Friend and Knee, 1969) present in the cell walls of the susceptible cultivar, King Edward. In later experiments it was found that there was a faster rate of deposition of lignin-like material in the walls of a resistant cultivar (Orion) compared with a susceptible cultivar (Majestic) inoculated with race 4 of *P. infestans*. Moreover there was a corresponding result when the levels of activity of PAL were measured (Friend, Reynolds and Aveyard, 1973). These and further experiments showed that a rapid, confined lignification and a correlated rise in PAL activity seem to be associated with the reaction of resistant rather than susceptible tuber slices and discs.

Later experiments (Henderson and Friend, 1979) showed that these correlations between lignification and PAL activity were a general feature of potato tuber discs, containing either the R_1 or R_3 genes, showing race-specific resistance to *P. infestans*. Another and possibly even more important mechanism of resistance of tuber discs to *P. infestans* is the accumulation of sesquiterpenoid phytoalexins (Varns, Currier and Kuć, 1971) but it seems likely that both phytoalexin accumulation and lignification may be involved in resistance, depending on cultivar, cultural conditions and physiological status of the tubers.

Potatoes may also show polygenic resistance to *P. infestans*. Recent experiments have indicated that such polygenic resistance in tubers is dependent upon deposition of phenolic compounds in cell walls, (Ampomah, 1983; Ampomah and Friend 1984 and 1988). This type of resistance occurs in potato cultivars which do not necessarily contain any major resistance genes. When tuber discs of two cultivars, one (Stormont Enterprise) resistant and the other (Majestic) susceptible to a complex race of *P. infestans* were inoculated 24 hours after cutting, the fungus was impeded faster in the resistant than in the susceptible cultivar. There was browning of the cell in the lesion; the formation of this dark material occurred at about the time that fungal growth was impeded. The walls of the cells in the dark region give a histochemical reaction with toluidine blue which indicated that they contained phenolic compounds. Tuber discs of the same two cultivars gave a similar response to the fungus which causes gangrene, *Phoma exigua* var. *foveata*. There were greater increases in the activities of PAL, peroxidase, especially ionically- and covalently-bound, and caffeic acid O-methyl transferase in the infected than in the uninfected tissue, whereas there was little difference in the increase of phenolase between infected and uninfected discs.

Treatment of the discs with amino-oxyacetic acid (AOA) before inoculation markedly inhibited browning, and stainability with toluidine blue, and permitted the fungi to grow further through the tuber tissue. AOA had no effect on the levels of the two sesquiterpenoid phytoalexins, rishitin and lubimin, in the tuber discs. These experiments therefore indicate that accumulation of phenolic compounds in cell walls is more important than phytoalexin accumulation in non-race specific resistance of tuber discs to *P. infestans* and in resistance to *P. exigua* (Ampomah, 1983; Ampomah and Friend, 1984 and 1988).

Changes in type and amounts of phenolic compounds in the discs after inoculation have also been determined. After freezing the discs in liquid nitrogen and extracting with chloroform:methanol:water (CMW), the residue was extracted with phenol:acetic acid:water (PAW). The CMW extracts contained chlorogenic acid; its level was lower in inoculated than in uninoculated tissue. Alkaline *minus* neutral difference spectra indicated that phenolic compounds were present both in the dialyzed PAW extract and in the PAW residue. There was always more phenolic material in both extract and residue from the inoculated discs; the levels were always higher in the resistant than in the susceptible discs.

The presence of quinic acid in the PAW extracts, but not in the PAW residue indicates that the extracts may well contain oxidized polymerized chlorogenic acid bound to protein (Pierpoint, Ireland and Carpenter, 1977; Davies, Newby and Synge, 1978). After saponification of the PAW residue the presence of *p*-coumaric and ferulic acids was shown by HPLC; these compounds were not detected in saponified PAW extracts. Ampomah and Friend (1984 and 1988) suggested that the acids were probably esterified to a cell wall component which could be lignin, suberin, carbohydrate or protein.

Hammerschmidt (1984) found that cell walls of potato tuber slices stained for lignin about 8-10 hours after inoculation with two fungi non-pathogenic on potatoes, *Cladosporium cucumerinum* and *Fusarium roseum* whereas tissue inoculated with the pathogen *F. roseum* f.sp. *sambucinum* did not give a positive lignin stain until 18-20 h. Lignin was verified by the isolation of a ligninthioglycollic acid and the presence of *p*-hydroxybenzaldehyde and vanillin from cupric oxide oxidation of alkali-treated tissue. Susceptibility of the tuber tissue was induced by pretreatment of the tuber tissue with either AOA or ABA, each of which reduced lignin deposition.

It remains difficult to draw distinctions between lignin and suberin in potato tuber tissue. Kolattukudy (1977 and 1981) has postulated that the aromatic domains in suberin contained *p*-coumaric and ferulic acids; the linkages proposed closely resemble those in most "classical" models of lignin structure. Nevertheless Hammerschmidt (pers. comm.) has found no indication of the characteristic suberin acids in potato discs aged for a short time and he has concluded that the brown material is lignin. However the hypothesis that hydroxycinnamoyl esters of galactan may be part of the insoluble phenolic material in potato reacting to pathogens (Friend, 1976) has further support from the recent work of Fry (1983) who has isolated feruloyl-arabinans and galactans and their corresponding *p*-coumaroyl derivatives from the walls of cultured spinach cells.

The phenolic material in the cell walls could act as either a chemical or a physical barrier (Friend, 1976). The hydroxycinnamoyl esterification of cell wall galactans would probably be a sufficient modification to make the galactan less degradable by the fungal galactanases and thus be a chemical barrier. Lignification (and suberization) may well impede the passage of fungal galactanases into the wall and thus be a physical barrier.

CONCLUSIONS

It is clear from the results reported here that fungal pathogens of potato degrade the galactan component of the host cell wall as part of their pathogenic mechanism. Mechanisms of resistance to these pathogens include cell wall modifications such as the deposition of phenolic compounds; it will be interesting to determine whether the apparent pre-formed resistance in the walls of certain tuber varieties also involves phenolic compounds.

REFERENCES

Ampomah YA (1983) Aspects of phenolic metabolism in the potato in response to challenge by *Phytophthora infestans* (Mont.) de Bary and *Phoma exigua* Desm. var *foveata* (Foister) Boerema. Ph.D. thesis, University of Hull

Ampomah YA, Friend J (1984). The role of lignification in the resistance of plants to attack by pathogens. App Biochem Biotechnol 9: 325-326

Ampomah YA, Friend J (1988) Insoluble phenolic compounds and resistance of potato tuber discs to *Phytophthora* and *Phoma*. Phytochemistry 27: 2533-2541

Cole AJ (1970) Pectic enzyme activity from *Phytophthora infestans*. Phytochemistry 9: 337-340

Davies AMC, Newby VK, Synge RLM (1978) Bound quinic acid as a measure of coupling of leaf and sunflower-seed proteins with chlorogenic acid congeners: loss of availability of lysine.

Friend J (1976) Lignification in infected tissue. In: Friend J, Threlfall DR (eds) Biochemical aspects of plant-parasite relationships. Academic Press, London and New York, pp 291-303

Friend J, Knee M (1969) Cell wall changes in potato tuber tissue infected with *Phytophthora infestans* (Mont.) de Bary. J Exp Bot 20: 763-775

Friend J, Reynolds SB, Aveyard MA (1973) Phenylalanine ammonia lyase, chlorogenic acid and lignin in potato tuber tissue inoculated with *Phytophthora infestans*. Physiol Plant Pathol 3: 495-507

Fry SC (1983) Feruloylated pectins from the primary cell wall: their structure and possible functions. Planta 157: 111-123

Hammerschmidt R (1984) Rapid deposition of lignin in potato tuber tissue as a response to fungi non-pathogenic on potato. Physiol Plant Pathol 23: 33-42

Henderson SJ, Friend J (1970) Increase in PAL and lignin-like compounds as race-specific responses of potato tubers to *Phytophthora infestans*. Phytopath Z 94: 323-334

Hirst EL, Jones JKN, Walder WO (1947) The pectic substances. VII Constitution of galactan from *Lupinus albus*. J. Chem. Soc. 1947: 1225-1229

Ishii S (1978) Analysis of the components released from potato tuber tissues during maceration by pectolytic enzymes. Plant Physiol 62: 586-589

Jarvis MC (1983) Personal communication

Jarvis MC, Hall MA, Threlfall DR, Friend J (1981) The polysaccharide structure of potato cell walls: chemical fractionation. Planta 152: 93-100

Jarvis MC, Threlfall DR, Friend J 91981) Potato cell wall polysaccharides: degradation with enzymes from *Phytophthora infestans*. J Exp Bot 32: 1309-1319

Keenan P (1984) The degradation of potato tuber cell walls by *Phoma exigua* var. *foveata*. Ph.D. thesis, University of Hull

Knee M (1970) The use of a new, rapid micro-method for analysing changes in the carbohydrate fractions of potato tuber tissue after invasion by *Phytophthora infestans*. Phytochemistry 9: 2075-2083

Knee, M, Fielding AH, Archer SA, Laborda F (1975) Enzymic analysis of cell wall structure in apple fruit cortical issue. Phytochemistry 24: 2213-2222

Knee M, Friend J (1968) Extracellular "galactanase" activity from *Phytophthora infestans* (Mont.) de Bary. Phytochemistry 7: 1289-91

Knee M, Friend J (1970) Some properties of the galactanase secreted by *Phytophthora infestans* (Mont.) de Bary. J Gen Microbiol 60: 23-30

Kolattukudy P (1977) Lipid polymers and associated phenols, their chemistry, biosynthesis and role in pathogenesis. In: Loewus FA, Runeckles VC (eds) The structure, biosynthesis and degradation of wood. Recent advances in phytochemistry. Plenum Press, New York, pp 185-246

Kolattukudy P (1981) Structure, biosynthesis and biodegradation of cutin and suberin. Annu Rev Plant Physiol 322: 539-567

Pierpoint WS, Ireland RJ, Carpenter JM (1977) Modification of proteins during the oxidation of leaf phenols: reaction of potato virus X with chlorogenoquinone. Phytochemistry 16: ,29-34

Sturdy ML (1973) Studies on the infection of potato tubers by *Fusarium caeruleum* (Lib.) Sacc. and *Phytophthora erythroseptica* Pethyb. Ph.D. thesis, University of Hull

Varns J, Currier W, Kuć J (1971) Specificity of rishitin and phytuberin accumulation by potato. Phytopathology 61: 968-971

Zucker M, Hankin L (1971) Physiological basis for a cycloheximide induced soft rot of potatoes by *Pseudomonas fluorescens*. Annals of Botany 34: 1047-1062

SECTION IV

Signal Specificity and Target Cell Status

INTERFERENCE BY ETHYLENE WITH THE ABSCISSION RETARDING EFFECTS OF AUXIN IN CITRUS LEAVES

R. Goren and J. Riov
Department of Horticulture
Hebrew University of Jerusalem
Rehovot 76100
Israel

INTRODUCTION

The ability of exogenously supplied ethylene to promote abscission of plant organs is a well-known phenomenon. Beyer and Morgan (1971) proposed a model for the role of ethylene in the regulation of abscission of intact leaves. The proposed regulatory system first involves a modification of the hormonal balance in the abscission zone achieved as ethylene reduces the indole-3-acetic acid (IAA) transport capacity of the petioles. After reducing auxin levels, ethylene acts directly on the abscission zone e.g. to stimulate synthesis of cell-wall degrading enzymes and the secretion of these enzymes into the cell-wall (Abeles and Leather, 1971). Also, ethylene has been demonstrated to reduce endogenous IAA levels (Ernest and Valdovinos, 1971; Beyer, 1975; Lieberman and Knegt, 1977) and there is some evidence that ethylene may achieve this in the abscission zone either by stimulating destruction (Ernest and Valdovinos, 1971; Gaspar *et al.*, 1978; Minato and Okazawa, 1978; Riov and Goren, 1979; Riov *et al.*, 1982; Riov, *et al.*, 1986;) or by inhibiting synthesis (Ernest and Valdovinos, 1971). Beyer (1975) demonstrated that the leaf blade is the initial target tissue for exogenously supplied ethylene, where some essential function of the hormone must first be performed before abscission can occur. There are data which indicate that this essential function of ethylene is to reduce the amount of auxin transported out of the blade, possibly by reducing auxin levels and inhibiting auxin transport in the veinal tissues (Beyer, 1975).

Conjugated forms of IAA are thought to be an important mechanism for regulating the level of free IAA in plant tissues (Epstein *et al.*, 1980). Effects of ethylene on IAA conjugation have been studied by several investigators, but so far no clear cut conclusions have emerged. Increased IAA conjugation by ethylene has been observed in various tissues (Ernest and Valdovinos, 1971; Gaspar *et al.*, 1978; Minato and Okazawa, 1978; Riov and Goren, 1979; Riov *et al.*, 1982; Riov *et al.*, 1986;), whereas in several other tissues no direct effect of ethylene on IAA conjugation could be found (Beyer and Morgan, 1970; Goren *et al.*, 1974; Walters and Osborne, 1979). Recently it has been reported that although ethylene increased IAA conjugation, the level of free IAA did not decline (Epstein, 1982; Wood, 1985).

NATO ASI Series, Vol. H 35
Cell Separation in Plants
Edited by D. J. Osborne and M. B. Jackson

In this chapter we present possible mechanisms by which ethylene interferes with the retarding effects of auxin on abscission of citrus leaves.

EFFECT OF ETHYLENE ON IAA TRANSPORT

Inhibition of auxin transport seems to be one of the main effects of ethylene at an early stage of the abscission process (Beyer and Morgan, 1971). We examined the effect of ethylene on IAA transport in midrib tissue of citrus and eucalyptus leaves as related to their abscission response (Riov and Goren, 1979). IAA transport was measured by the classic donor-receiver agar cylinder technique. The ability of ethylene to promote leaf abscission differed greatly among these species. Whereas citrus leaves reached 100% abscission after 48 hours of ethylene treatment (Fig. 1), abscission of eucalyptus leaves started between 96 and 120 hours after the beginning of treatment and reached about 70% after 168 hours (Fig. 2). In both species, abscission of ethylene-treated leaves showed a kinetic trend similar to that of the inhibition of auxin transport (Figs. 1 and 2). As might be expected, the inhibition of auxin transport preceded abscission. In citrus leaves, ethylene induced a rapid and marked reduction in IAA transport, reaching about 80% inhibition after 24 hours of exposure (Fig. 1). The reduction of IAA transport in eucalyptus midrib sections was much slower and less pronounced (Fig. 2). A significant inhibition of IAA transport was observed only after a 96 hours ethylene pretreatment, and reached 50% after 144 hours. The data suggest that as long as the basipetal flow of IAA from the leaf blade continues unaltered, ethylene cannot induce abscission. The ability of ethylene to inhibit auxin transport may therefore be an important factor in determining the sensitivity to ethylene.

Ethylene and 2,5-norbornadiene (2,5-NBD), a competitive inhibitor of ethylene binding (Sisler and Pian, 1973), have no significant effect on IAA transport if applied during the running time of IAA transport (Sisler *et al.*, 1985). This suggests that ethylene and 2,5-NBD have no direct effect on IAA transport. Ethylene (2 $\mu l\ l^{-1}$) pretreatment for 24 hour reduced IAA transport by 65% in citrus leaf explants (Table 1). If 2,000 $\mu l\ l^{-1}$ of 2,5-NBD were present, IAA transport was not reduced. If both 2,000 $\mu l\ l^{-1}$ of 2,5-NBD and 2 $\mu l\ l^{-1}$ of ethylene were present, most of the effect of ethylene appeared to be overcome. This suggests that a small effect of ethylene did occur during the 24-hour incubation, and that 2,000 $\mu l\ l^{-1}$ are slightly less than the amount necessary to completely overcome 2 $\mu l\ l^{-1}$ of ethylene. The data indicate that inhibition of auxin transport by ethylene is specific via binding to receptors.

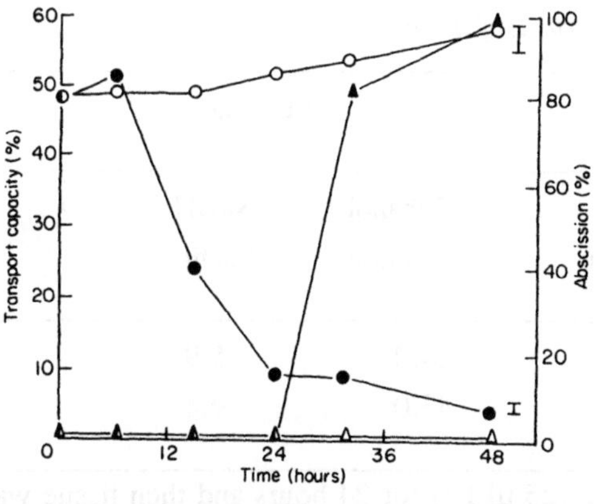

Figure 1. Effect of ethylene on abscission and IAA transport in midrib sections of citrus leaves. Air: (Δ) abscission; (O) transport capacity. Ethylene: (▲) abscission; (●) transport capacity. (From Riov and Goren (1979) with permission).

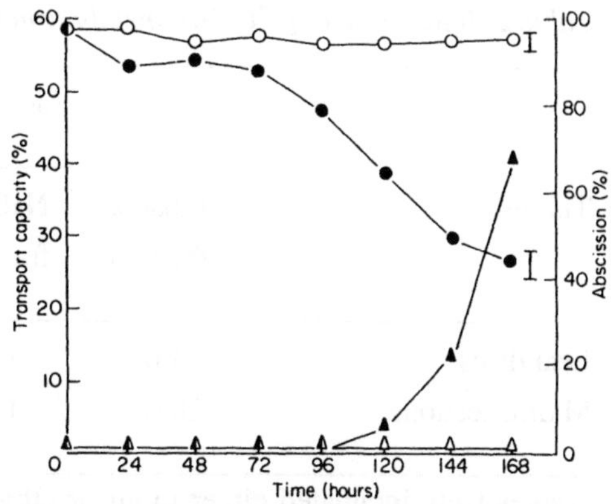

Figure 2. Effect of ethylene on abscission and IAA transport in midrib sections of eucalyptus leaves. Key as in Figure 1 (From Riov and Goren (1979) with permission).

Table 1. Effect of 24 hour preincubation in ethylene (2 μl l⁻¹) and 2,5-norbornadiene (2,000 μl l⁻¹) on IAA transport in midrib sections of citrus leaves. Values are % transport capacity ± SE

Transport	Transport capacity (%)
Air	25.2±1.4
Ethylene	8.9±2.1
2,5-NBD	28.6±2.9
Ethylene + 2,5-NBD	20.1±0.8

EFFECT OF ETHYLENE ON [^{14}C] IAA METABOLISM

Metabolism of IAA was studied after a short incubation period (1 hour) in IAA as previously suggested (Riov *et al.*, 1982). Treatment of intact leaves with 25 μl l^{-1} ethylene induced abscission which reached 100% 48 hours after the initiation of exposure to ethylene. Concomitantly with the onset of abscission, levels of both low (ethanol fraction) and high molecular weight conjugates (NaOH fraction) increased significantly in ethylene pretreated tissue (Table 2).

Table 2. Initial rate of [^{14}C]IAA metabolism in citrus leaf tissues.

Tissue	Air		Ethylene	
	Ethanol fraction	NaOH fraction	Ethanol fraction	NaOH fraction
Leaf discs	14.0	1.4	24.0	5.9
Midrib sections	28.0	1.9	43.0	9.4

Leaves were incubated either in air or ethylene (25 μl l^{-1}) for 24 hours and then tissue was excised and incubated in labelled IAA for 1 h. Bound IAA is expressed as percentage of total uptake.

In untreated citrus leaves the major low molecular weight conjugate was identified by GC-MS as indole-3-acetylaspartic acid (IAAsp). Following ethylene pretreament, a substantial accumulation of a neutral conjugate was detected, with a concomitant decrease in IAAsp level. The conjugate was purified by PVPP column chromatography, preparative TLC and HPLC. Analysis by GS-MS indicated that the conjugate is indole-3-carboxylic-ß-D-glucose (ICAglu).

The indole-3-carboxylic acid (ICA) identified in the present study originated from externally applied IAA after exposure of leaves to ethylene. It has been claimed that ICA may be an artifact, originating from IAA during extraction (Sandberg *et al.*, 1984). This claim does not apply to our study for two reasons: a) no free labelled IAA was present in the tissue during the extraction since all the IAA applied was metabolized before extraction, and b) the ICA was detected in the form of a glucose ester which could be formed only *in vivo*. The present data, as well as previous reports showing that indole-3-methanol (IM) and ICA are natural constituents (Sandberg *et al.*, 1984; Brown *et al.*, 1986) suggest that ICA is a major catabolite of IAA in plants.

Ethylene treatment of citrus leaves resulted in lower levels of [^{14}C]IAA. The decline in free IAA level emphasizes the importance of studying changes in IAA metabolites following ethylene treatment as a possible mechanism for removal of free IAA. Therefore we investigated further the effect of ethylene on the formation of ICAglu in relation to its effect on abscission. ICAglu was assayed by applying [^{14}C]IAA to midrib tissue and then analyzing its level by TLC.

The effect of the duration of ethylene pretreatment on the formation of ICAglu was studied by monitoring the abscission of the same leaves at different times prior to incubation in [^{14}C]IAA. A slight, non-significant accumulation of ICAglu was observed following 8 hours of ethylene pretreatment. After 24 hours of ethylene pretreatment, a significant increase in the level of the catabolite was detected. At that time abscission was only 20%. ICAglu reached its maximum level after 72 hours, when 100% abscission was obtained. Level of ICAglu in untreated leaves remained constant during the entire experimental period. These leaves did not abscise during this period.

When leaves were subjected to various concentrations of ethylene, a positive correlation between abscission and the increase in ethylene concentration above 1 $\mu l\ l^{-1}$ was observed. The pattern of ICAglu formation in the same leaves, due to increase in ethylene concentration, was similar to that of abscission. Formation of ICAglu increased with the rise in ethylene concentration above 1 $\mu l\ l^{-1}$ (Table 3).

Table 3. Effect of ethylene concentration on abscission and [^{14}C]IAA metabolism in citrus leaves

Ethylene ($\mu l\ l^{-1}$)	Abscission (%)	Radioactivity (% of total)	
		IAAsp	ICAglu
0.12	0	40 ± 10	8 ± 5
1	3	38 ± 11	13 ± 5
12	82	15 ± 6	56 ± 9
100	100	10 ± 5	58 ± 9

The data just described indicate that formation of ICAglu is induced by ethylene. The question whether exogenous IAA will by itself induce formation of ICAglu in ethylene-pretreated tissue was studied. Incubation in cold IAA up to 8 hours did not induce any additional formation of ICAglu. Furthermore, addition of cycloheximide to the incubation medium had no effect on formation of ICAglu. This is the first study showing that a plant hormone, namely ethylene, is capable of inducing the oxidative decarboxylated catabolism of IAA. It still remains to be shown that ethylene is capable of inducing the formation of ICAglu from endogenous IAA.

Chromatography of the NaOH extract on a Sephadex G-25 column, indicated that the molecular weight of the IAA conjugate(s) present in this fraction is higher than 5,000 (Fig. 3). Differential centrifugation of the ethanol- and water-insoluble material (before NaOH extraction), revealed that most radioactivity was associated in the 5,000 g pellet, indicating that it was associated with the cell-wall. Further analysis suggested that these conjugates are probably bound to different cell-wall components (Riov *et al.*, 1986).

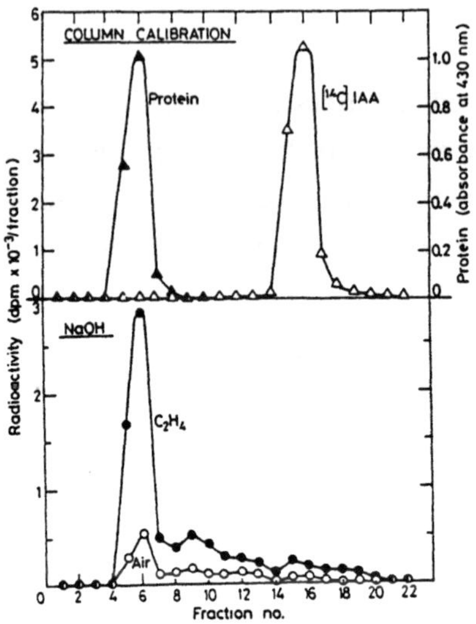

Figure 3. Sephadex G-25 chromatography of NaOH-extractable IAA conjugates of citrus leaf discs from control and ethylene-pretreated leaves. (From Riov et al., 1982 with permission).

Table 4. IAA metabolism in naturally senescing and ready to abscise citrus leaves

Leaf status	ICAglu (% of total uptake)	NaOH fraction (dpm)
Mature	7.3	540
Senescing	23.1	1,150

Senescing and ready to abcise leaves produced significant amounts of ethylene as compared to mature leaves (22.6 and 0.2 nl g^{-1} h^{-1}, respectively). It was thus interesting to study whether this increased ethylene production was accompanied by an increased metabolism of IAA. When tissue excised from senescing leaves was incubated in labelled IAA, formation of ICAglu and high molecular weight conjugates was significantly higher than in mature leaves (Table 4). When leaves at different stages of senescence (green to yellow colour) were examined, a gradual increase in ethylene production with the advancement of senescence was observed. This increase in ethylene production was accompanied by an increase in IAA conjugates.

LEVELS OF ENDOGENOUS IAA IN SENESCING AND ETHYLENE-TREATED LEAVES

Endogenous IAA was determined by enzyme-immunoassay (Sagee *et al.* 1986) in the petioles of senescing leaves at abscission. Non-abscising mature leaves were used for comparison. Whereas mature leaves contained 74.6 ng g^{-1} IAA, senescing leaves contained 32.8 ng g^{-1} (Table 5). This difference in IAA content was found to be statistically highly significant.

In midrib tissue of mature leaves aged in air, IAA concentration ranged from 58.6 ng g^{-1} at zero time to 67.3 ng g^{-1} after 32 hours, the difference being insignificant (Table 5). Exposure to ethylene somewhat reduced IAA levels but the reduction was not statistically significant.

Table 5. Endogenous free IAA levels in mature and senescing citrus leaves and in mature leaves treated with ethylene (25 $\mu l\ l^{-1}$) for 32 hours

Experiment	Leaf status	IAA content (ng g^{-1} fresh weight)
1	Mature	74.6±14.9
	Senescing	32.8±8.1
2	Mature, zero time	58.6±17.8
	Mature, 32 h air	67.3±16.9
	Mature, 32 h ethylene	51.4±13.9

CHANGES IN IAA-INDUCED H^+ EFFLUX AND IAA BINDING DURING ABSCISSION

According to the general concept, if auxin decreases in the abscission zone during the first 24 hours, then the ability of the abscission zone to respond to auxin may also decrease. To monitor this possibility, two auxin responses were examined: auxin-induced H^+ efflux and auxin-induced retardation of abscission (Jaffee and Goren, 1979). Specific IAA binding to particulate fractions of abscission zones was also determined.

The time course of IAA-induced H^+ efflux from isolated abscission zones showed a 70% decrease during the first 12 hours and then a levelling off (Fig. 4A). In reaching a minimum by 12 hours, auxin-induced H^+ efflux capacity followed a similar time course as that of specific auxin binding. Parallel to these changes, the retarding effect of 2,4-D on abscission was reduced (Fig. 4B).

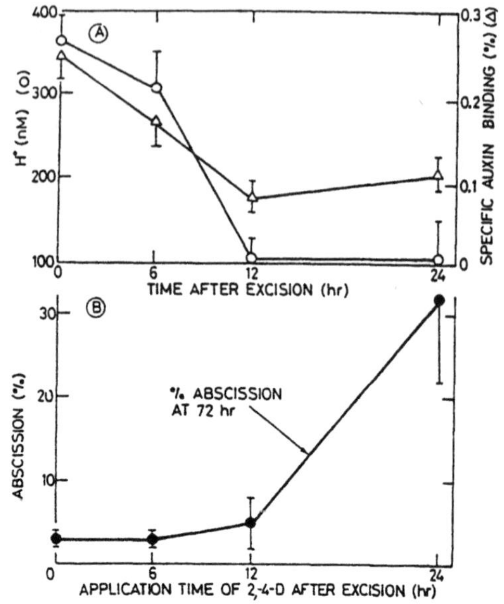

Figure 4. Time course of (A) IAA (4.4 x 10^{-6} M)–induced H^+ efflux from isolated abscission zones and the specific IAA binding capacity of the particulate fraction of abscission zones, and (B) the retarding effect of 2.3 x 10^{-5} M 2,4-D on abscission (applied as a 15 minute soak at different intervals after excision). The vertical bars represent standard errors (From Jaffe and Goren, 1979 with permission).

The loss of auxin-induced H^+ efflux from the abscission zone, the reduction in the ability of microsomal fractions to bind auxin specifically, and the loss of the ability of the tissue to respond physiologically or biochemically to auxin may provide an explanation for the well-known phenomenon in which exogenous auxin will retard abscission only when applied during the few hours after excision. We suggest that auxin loses its effect later simply because the machinery required to respond to it is no longer functional. In two other plants, in which auxin binding has been related to physiological changes, auxin-binding capacity decreases along with the ability to respond to auxin (Jablanovic and Noodén, 1974; Jaffee, 1975).

CONCLUSIONS

The accepted model of ethylene-induced abscission of intact leaves is based on the hypothesis that the reduction of IAA content in the abscission zone is a pre-requisite for ethylene action in accelerating abscission (Beyer and Morgan, 1971). Reduction in IAA level in the abscission zone of ethylene-treated leaves has been attributed to the inhibition of auxin transport (Beyer and Morgan, 1971). Whether IAA conjugation is involved in the reduction of IAA level in ethylene-treated leaves is still unclear.

In the present study, we demonstrated that the inhibition of auxin transport by exogenous ethylene is indeed causually related to the induction of abscission by the applied ethylene (Figs. 1 and 2). Moreover, we were able to show that ethylene increases the catabolism of IAA as well as its conjugation to high molecular weight substances (Tables 2, 3). Similar changes in auxin metabolism were observed in naturally senescing leaves which produce high levels of endogenous ethylene.

Based on the effect of ethylene on IAA transport and metabolism, which were observed in the course of the present study, it was expected that the level of free IAA in the abscission zone will decline following ethylene treatment. Surprisingly, no significant reduction in endogenous level of free IAA could be found in ethylene-treated leaves (Table 3). Similar results were recently obtained in other species (Epstein, 1982; Wood, 1985). In contrast to ethylene-treated leaves, endogenous IAA levels declined significantly as leaves senesce (Table 3).

The interaction between ethylene and IAA in the control of abscission seems to be more complex than is usually believed. In nature, the level of ethylene in leaves is relatively low, compared to the concentrations of ethylene usually applied experimentally, therefore the decrease of IAA level may be critical in order for ethylene to induce abscission. In experiments with exogenous ethylene, apparently the relatively high concentration of ethylene applied induces abscission, although IAA content does not decline. This does not rule out the possibility that when IAA content decreases, the tissue becomes more sensitive to ethylene action. Why endogenous level of IAA does not change when ethylene induces increased IAA conjugation remains to be clarified.

REFERENCES

Abeles FB, Leather GR (1971) Abscission: control of cellulase secretion by ethylene. Planta 97:87–91

Beyer EM Jr (1975) Abscission: the initial effect of ethylene is in the leaf blade. Plant Physiol 64:971–974

Beyer EM Jr, Morgan PW (1970) Effect of ethylene on the uptake distribution, and metabolism of indole-acetic acid-1-^{14}C and -2-^{14}C and naphthaleneacetic acid-1-^{14}C. Plant Physiol 46:157–162

Beyer EM Jr, Morgan PW (1971) Abscission: the role of ethylene modification of auxin transport. Plant Physiol 48:208–212

Brown BH, Crozier A, Sandberg G (1986) Catabolism of indole-3-acetic acid in chloroplast fractions from light-grown *Pisum sativum* L. seedlings. Plant Cell Environ 9:527–534

Epstein E (1982) Level of free and conjugated indole-3-acetic acid in ethylene-treated leaves and callus of olive. Physiol Plant 56:371–373
Epstein E, Cohen JD, Bandurski RS (1980) Concentration and metabolic turnover of indoles in germinating kernels of *Zea mays* L. Plant Physiol 65:415–421
Ernest LC, Valdovinos JG (1971) Regulation of auxin levels in *Coleus blumei* by ethylene. Plant Physiol 48:402–406
Gaspar T, Goren R, Huberman M, Dubucq M (1978) Citrus leaf abscission: auxin and ethylene regulatory role on peroxidase and endogenous growth substances. Plant Cell Environ 1:225–230
Goren R, Bukovac JM, Flore JA (1974) Mechanism of indole-3-acetic acid conjugation. No induction by ethylene. Plant Physiol 53:164–166
Jablanovic M, Noodén LD (1974) Changes in compatible IAA binding in relation to bud development in pea seedlings. Plant Cell Physiol 15:687–692
Jaffe MJ (1975) The role of auxin in the early events of the contact coiling of tendrils. Plant Sci Letters 5:217–225
Jaffe MJ, Goren R (1979) Auxin and early stages of the abscission process of citrus leaf explants. Bot Gaz 140:378–383
Lieberman M, Knegt E (1977) Influence of ethylene on indole-3-acetic acid concentration in etiolated pea epicotyl tissue. Plant Physiol 60:475–477
Minato T, Okazawa Y (1978) Effect of ethylene treatment on auxin metabolism of potato tubers. J Fac Agric Hokkaido Univ 58:535–547
Riov J, Dror N, Goren R (1982) Effects of ethylene on ^{14}C-indole-3-acetic acid metabolism in leaf tissues of woody plants. Plant Physiol 70:1265–1270
Riov J, Sagee O, Goren R (1986) Ethylene-induced changes in indole-3-acetic acid metabolism in citrus leaf tissues during abscission and senescence. Acta Hort 179:613–620
Riov J, Goren R (1979) Effects of ethylene on auxin transport and metabolism in midrib sections in relation to leaf abscission of woody plants. Plant Cell Environ 2:83–89
Sagee O, Maoz A, Mertens R, Goren R and Riov J (1986) Comparison of different enzyme immunoassays for measuring IAA in vegetative citrus tissues. Physiol Plant 68:265–270
Sandberg G, Jensen E and Crozier A (1984) Analysis of 3-indole carboxylic acid in *Pinus sylvestris* needles. Phytochem 23:99–102
Sisler EC, Goren R, Huberman M (1985) Effect of 2,5-norbornadiene on abscission and ethylene production in citrus leaf explants. Physiol Plant 63:114–120
Sisler EC, Pian A (1973) Effect of ethylene and cyclic olefins on tobacco leaves. Tobacco Sci 17:68–72
Walters J, Osborne DJ (1979) Ethylene and auxin-induced cell growth in relation to auxin transport and metabolism and ethylene production in the semi aquatic plant *Regnellidium diphyllum*. Planta 146:309–317
Wood BW (1985) Effect of ethephon on IAA transport, IAA conjugation and antidotal action of NAA in relation to leaf abscission of pecan. J Amer Soc Hort Sci 110:340–343

IDENTIFICATION OF LEAF ABSCISSION ZONES AS A SPECIFIC CLASS OF TARGET CELLS FOR ETHYLENE

Michael T. McManus and Daphne J. Osborne[1]
Department of Biochemistry
Royal Holloway and Bedford New College
University of London
Egham
Surrey, TW20 OEX
U.K.

INTRODUCTION

The anatomy of leaf abscission zones from many species has been well described. In many plants, the fully differentiated abscission zone can be discerned as a single layer, or several layers of cells which are morphologically distinct from neighbouring tissue (Addicott, 1982). The stimulus for shedding in a large number of plants is ethylene. For example, in the initiation of cell separation at the distal pulvinus:petiole leaf abscission zone of Phaseolus vulgaris, ethylene is produced just distally to the zone as part of a programmed series of events associated with leaf senescence (Jackson and Osborne, 1970). The cell separation event is preceded by the enlargement of the one or two rows of the cells that comprise the zone and by an increase in the activity in these cells of a carboxymethyl-β-1:4-glucanase. These responses are not observed in neighbouring tissue (Wright and Osborne, 1974). In another plant, Sambucus nigra, an abscission zone traverses the rachis of the compound leaf and is comprised of cells which are also distinct morphologically from their neighbours. On treatment with ethylene, these zone cells undergo major changes in ultrastructure, suggestive of active protein synthesis and secretion, while neighbouring cells remain comparatively quiescent (Osborne and Sargent, 1976).

Abscission cells, therefore, have been proposed as an example of a class of target cells for ethylene, in which the hormone can induce

[1]Department of Plant Sciences
University of Oxford
Oxford, OX1 3RA
U.K.

NATO ASI Series, Vol. H 35
Cell Separation in Plants
Edited by D.J. Osborne and M.B. Jackson

specific events which do not occur in cells of neighbouring (non-target) tissue (Osborne, 1984). Since cells comprising the abscission zone are structurally and functionally distinct from their non-target neighbours we have predicted that they should, as a specific target cell class, also express a specific protein complement (Osborne and McManus, 1986).

In this presentation, we review evidence that the positionally differentiated leaf abscission zone cells of S. nigra and P. vulgaris do contain specific polypeptides which are either wholly, or preferentially expressed in the abscission zone and furthermore, that quantitative differences exist in immunologically-recognized components of the cell walls of target and neighbouring non-target cells of P. vulgaris.

COMPARISON OF THE SPECTRUM OF POLYPEPTIDES PRESENT IN TARGET AND NON-TARGET TISSUES

To examine the spectrum of polypeptides contained within each tissue type, total cell protein is conveniently separated into soluble and bound fractions. The soluble fraction is supernatant obtained after homogenization of the tissue in water containing PMSF and β-mercaptoethanol and subsequent centrifugation at 700 x g. The bound fraction is supernatant obtained from extraction of this pellet (after washing with water) with salt (1.0 M NaCl, 3.0 M LiCl) followed by centrifugation at 10,000 x g.

The spectrum of polypeptides obtained after fractionation of the soluble fraction from P. vulgaris by sodium dodecyl sulphate-polyacrylamide gel electrophoresis (SDS-PAGE) and visualisation by silver staining reveals that, apart from concentration differences, the majority of the polypeptides detected are common to all three tissues (Fig. 1A). Examination of the profile of polypeptides extracted with high salt (the bound fraction) reveals a similar pattern (Fig. 1B). Again, apart from concentration differences, most notably a polypeptide of ca. 36 kDa (see arrow) which is expressed strongly in the pulvinus and the petiole but less so in the abscission zone, the majority of polypeptides visualized are present in all three tissue types.

To ascertain if there are proteins which are wholly abscission cell-specific it was necessary to improve the resolution of polypeptide differences between the three tissue types. To achieve this, polyclonal antibodies were raised in rabbits against pooled extracts of pulvinus and petiole tissue. The IgG was coupled to cyanogen bromide (CNBr)-activated Sepharose 4B and used as an immuno-affinity column to bind proteins common to all three tissue types. Proteins which are not recognized by the anti-pulvinus+petiole IgG (i.e. those which elute) are then subjected to gel electrophoresis. Examination of the spectrum of polypeptides obtained following this affinity column procedure reveals a major staining component of ca. 68 kDa (see arrow) in the abscission zone eluate which is almost entirely absent from the pulvinus and petiole eluates (Fig. 1C).

In S. nigra, a similar approach to specifically enrich the abscission zone protein extract has been used. To do this, total (soluble and bound) abscission zone and mid-rachis extracts were passed through an affinity column comprised of IgG purified from antiserum raised against mid-rachis

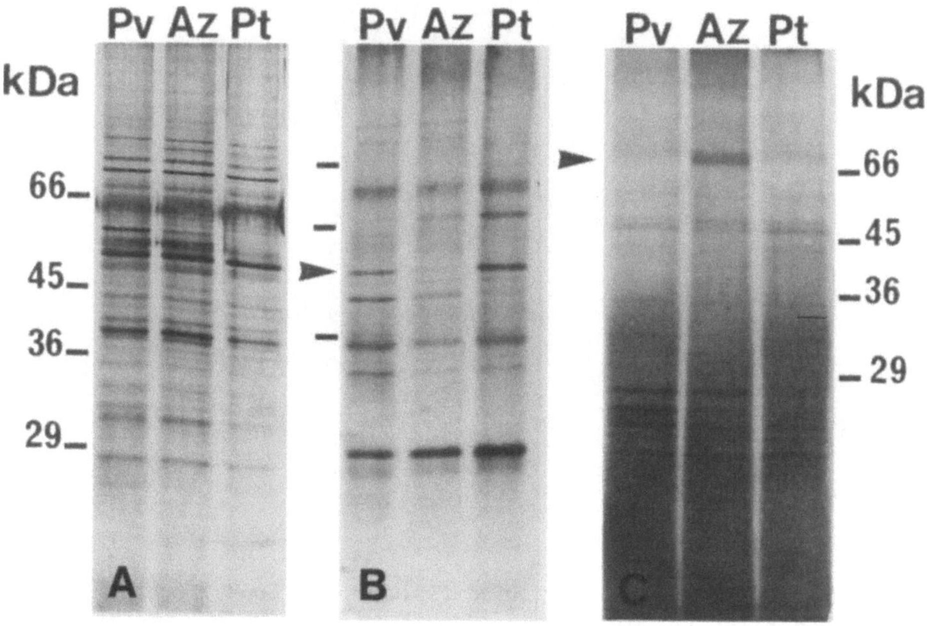

Figure 1A,B,C. Silver stain of proteins extracted as either the soluble fraction (A), or the bound fraction (B) from P. vulgaris, and fractionated through a 12.5% SDS-polyacrylamide gel, or of proteins extracted as a soluble fraction, chromatographed through an anti-pulvinus+petiole IgG affinity column and the eluates (i.e. non-recognized proteins) fractionated through a 12.5% SDS-polyacrylamide gel (C). Pv = pulvinus, Az = abscission zone, Pt = petiole (from McManus and Osborne, 1988a)

protein extracts (anti-Mr IgG), coupled to Sepharose. Two distinct peaks were observed as each extract eluted (Fig. 2A) and fractions comprising each peak were pooled and subjected to SDS-PAGE. A polypeptide of ca. 34 kDa (see arrow) appears as a prominent staining band in peak 1 of the abscission zone eluate and is present, but at a very low concentration, in the mid-rachis eluate. The majority of proteins elute as the second peak in both tissues and the spectrum of polypeptides is similar in both.

In a second immuno-affinity column approach, abscission zone and mid-rachis extracts were passed through a column comprised of IgG purified from antiserum raised against an abscission zone extract, the antiserum being first competed with protein extracts from mid-rachis. Proteins which

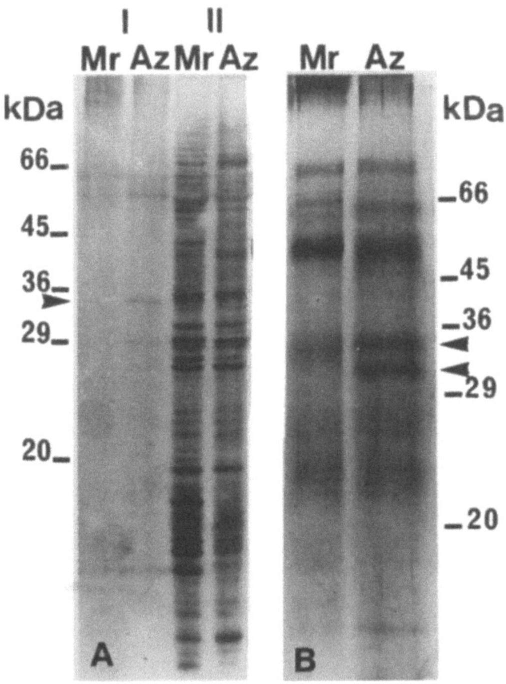

Figure 2A,B. A. Silver stain of pooled (soluble and bound) protein extracts from S. nigra after chromatography through an anti-Mr IgG affinity column and fractionation of the eluates (i.e. non-recognized proteins) through a 12.5% SDS-polyacrylamide gel; I = proteins eluting as the first peak, II = proteins eluting as the second peak. B. Silver stain of extracts (as in A) which were chromatographed through an Mr-competed anti-Az IgG affinity column, the proteins which bound (putatively Az-specific) eluted with pH 11.0 and pH 2.5 and fractionated through a 12.5% SDS-polyacrylamide gel. Mr = mid-rachis, Az = abscission zone (from McManus and Osborne, 1988b)

bound to this column (putatively abscission cell-specific) were eluted using a combination of high and low pH, and separated using SDS-PAGE (Fig. 2B). Using this technique, two major staining polypeptides, of ca. 34 kDa and ca. 32 kDa (see arrows), are detected in the abscission zone extract but not in extracts of mid-rachis.

Both of these immuno-affinity column procedures revealed an abscission cell-specific polypeptide of ca. 34 kDa. An antibody was raised against this polypeptide by direct immunization of the peptide after excision from the polyacrylamide gel. Using Western blotting, this antibody recognises a ca. 34 kDa (see arrow) polypeptide which is present both in abscission zone and ethylene-treated abscission zone extracts, but is absent from mid-rachis extracts (Fig. 3). The presence of this polypeptide in ethylene-treated abscission zone tissue confirms its abscission cell-specific location, since this extract is prepared from a homogeneous population of fully separated cells which contain no detectable mid-rachis contamination.

THE USE OF A MONOCLONAL ANTIBODY DIRECTED AGAINST A GLYCOSYL EPITOPE TO IDENTIFY ABSCISSION CELL-SPECIFIC GLYCOPROTEINS

A monoclonal antibody, YZ1/2.23, raised against a total abscission zone extract from S. nigra and which recognises an epitope contained solely within the N-linked oligosaccharide structure Manα3(Manα6)(Xylβ2)GlcNAcβ4 (Fucα3)GlcNAc (McManus *et al.*, 1988), has been used to probe for differences in the spectrum of glycoproteins present in target and non-target tissue.

The monoclonal antibody recognises many common polypeptides in soluble extracts of the abscission zone, pulvinus and petiole tissue of P. vulgaris (Fig. 4) although most notably a polypeptide of ca. 36 kDa (see arrow) which is always particularly prominent in the abscission zone.

YZ1/2.23 binds to several polypeptides which are present in the high salt (bound) extracts of both target and non-target tissues from P. vulgaris (Fig. 5A), and with the exception of concentration differences, there is no evidence for any abscission cell-specific polypeptides.

However, when these extracts are chromatographed through an affinity column of anti-pulvinus+petiole rabbit IgG, the eluting (i.e. non-recognized) proteins fractionated through an acrylamide gel and challenged with

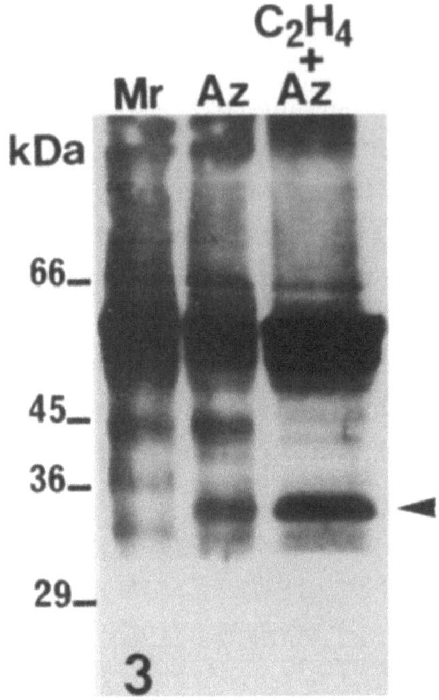

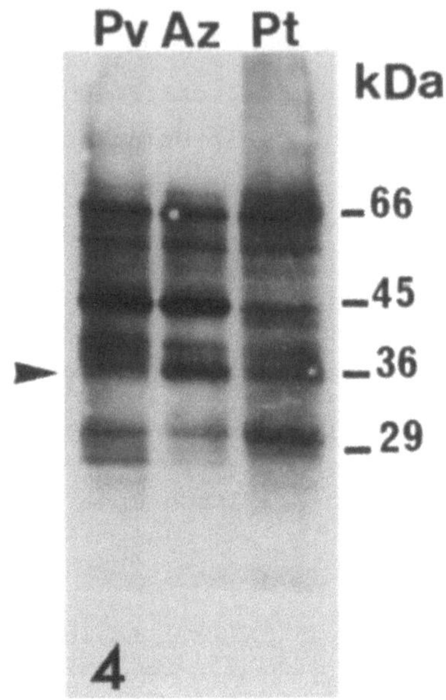

Figure 3. Autoradiography after Western blotting onto nitrocellulose of pooled (soluble and bound) proteins extracted from S. nigra, and fractionation through a 12.5% SDS-polyacrylamide gel; the nitrocellulose blot was first treated with a rat antibody (raised against a ca. 34 kDa polypeptide identified in abscission zone extracts after immuno-affinity column chromatography), and then with the ^{125}I-labelled F(ab')2 fragment of sheep anti-rat Ig. Mr = mid-rachis, Az = abscission zone, C_2H_4 + Az = ethylene-treated abscission zone (from McManus and Osborne, 1988b).

Figure 4. Autoradiography after Western blotting onto nitrocellulose of proteins extracted as the soluble fraction from P. vulgaris and fractionated through a 12.5% SDS-polyacrylamide gel; the nitrocellulose blot was treated with YZ1/2.23 monoclonal antibody and then with the ^{125}I-labelled F(ab')2 fragment of sheep anti-rat Ig. Pv = pulvinus, Az = abscission zone, Pt = petiole (from McManus and Osborne, 1988a).

YZ1/2.23, then differences in the intensity of monoclonal recognition are evident (Fig. 5B). A polypeptide of ca. 46 kDa (see arrow), which appears as a minor component recognised by YZ1/2.23 in non-competed bound extracts from all three tissues (in Fig. 5A), now appears as a major component but predominantly in the abscission zone (Fig. 5B).

EVIDENCE FOR DIFFERENCES IN CELL WALL COMPONENTS OF TARGET AND NON-TARGET CELLS OF PHASEOLUS VULGARIS

Antibodies have been raised by fractionating a total abscission zone extract through an SDS-polyacrylamide gel, slicing the gel into 2 cm slices and then extracting each slice separately with phosphate buffer, pH 7.4, containing 1.0 M NaCl. The slurry was then filtered, each filtrate chromatographed through affinity columns of anti-pulvinus+petiole IgG immobilized onto Sepharose and eluates immunized into separate laboratory-strain rats. These antibodies were used to probe recognition sites in sections of fixed tissue cut to include the pulvinus, abscission zone and petiole. The binding of one of these polyclonals, raised against polypeptides of molecular weight range 30-45 kDa (and designated PcAb 6), has been determined using gold-labelled anti-rat IgG and electron

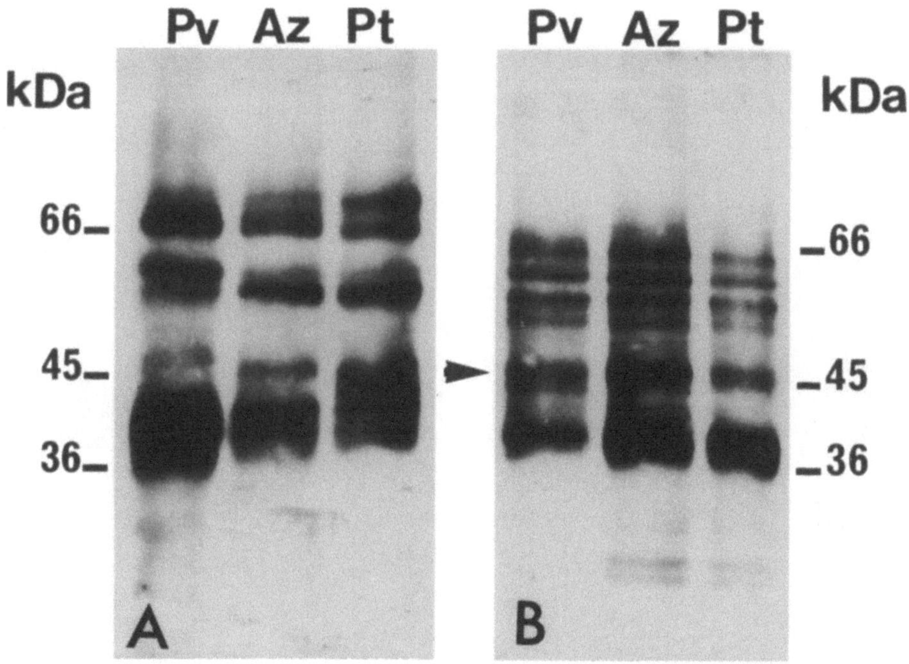

Figure 5A,B. Autoradiography after Western blotting onto nitrocellulose of proteins extracted from P. vulgaris as the bound fraction and fractionated through a 12.5% SDS-polyacrylamide gel (A), or chromatographed through an anti-pulvinus+petiole IgG affinity column and the eluates (i.e. non-recognized proteins) separated through a 12.5% SDS-polyacrylamide gel (B); the nitrocellulose was treated with YZ1/2/23 monoclonal antibody and then with ^{125}I-labelled F(ab')2 fragment of sheep anti-rat Ig. Pv = pulvinus, Az = abscission zone, Pt = petiole (from McManus and Osborne, 1988a).

microscopy. This polyclonal antibody binds specifically to cell walls and shows a high affinity for the walls of abscission zone cells (Fig. 6B) compared with those of adjacent pulvinus tissue (Fig. 6A) or immediately adjacent petiole cells. In each tissue, deposition is predominantly (but not exclusively) at junctions of several cells in the corner regions.

DISCUSSION

The results of the present study extend the concept that higher plants possess cells (target cells) which can respond in a specific way to hormonal stimuli. We show that several polypeptides are expressed preferentially in the pulvinus:petiole abscission zone of P. vulgaris and the leaflet and rachis abscission zone of S. nigra. The demonstration of an accumulation of some specific cell wall component in zone cells of

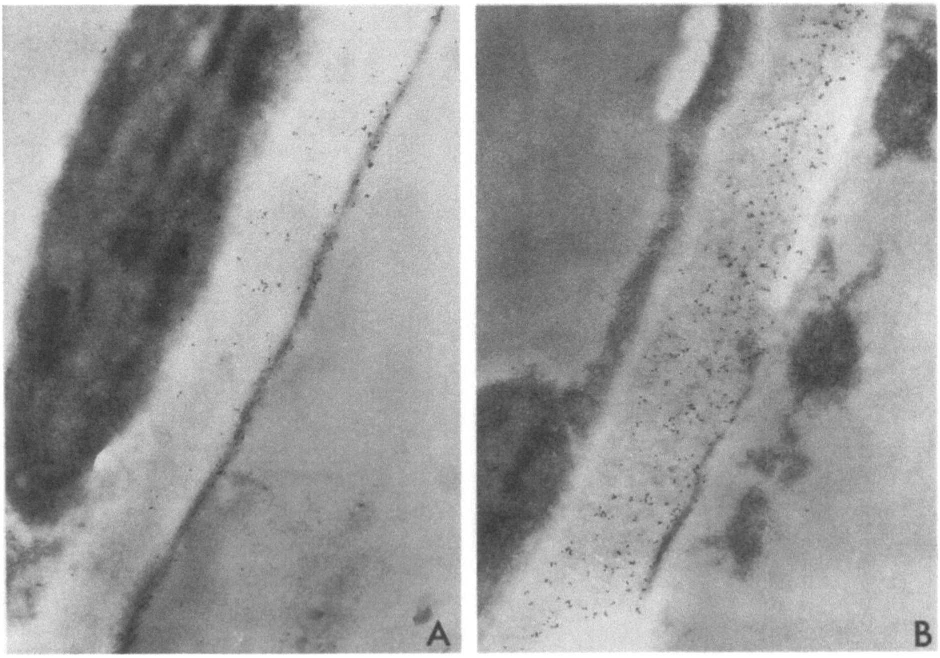

Figure 6A,B. Immuno-gold electron micrographs recording the binding of gold-labelled anti-rat IgG to PcAb 6 on cells from either pulvinus (A), or abscission zone (B) tissue prepared by fixation of sections in 2.5% (v/v) glutaraldehyde in sodium cacodylate buffer, pH 7.2, and embedding at -40°C in Lowicryl K4M resin. In collaboration with B.J. Wells, John Innes Institute, Norwich, U.K. (Wells, 1985).

P. vulgaris compared with that in the neighbouring non-zone cells of pulvinus and petiole is further evidence for the molecular specificity of cell differentiation during development.

In addition, we show that certain of these zone-abundant peptides are glycosylated. Two polypeptides of ca. 36 kDa (in the soluble fraction) and ca. 46kDa (in the bound fraction) in P. vulgaris which are expressed preferentially in the abscission zone have been discerned by the monoclonal antibody YZ1/2.23. This antibody recognizes an epitope solely within a xylose/fucose-containing modified oligosaccharide structure (McManus *et al.*, 1988), and so these polypeptides must be glycosylated.

The presence in plants of tissue-specific and developmentally-regulated polypeptides, particularly those known to be glycosylated, has important parallels with animal cell differentiation and development. The use of oligosaccharide-specific antibodies is becoming increasingly important in characterizing the changing profiles of cell-surface glycoproteins which accompany cell differentiation during development in animals (Feizi and Childs, 1987). It is possible that changes in the spectrum of glycoproteins will prove to be an important component of plant cell differentiation. For example, the expression of a hydroxyproline-rich glycoprotein in the cell walls of soybean seed coats is known to be developmentally-regulated (Cassab *et al.*, 1986).

Until the present study, the identification of specific, or preferentially expressed, peptides in functionally differentiated cell types *before* they respond to a specific hormonal signal has not been demonstrated in higher plants. We contend that such peptides should serve as "markers" to delineate the time-course of cell differentiation *in vivo*. Using polyclonal and monoclonal antibodies raised against abscission cell-specific components we can investigate which hormonal or environmental factors are important to the development of abscission zone cells as a distinct, positionally differentiated ethylene-responsive target class.

ACKNOWLEDGEMENTS

We wish to thank Mrs. Ann Edwards for help with the preparation of the manuscript and Mr. Roy Davies for the photography.

REFERENCES

Addicott FT (1982) Abscission. University of California Press, Berkeley Los Angeles London

Cassab GI, Nieto-Sotelo J, Cooper JB, van Holst G-J, Varner JE (1985) A developmentally regulated hydroxyproline-rich glycoprotein from the cell walls of soybean seed coats. Plant Physiol 77:532-535

Feizi T, Childs RA (1987) Carbohydrates as antigenic determinants of glycoproteins. Biochem J. 245:1-11

Jackson MB, Osborne DJ (1970) Ethylene, the natural regulator of abscission. Nature 225:1019-1022

McManus MT, Osborne DJ (1988a) Evidence for the preferential expression of specific polypeptides in leaf abscission zones of the bean Phaseolus vulgaris. (submitted for publication)

McManus MT, Osborne DJ (1988b) Identification of polypeptides specific to rachis abscission zone cells of Sambucus nigra L. (submitted for publication)

McManus MT, McKeating J, Secher DS, Osborne DJ, Ashford D, Dwek RA, Rademacher TW (1988) Identification of a monoclonal antibody to abscission tissue that recognizes xylose/fucose containing N-linked oligosaccharides from higher plants. Planta 175:506-512

Osborne DJ, (1984) Concepts of target cells in plant differentiation. Cell Diff 14:161-169

Osborne DJ, Sargent JA (1976) The positional differentiation of ethylene-responsive cells in rachis abscission zones in leaves of Sambucus nigra and their growth and ultrastructural changes at senescence and separation. Planta 130:203-210

Osborne DJ, McManus MT (1986) Flexibility and commitment in plant cells during development. Current Topics in Develop Biol 20:383-396

Wells B (1985) Low temperature box and tissue handling device for embedding biological tissue for immuno-staining in electron microscopy. Micron and Microscopica Acta 16:49-53

Wright M, Osborne DJ (1974) Abscission in Phaseolus vulgaris. The positional differentiation and ethylene-induced expansion growth of specialised cells. Planta 120:163-170

IAA MEDIATED CELL SEPARATION IN THE SEPAL SPLITTING OF OENOTHERA LAMARKIANA FLOWER BUDS

T. Yamaki and K. Takeda
Department of Biology
University of Occupational and Environmental Health
Yahatabishu-ku
Kitaktushu-shi 807
Japan

INTRODUCTION

The flower bud of *Oenothera lamarkiana* has four petals and four sepals. When the bud starts to open, the turgor pressure of petals increases sufficiently to unfold the petals, however, the sepals remain sutured to each other. When light intensity decreases to below a certain value, the sutured tissue of two adjacent sepals start to separate from the tip to the base (1st split), and a little later, the sutured tissue on the opposite side of the bud begin to separate from the tip to the middle (2nd split). Several minutes later, the basal parts of the three still sutured sepal margins, including the one that had not yet begun to separate, start to split (3rd split). A little later still, the petals push the four separating sepals outwards and the flower opens (flower open). The main object of the present study is the mechanism of cell separation of sutured cells in the epidermal layers of adjacent sepal margins.

MORPHOLOGICAL CHANGES

Electron microscopic observations of the epidermal layers of the adjacent sutured sepal margins indicate the following: 1) The cells of adjacent sepal margins are sutured very tightly, rather like a zip fastener. 2) When the time comes, the substance(s) which cement the cell walls of marginal epidermal cells of adjacent sepals starts to melt and the cells begin to separate. 3) Accordingly, cell separation is the first step of sepal splitting, and sepal splitting is a necessary step for bud opening together with the outwardly pushing forces of the petals.

INFLUENCE OF HABITUATED RHYTHMS AND ENVIRONMENTAL FACTORS ON CELL SEPARATION

1) *Seasonal changes of flower opening time outdoors.* The time of flower opening is closely related to cell separation at the sepal margins and to sepal splitting. Flower opening time between May and December is closely

NATO ASI Series, Vol. H 35
Cell Separation in Plants
Edited by D. J. Osborne and M. B. Jackson

linked to the time of sunset, but shows little fluctuation according to the cloudiness of the day or the day before. Other meteorological factors, like temperature also have very little affect (Saito and Yamaki, 1967).

2) *Habituation induced by light-dark periodicity.* When outdoor grown plants are moved into a dark room every day at 12:00 h, the time at which 50% of the matured flower buds open is brought forward a little each day, on the first day flowers open at 19:00 h (the opening time under natural conditions out of doors) but by the seventh day flower opening occurs at approximately 12:25 - 12:30 h. Thereafter, flowers open 25-30 minutes after the start of the dark period. When plants with flowers opening at 19:00 h every day are moved into a continuously dark room, flowers continue to open at around 19:00 h for two or three successive days. When the plants habituated to open their flowers at 16:00 h are moved into a room, with the following dark and light periodicity (-22:00 h--dark--08:00 h--light--22:00 h--dark), they open their flowers at 20:30 h on the first day and at 22:15 h on the second day. The above three kinds of experiments indicate the existence of habituated flowering rhythms, as well as an inhibitory effect of light and promoting effect of darkness on cell separation in sepal margins, and consequently on flower opening.

3) *Length of dark period for flower opening (cell separation and sepal splitting), in the case of habituated plants in controlled environment cabinets.* Plants were grown for 16 days in a growth cabinet illuminated from 02:00-14:00 h each day. The time of flower opening from the start of dark period (14:00 h) decreased every day until the 10th day, and later a 25 minute dark period was enough to induce sepal splitting and flower opening. Next, a 12 minute dark period followed by 15 minutes of 1,700 lux white light was given to these habituated plants at 14:00 h. In this case flowers opened at 14:15 h. This result seems to indicate that the light sensitive reaction(s) is completed within the first 12 minutes of darkness.

4) *Inhibitory effect of different light intensities.* An effect on flower opening was observed using habituated plants grown in a cabinet providing 7.5 hours darkness from 09:30 h and 16.5 hours light from 17:00 h in each 24 hours. Plants were illuminated with different intensties of white light at the end of the light period and the time of flower opening was measured. Light above 15,000 lux inhibited flower opening almost completely, however, light below 13,000 lux was somewhat less effective. In these cases the

lower the light intensity, the earlier the flowers opened.

5) *Effect of light of different wavelengths on flower opening.* These effects were observed using the spectrograph of the Okazaki National Institute for Basic Biology with Dr M. Watanabe. Significant inhibition of flower opening occurred at wavelengths 450, 475, 495, 640, 660 nm and between 700 and 750 nm. In addition, a weak but definite inhibition was observed at 510 and 535 nm. We are not acquainted with any biological photoresponse to wavelengths in the green region, except in our previous experiment using *Oenothera biennis* (Yamaki, Saito and Gordon, 1966). These two inhibition peaks correspond exactly to the green-absorbing peaks of the protein binding phaeophytins.

LIGHT SENSITIVE ORGANS AND TISSUES

To test which parts of the plant are light sensitive, either the basal part of the plant (having a large number of leaves) or the apical part of plant (having many mature flower buds) were covered at 10:00 h with a black sheet, out of doors. In the former case, flower buds opened very close to sunset (19:00 h), but in the latter case, flowers opened around 14:30 h. These results indicate that the bud itself is the light sensitive organ. In other tests, apical or basal halves of the plant or the entire surface of flower buds were covered with black-lanolin at 14:00 h. At the same time, the entire surface of several buds was covered with clear water-lanolin to serve as controls. In the next experiment, the apical half, basal half or the entire length of the sutured tissue of adjacent sepals of buds was covered with black-lanolin. The buds whose basal halves of the sutured tissues were covered with black-lanolin were found to open earlier than the others. These results indicate that the basal half of the sutured tissue is the most sensitive part to the wavelength of light responsible for inhibiting flower opening.

INCREASES IN ENDOGENOUS IAA IN THE SEPAL TISSUE DURING SEPAL SPLITTING

In a preliminary experiment, the endogenous IAA in the sepal tissue of a plant grown out of doors was measured using a new Shimadzu HPLC method (Kabori *et al.*, 1983), capable of detecting below 0.01 ng of IAA. This method used two preparative columns packed with Shim-pack 5PC-RPI and an analytical column of Shim-pack CLC-ODS. IAA in the eluate was detected by fluorescence emission at 360 nm. It was found that IAA increased quickly just before opening. To determine the IAA increases more accurately, IAA in sepals was measured on three different days, using outdoor grown plants (Fig. 1).

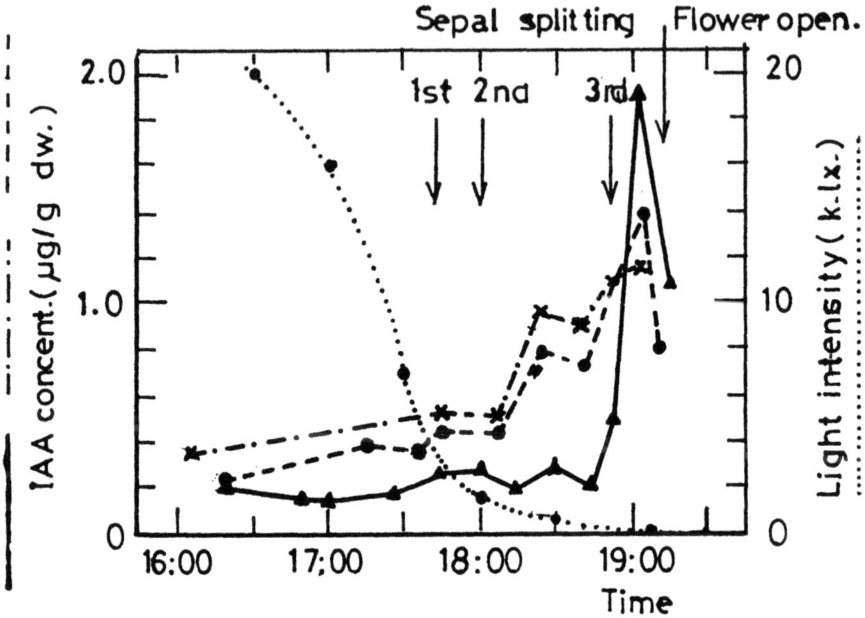

FIGURE 1. IAA concentrations in sepals (out of doors).

Buds were collected several times from 16:00 h to 19:10 h just after the completion of flower opening. The results of the three series of experiments are illustrated in Fig. 1, together with changes in light intensity. The concentration of IAA increased at the time of 1st, 2nd and 3rd sepal splittings. These phenomena seem to indicate that decreasing light intensity could accelerate the production of endogenous IAA, and in turn, IAA accelerates cell separation of the sutured epidermal cells. The IAA increase was far more rapid in the sutured tissue than in other parts of the sepal. This result also seems to support the above hypothesis.

EFFECT OF EXOGENOUS APPLIED IAA AND PRECURSORS ON SEPAL SPLITTING AND FLOWER OPENING

Results of applying, lanolin paste with different concentrations of IAA (from 0.06 to 120 x 10^{-4}M) to the whole surface of flower buds of outdoor-grown plants at 14:00 h, show that IAA accelerates flower opening at all the concentrations tested, the most effective being between 7.0 and 20 x 10^{-4}M.

Application of IAA lanolin paste on various parts of sepals. As mentioned earlier, the sutured epidermal tissues of adjacent sepals are highly sensitive to inhibition by light. So we tried to find whether exogenous IAA applied to these tissues induces cell separation. We applied IAA (5 x 10^{-4}M) in lanolin to the entire surface of the bud, to the basal half of the bud, or to its apical half. We also placed IAA in lanolin to the whole suture line, to its basal half or to its apical half at approximately 10:00 h and left the test plants out of doors. When IAA in lanolin covered the sutured epidermal tissues of adjacent sepals, especially their basal half, IAA accelerated cell separation, an effect similar to that obtained with black lanolin. These results seem to indicate that light inhibition of cell separation is caused by an inhibition of IAA synthesis in the same tissues.

Effect of IAA precursors on sepal splitting outdoors. Test substances in lanolin were given to the entire surface of buds in lanolin paste. Water in lanolin paste was used as a control. Tryptamine (5 x 10^{-4}M), indole acetaldehyde (5 x 10^{-5}M) and indole pyruvic acid (5 x 10^{-5}M) all accelerated sepal splitting to the same extent as equimolar concentrations of IAA. However, tryptophane (5 x 10^{-4}M), indole acetonitrile (5 x 10^{-4}M), indole lactic acid (5 x 10^{-5}M) and naphthalene acetamide (1 x 10^{-4}M, substituting for indole acetamide), had no accelerating effect. From these experiments, it is clear that IAA is synthesized from tryptophane through the following steps:

tryptophane → indole pyruvic acid → indole acetaldehyde → IAA

tryptophane → tryptamine → indole acetaldehyde → IAA

These results also seem to indicate that light inhibits the synthesis of indolepyruvic acid or tryptamine from tryptophane.

EFFECT OF SEVERAL GROWTH SUBSTANCES AND AN ANTAGONIST OF IAA ON SEPAL SPLITTING AND FLOWER OPENING

Lanolin containing one of several compounds was applied to the entire surface of flower buds at approximately 13:00 h, out of doors. Naphthalene acetic acid (7-10 x 10^{-3}M), 2,4-dichlorophenoxyacetic acid (4-12 x 10^{-3}M), gibberellic acid (0.1-64 x 10^{-4}M) kinetin (0.2-20 x 10^{-4}M), abscisic acid (0.2-20 x 10^{-3}M) were tested. NAA and 2,4-D accelerated sepal splitting. However, other compounds had very little effect. These results indicate that only auxins accelerate sepal splitting. Parachlorophenoxy iso-butyric acid (PCIB), an IAA antagonist, was applied to the bud at approximately 16:00 h in lanolin at concentrations between 1.4 and 16 x 10^{-2}M. Concentrations above 5 x 10^{-2}M, PCIB inhibited sepal splitting remarkably, even in the dark. This result also indicates that endogenous IAA plays a strong promoting role in sepal splitting.

EFFECT OF RNA SYNTHESIS INHIBITORS ON SEPAL SPLITTING AND FLOWER OPENING

Actinomycin-D at 1.0 mg ml^{-1} in water was mixed with the same volume of lanolin and applied all over the surface of flower buds at approximately 14:00 h out of doors. This compound accelerated sepal splitting slightly

TABLE 1. Effect of IAA on the separation of sutured epidermal cells of adjacent sepal margins

Stage of sepal splitting of 50% of buds	Time of splitting [clock time]: Water paste	Time of splitting [clock time]: IAA-paste	Acceleration by IAA (minutes)
1st split	17:20	15:27	113
2nd split	18:31	16:49	102
3rd split	18:56	18:01	55
Flower open	19:35	18:41	54

Applied at 12:30 h. The IAA concentration was 5 x 10^{-4}M but had almost no effect on flower opening. If buds were covered with 6.3 x 10^{-3}M IAA lanolin paste 4 hours after actinomycin-D pre-treatment the accelerating action of IAA remained very marked, suggesting that newly synthesized RNA is not necessary for the action of IAA on sepal splitting. Amanitin was also used for the same purpose. This compound accelerated sepal splitting, and had almost no inhibiting effect on flower opening. IAA given to the bud 4 hours after amanitin treatment accelerated sepal splitting and flower opening remarkably. The result of this experiment supports the conclusion drawn from the actinomycin-D experiment.

EFFECT OF INHIBITORS OF PROTEIN SYNTHESIS

Chloramphenicol at 3 mg ml^{-1}, 13 mg ml^{-1} or 20 mg ml^{-1} in water was mixed with the same volume of lanolin and spread over the surface of buds at around 10:00 h. This compound strongly inhibited sepal splitting and flower opening at higher concentrations. However, at lower concentrations it promoted 2nd sepal splitting slightly. Cycloheximide applied at 0.88-0.9 mg ml^{-1} accelerated 1st sepal splitting, but inhibited 2nd splitting and flower opening remarkably. When IAA (7.5 x 10^{-3}M) or NAA (6.6 x 10^{-3}M)lanolin was applied to cycloheximide-pretreated buds, auxin promoted only 1st sepal splitting, and 2nd splitting was inhibited considerably. Where a high concentrations (4 mg/ml) of cycloheximide were applied, sepal splitting and flower opening was inhibited completely. These results seem to indicate that one or several kinds of newly synthesized protein might be necessary for the advance of cell separation represented by 1st and 2nd sepal splitting.

EFFECT OF SPLIT SEPAL EXTRACT ON CELL SEPARATION

Splitting and split sepals of 400 buds were washed twice with 900 ml water by ultrasonification for 15 minutes. The washings were freeze-dried, dissolved in 20 ml of water and dialysed. This dialysate was freeze-dried again to give 300 mg of greyish residue. In another experiment, splitting sepals from 250 or 700 buds were washed with water and high molecular substances were obtained by freeze-drying or by distillation under reduced pressure prior to dialysis. About 40 mg of the dialysate was dissolved in 1 ml of water and mixed with the same volume of lanolin. The lanolin paste was spread on the sutured epidermal tissue of four adjacent sepals on five to ten buds. Water in lanolin was used as the control treatment. The results are shown in Table 2. They indicate that the extract contains one or more high molecular weight substances with the ability to separate sutured epidermal cells of adjacent sepals. We tried to confirm whether the cell separating action of the extract was due to the high molecular substance(s) itself or contamination by traces of IAA. However, the amount of contaminating endogenous hormone was very small; the concentration of IAA in the test solution of the extract being about 3×10^{-6}M. This concentration of IAA has no accelerating effect on sepal splitting. When the extract was applied to the bud with PCIB, PCIB had no inhibitory effect on cell separation. The results of these experiments indicate that one or more high molecular weight substances in the extract promote cell separation by softening the cement between the sutured marginal cells of adjacent sepals.

TABLE 2.. Effect of water extract from split sepals on the separation of sutured epidermal cells of adjacent sepal margins

Stage of sepal splitting of 50% of buds	Time of splitting: Water paste [clock time]	Time of splitting: Paste of sepal ext. [clock time]	Acceleration by sepal extract (minutes)
1st split	17:05	16:42	23
2nd split	18:09	17:07	62
3rd split	18:48	18:36	12
Flower opening	18:57	18:55	2

Applied at 13:00 h

Cycloheximide, in concentrations high enough to inhibit sepal splitting had no inhibitory effect on the cell separating action of the extract. So the high molecular substance(s) extracted by water from the splitting sepals or excreted by the sutured epidermal cells of the adjacent sepals could be an enzyme(s).

PRELIMINARY EXPERIMENT TO FIND THE ENZYME(S) WHICH IS SYNTHESIZED IN THE SUTURED TISSUES OF ADJACENT SEPALS

Aqueous solutions of cellulase RS (Onozuka), macerozyme R-10, pectolyase Y-23 and several other enzymes commonly used for the preparation of protoplasts were tested. Lanolin paste containing these enzymes was applied to the sutured part of adjacent sepals of buds, out of doors. None of the enzyme preparations accelerated the separation of sutured epidermal cells, but they did degrade the cell wall of epidermal and mesophyll cells. In contrast, chitinase T-1 (Asahi Industries Co. Ltd) (10 mg ml^{-1}) effectively separated the sutured epidermal cells of adjacent sepals without affecting the structure of other cells. In purifying chitinase, cultured solutions of *Trichoderma* sp. were ultra- filtrated to remove small molecular substances. So, the resulting enzyme sample was unlikely to contain any IAA. However, there still remained an anxiety that trace amounts of IAA were present. So, PCIB was given to buds together with the purified chitinase to minimise the effect of any IAA which might be present in the enzyme preparation. We found that the PCIB had almost no inhibitory effect on the action of chitinase. Thus, the effect of the enzyme preparation cannot be attributed to traces of IAA.

TABLE 3. Effect of "chitinase" on the separation of sutured epidermal cells of adjacent sepal margins

Stage of sepal splitting of 50% of buds	Time of splitting [clock time]		Acceleration by "Chitinase" (minutes)
	Water paste	"Chitinase"-paste	
1st split	17:14	15:26	108
2nd split	18:24	16:47	97
3rd split	18:47	18:03	45
Flower opening	19:20	18:46	34

Pasted at around 12:30 Chitinase: 3% solution

CONCLUSIONS

From the results of our present experiments on the separation of sutured epidermal cells of adjacent sepals, the physiological steps towards cell separation may be as follows:

1) The action of IAA synthesizing enzyme(s) is suppressed by blue and red light and very characteristically, by green light.

2) This enzyme(s), located mainly in the sutured epidermal cells of adjacent sepals, is activated when the light intensity decreases.

3) In the case of habituated plants, the activity of IAA synthesizing enzyme(s) is accelerated during the first 12 minutes of darkness, or even sooner.

4) IAA synthesized in these tissues by the above-mentioned enzyme(s) catalyses the synthesis of a cell separating enzyme(s) in these same tissues.

5) The activity of this cell separating enzyme can be observed within 15 minutes of the start of IAA synthesis.

6) The cell separating enzyme(s) might be a chitinase dissolving only the cementing substance(s) which binds the epidermal cells of adjacent sepals margins and thereby starting cell separation.

7) The sepals are pushed outward by the expanding petals and adjacent sepal start to separate, marking the beginning of sepal splitting.

8) This cell separation extends to the remaining sutured parts of sepals, and finally all four sepals separate and the flower opens.

REFERENCES

Kobori K, Sakakibara H, Maruyama K, Kobayashi T, Yamaki T (1983) A rapid method for determining urinary indoleacetic acid concentration and its clinical significance as the tumour-marker in the diagnosis of malignant diseases. J. UOEH 5: 213-220

Saito M, Yamaki T (1967) Retardation of flower opening in *Oenothera lamarkiana* caused by blue and green light. Nature 214: 1027

Yamaki T, Saito M, Gordon SA (1966) On the inhibition of green light of flower bud opening in *Oenothera*. Argonne Nat. Lab. Annual Report, ANL-7278: 289-290

POLLINATION-INDUCED COROLLA ABSCISSION AND SENESCENCE AND THE ROLE OF SHORT-CHAIN FATTY ACIDS IN THE PROCESS

A.H. Halevy and C.S. Whitehead[1]
The Hebrew University of Jerusalem
Department of Horticulture
PO Box 12
Rehovot 76-100
Israel

INTRODUCTION

Pollination causes distinct and sometimes dramatic changes in the metabolism and development of various flower parts. In most flowers, pollination results in considerable acceleration of senescence. Some orchid flowers may persist for months but fade within days or hours after pollination. Pollination-induced senescence may be manifested differently in different flowers. It may cause wilting and fading, as in carnation and petunia; promote anthocyanin synthesis, as in *Cymbidium* (Arditti and Flick, 1976) and *Lantana* (Mohan Ram and Mathur, 1984) or accelerate corolla abscission, as in *Nicotiana* (Kendall, 1918), sweet peas, snapdragons, *Pelargonium* (Wallner *et al.*, 1979), *Linum* (Addicott, 1982; Halevy and Mayak, 1981) and *Digitalis* (Stead and Moore, 1979, 1983). Pollination promotes abscission of the corolla or of individual petals, while in unpollinated flowers the entire flower often abscises at the base of the peduncle.

It is generally accepted that ethylene production is increased during flower senescence and that ethylene accelerates flower senescence. It has also been reported for a number of flowers that sensitivity to applied ethylene increases with flower age (Halevy and Mayak, 1981). However, these generalized assumptions may not apply to all flowers. It seems that from the senescence point of view, flowers may be generally divided in two ways: (1) as to their relation to ethylene and (2) as to the cause of the termination of their life.

As to ethylene, one can distinguish two groups (a) flowers which show a climacteric rise in ethylene production during senescence and are

[1] Rand Africaans University, Dept of Botany, PO Box 524, Johannesburg 2000, South Africa

NATO ASI Series, Vol. H 35
Cell Separation in Plants
Edited by D.J. Osborne and M.B. Jackson

sensitive to ethylene (e.g. carnation, morning glory), (b) those in which ethylene production does not increase substantially during senescence and are not sensitive to ethylene (e.g. *Chrysanthemum, Gerbera*). As to the cause of corolla death, there seems to exist two distinct groups: (a) Those showing a gradual change in composition of the corolla with age, a loss of turgor and final wilting (e.g. carnation, petunia, morning glory) - the "wilting type". (b) Those in which the life of the corolla is terminated by abscission, when it is still turgid (e.g. *Pelargonium, Digitalis*) - the "abscissing type". Ethylene may accelerate senescence whether it is manifested by gradual wilting or by abrupt abscission.

WHERE AND WHEN IS THE POLLINATION SIGNAL(S) PRODUCED?

The pollination process can be generally divided into three stages (a) contact of pollen with the stigma, (b) passage of pollen tubes in the style and (c) fertilization of ovules. Since the pollination process operates in the pistil and the senescence symptoms appear in the petals, senescence signal(s) are apparently transmitted from the gynoecium to the corolla. Gilissen and Hoekstra (1984) have indeed found that eluates collected from pollinated petunia gynoecia, induced corolla wilting.

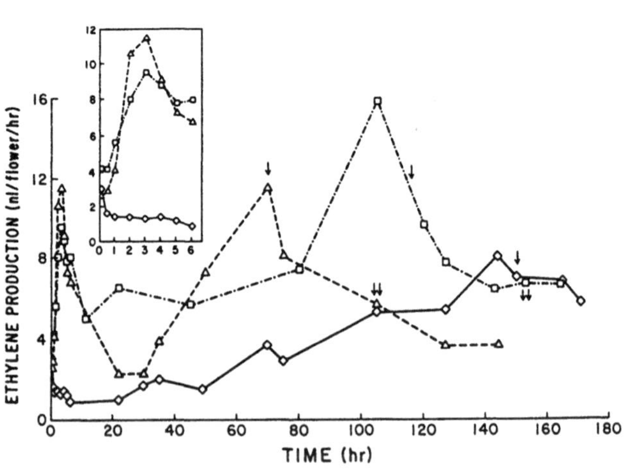

FIGURE 1. Ethylene production by senescing petunia flowers. Freshly harvested flowers (just fully open) were placed in ventilated jars immediately (diamonds) or after pollination (triangles) or wounding the pistil (squares). The inset details ethylene production during the first 6 hours. Single arrows indicate the time of visible blueing, double arrows the onset of wilting. Means of four replicate flowers per treatment. (From Whitehead et al., 1984, with permission)

The question arises as to which of the three stages produces the pollination signal(s). Soon after pollination there is a substantial transient rise in ethylene production in petunia flowers (Fig. 1). Pollen of petunia, carnation and other flowers contains substantial concentrations of ACC (Whitehead *et al.*, 1983), and application of ACC to the stigma of petunia (Whitehead *et al.*, 1984) and carnation (Reid *et al.*, 1984) resulted in a dramatic transient rise in ethylene production by the gynoecium. The initial peak of ethylene production by pollinated petunia flowers may be the result of oxidation of pollen ACC, since the germination of the pollen starts only after 2 hours. However, when the stigma is removed soon after this first burst of ethylene, no stimulation of senescence occurs. This indicates that no senescence-pollination signal is produced during the contact of the pollen and the stigma in the first 2-3 hours after pollination (see also Hoekstra and Weges, 1986; Pêch *et al.*, 1987).

Time-course studies of pollination-induced corolla wilting (Gilissen, 1977) and abscission (Stead and Moore, 1979), have shown that senescence symptoms in the corolla or the weakening of the abscission zone, begin several hours before the pollen tubes reach the ovary. This eliminates fertilization as a senescence-inducing stage. Further support for this conclusion comes from the findings that corolla senescence can also be induced by self-pollination in self-incompatible petunias (Gilissen, 1977; Halevy, unpublished).

Another aspect related to pollination and fertilization is the hydrolysis of macromolecules in the corolla and the mobilization of the breakdown products to the ovary and the developing fruit (Halevy and Mayak, 1979, 1981), implicating a senescence signal from the fertilized ovary to the corolla. However, when pollinated tomato flowers were sprayed with silver thiosulphate (STS) no petal abscission occurred, the fruits developed normally with the petals attached to them (R. Nichols, personal communication). This indicates that mobilization from the petals to the ovary is a consequence and not the cause of petal senescence and abscission.

Pollination with pollen killed by x-ray irradiation, does not accelerate wilting. Gilissen (1977) also reported that about 800 viable pollen grains are necessary to initiate maximal wilting of petunia flowers, which means that a fifth of the stigmatic surface must be covered with living pollen grains. He suggested that the pollen tube penetration and injury of the style are the causes of the pollination-induced corolla senescence (Gilissen, 1976). We have indeed demonstrated that wounding of the pistil by piercing it with a thin copper needle accelerates ethylene production and corolla wilting (Whitehead *et al.*, 1984; Fig. 1, Plate 1). This indicates that the pollination signal(s) are produced in the style in reaction to the passage of the pollen tubes through it. Seventy years ago, Kendall (1918) reached the same conclusion stating "it is the stimulation of the stylar tissues caused by the growth of the pollen tubes which shortens the time between anthesis and the abscission of the corolla".

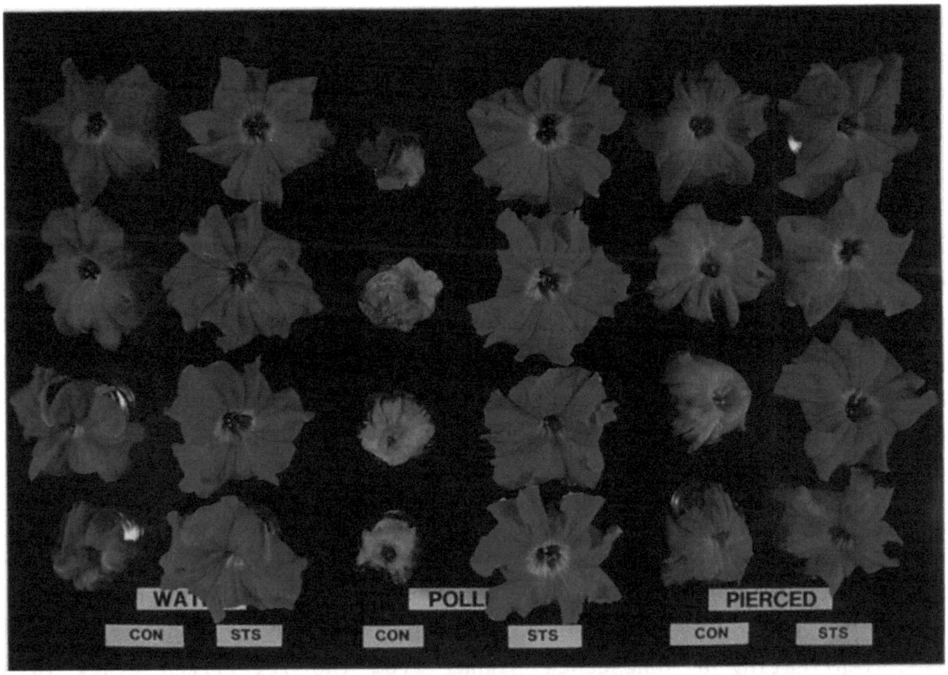

Plate 1. Effects of silver thiosulphate (STS) pre-treatment on the response of petunia flowers to pollination or piercing the pistil. Freshly harvested flowers (just fully opened) were pre-treated with STS (4 mM for 1 hour), then pollinated, pierced, or left untreated. The photograph was taken on day 7 with a red filter to accentuate the blue senescent corollas (darker shades). (From Whitehead et al., 1984, with permission)

THE POLLINATION SIGNAL(S)

The dramatic effects of pollination on flower senescence were attributed by Burg and Dijkman (1967) to auxin, which was known to be present in pollen. Auxin is known to promote ethylene production in several tissues (Yang and Hoffman, 1984). It was suggested that pollination caused transfer of auxin to the stigma, that moved to the pistil and induced further ethylene production and senescence. Indeed, in orchids, application of IAA caused a sequence of events similar to that resulting from pollination. However, studies with labelled IAA applied to the stigma have shown that its movement is very slow both in orchids (Strauss and Arditti, 1982) and in carnations (Reid *et al.*, 1984). Furthermore, application of IAA to the stigma (Lovell *et al.*, 1987) or injection into the ovaries of petunia flowers (Whitehead and Halevy, 1989) did not enhance corolla senescence, and application of IAA to the stigma of *Digitalis* did not cause rapid corolla abscission (Stead and Moore, 1983). Therefore, auxin does not seem a likely candidate for the pollination signal.

Pollination-induced senescence is preceded or accompanied by enhancement of ethylene production by the flowers. This has been established in some orchids (Burg and Dijkman, 1967), carnations (Nichols, 1977), Petunia (Whitehead *et al.*, 1984) and other flowers (Hall and Forsyth, 1967; Wallner *et al.*, 1979). However, it seems that there is a difference in principle between the "wilting type" flowers and the "abscissing type" i.e. the sites of ethylene production are different. In both types initially, almost all the ethylene is produced by the stigma, followed by a sequential increase in ethylene production, from the upper portion of the style to the lower portion of the style, ovary and receptacle or calyx (Nichols *et al.*, 1983; Stead and Moore, 1983). However, in the "wilting type" flowers such as carnation and petunia, most of the ethylene is finally produced by the

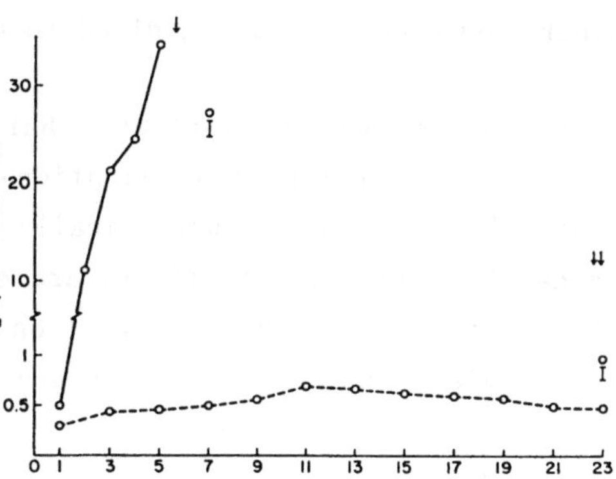

FIGURE 2. Ethylene production by pollinated (solid line) and unpollinated (dashed line) cyclamen flowers of different ages. Intact flowers were left unpollinated or pollinated on opening day. Flowers were harvested daily for ethylene production measurements. Means of five replications. Bars indicated the SE of the means. One arrow indicates the time of corolla abscission and two arrows the time of wilting. (From Halevy et al., 1984, with permission)

corolla (Nichols *et al.*, 1983; Whitehead *et al.*, 1984). While in the "abscissing type" flowers, the corolla of pollinated flowers produces little or no ethylene before abscissing, as found in *Digitalis* (Stead and Moore, 1983), and other flowers (Hall and Forsyth, 1967).

An interesting case is the cyclamen (Halevy *et al.*, 1984). Non-pollinated cyclamen flowers are long-lived and their normal senescence is characterized by gradual discoloration, loss of turgor and final wilting. They are insensitive even to high ethylene concentrations and their sensitivity does not change with age. Non-pollinated flowers produce very little ethylene and no change in production is observed during the whole life span of the flower until death. However, pollination induced a dramatic increase in ethylene evolution and caused abscission of the fully turgid corolla four days after pollination (Fig. 2). The pollination-induced ethylene production is almost exclusively from the gynoecium and not from the corolla (Halevy, unpublished). Non-pollinated cyclamen flowers thus belong to the non-ethylene producing "wilting type" while the pollinated flowers belong to the ethylene producing "abscissing type". In the "wilting type" flowers the senescence symptoms appear first in the distal portion of the petals or the corolla. These portions are also the major production sites of ethylene production at senescence. However, when the upper part of carnation petals and petunia corollas are detached from their base, their senescence is considerably delayed and they produce very little ethylene (Mor *et al.*, 1985). It was found that most ACC-synthase activity is located in the basal portion of the corolla. ACC is translocated to the upper parts where it is oxidised to ethylene.

The results presented above seem to indicate that ethylene may be a candidate for the co-ordinating signal of pollination-induced senescence. Pollination as well as pistil wounding dramatically increased ethylene production and senescence (Fig. 1). These effects are counteracted by STS, the inhibitor of ethylene action in petunia, carnation and orchids (Goh *et al.*, 1985). However, ethylene production and senescence of pollinated carnation petals are not prevented by removal of ethylene produced by the gynoecium (Reid *et al.*, 1984). This suggests that pollination-induced corolla senescence is a response to movement from the gynoecium of some stimulus other than ethylene.

Stead and Moore (1979) indicated that the pollination stimulus passes along the style at a rate which could be determined by stylar excision experiments. We have recently found a sequential increase in ACC-synthase activity and ACC levels down the style following pollination of petunias. Very low ethylene-forming-enzyme (EFE) activity is present in the style. The ACC thus produced in the style is transported to the upper portion of the corolla which has very low ACC-synthase activity. These results may indicate that ACC is the transmitted pollination-induced senescence signal. In the "wilting type" flowers, ACC moves up to the corolla, while in the "abscissing type" flowers it moves only to the abscission zone.

Results with cyclamen flowers demonstrate however, that ACC may not be the only pollination signal. Fig. 2 shows a dramatic increase in ethylene production by pollinated flowers, followed by corolla abscission. However, when unpollinated flowers were exposed for 48 hours to ethylene even at the very high concentrations of 50 µl/l, or were treated with ACC that increased endogenous ethylene production more than 100-fold and to a rate 5 times faster than in pollinated flowers, there was no effect on flower senescence or on corolla drop (Halevy *et al.*, 1984). However, when ACC was applied to *pollinated flowers* it enhanced corolla abscission. STS counteracted the effect of pollination, prevented corolla abscission and turned the senescence of pollinated cyclamen flowers from the "abscissing type" to the "wilting type" of flower senescence.

These results indicate that the promotive effect of pollination on corolla abscission cannot be caused merely by promotion of ACC synthesis and stimulation of ethylene production. It is obvious that apart from the promotion of ethylene evolution, pollination also renders the tissue *sensitive* to ethylene. This means that ethylene is necessary but not itself sufficient to induce the pollination-promoted senescence syndrome. Therefore, it seems that pollination induces at least two signals, one is ACC and the other is a "sensitivity factor" which renders the tissue sensitive to ethylene. We have followed the time-course of the increase in sensitivity to ethylene in pollinated flowers, by exposing the upper portion of petunia corolla to ethylene at various times after pollination. It was found that the "sensitivity factor" is the first signal moving into the corolla. Increase in sensitivity to ethylene can be detected 7 hours after pollination (Fig. 3), while a significant increase in ACC is found only 16 hours after pollination.

WHAT IS THE "SENSITIVITY FACTOR" TO ETHYLENE?

Results presented above clearly demonstrate that pollination not only increases the sensitivity of the corolla to ethylene, but that following pollination there is production of a *transmissible* factor that renders the corolla or the abscission layer sensitive to ethylene. The identity of this factor has been sought.

When petunia flowers were treated with AVG prior to pollination there was no increase in ACC and ethylene in the corolla, but there was a marked increase in sensitivity to ethylene (Fig. 3). This demonstrates that the "sensitivity factor" is not directly related to ethylene biosynthesis.

Our first candidate for the sensitivity factor was ABA (Kao and Yang, 1983). Indeed ABA does increase the sensitivity of detached unpollinated corolla to ethylene and furthermore, an increase in ABA was found in the pistil following pollintation. However, we were unable to find any change in ABA content in the corolla after pollination, and injection of ABA into the ovary did not promote senescence or the sensitivity to ethylene. These results indicate that ABA is not transported from the pistil to the corolla (Whitehead and Halevy, 1989a).

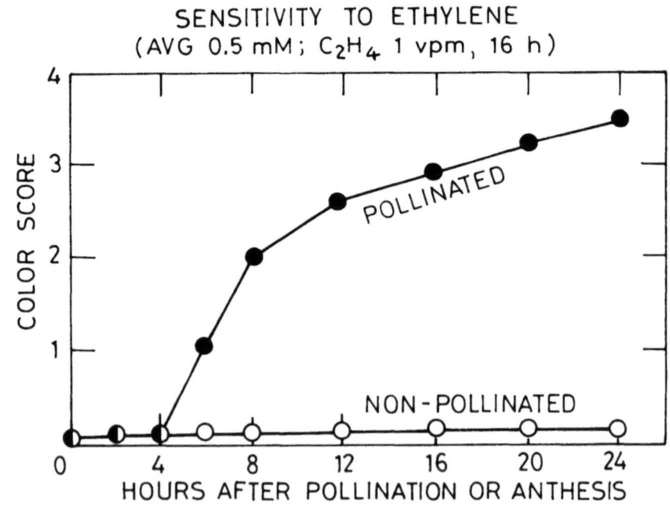

FIGURE 3. The effect of pollination of petunia flowers on sensitivity of the corolla to ethylene. Corollas were pretreated with AVG (0.5 mM) detached at various times after pollination or anthesis and exposed to ethylene (1 vpm for 16 hours). Sensitivity to ethylene was recorded as change in colouration and turgidity (0 - pink and turgid; 4 - blue and wilting)

When stylar exudates were collected from pollinated and nonpollinated petunia flowers and applied to the stigmas of nonpollinated flowers, only the flowers treated with exudate from pollinated styles senesced at an accelerated rate, while the exudate from unpollinated styles had no effect (Whitehead and Halevy, 1989b). In an attempt to identify compounds in the exudate that might be involved in the acceleration of the sensitivity of pollinated flowers to ethylene, the exudates from three batches of 25 pollinated or unpollinated styles were pooled and freeze dried. Trimethylsilyl (TMS) ethers of the resulting residues were prepared by dissolving the residues in a mixture containing bis-(trimethylsilyl) trifluoroacetamide and trimethylchlorosilane (5:1). The mixture was then heated at 70°C for 30 min. TMS ethers were separated by capillary gas chromatography on a gas chromatograph equipped with a flame ionization detector and a 20 m x 0.32 mm SE 30 fused silica capillary column. The different ethers eluting from the capillary column were positively identified by GC-MS analysis (El, 70 eV). Analysis of the TMS ethers synthesized from these exudates revealed the presence of large quantities of two compounds in the exudate from pollinated styles which could not be detected in the exudates from unpollinated styles. These compounds were identified by GC-MS analysis as the short-chain saturated fatty acids octanoic acid and decanoic acid. Application of the two fatty acids to the stigmas of unpollinated flowers resulted in an acceleration of senescence similar to that brought about by pollination. Exposure of flowers to ethylene after application of these acids to their stigmas also resulted in an increase in the sensitivity of the corollas to ethylene (Whitehead and Halevy, 1989a). When ^{14}C-labelled octanoic and decanoic acids were applied to petunia stigmas, they were recovered in the styles, ovaries and corollas. Furthermore, when deuterated acetate was applied to pollinated stigmas, the deuterated label was recovered in octanoic and decanoic acid of the stylar exudates.

These and other results that will be presented elsewhere (Whitehead and Halevy, 1989b) indicate that the short-chain saturated fatty acids are the pollination-induced signals inducing sensitivity to ethylene in petunia flowers.

We now have preliminary results indicating that these fatty acids also increase the sensitivity to ethylene in a few other plant systems, and may, thus, have a more general role in regulating the sensitivity of plant tissues to ethylene.

Short-chain fatty acids have been demonstrated to inhibit several plant processes such as germination, cell elongation and endosperm amylolysis (Babiano *et al.*, 1984; Buller *et al.*, 1976; Stewart and Barrie, 1979; Ulbright *et al.*, 1982a, b). They have also been shown to be involved in inducing dormancy in some seeds and bulbs (Ando and Tsukamoto, 1971; 1974; Barrie *et al.*, 1975, 1979). These effects resemble the action of ABA which also increases the sensitivity of petunia corollas to ethylene (see above). It is interesting to note that as in petunia, fatty acids in the range of C_8 to C_{10} were found to be the most active in the above system.

The mode of action of fatty acids in increasing the sensitivity to ethylene is still unknown. We have found that they increase the membrane permeability of petunia corollas as was reported earlier for other plant tissues (Jackson and Taylor, 1970; Wilmer *et al.*, 1978; Metzger and Sebasta, 1982). It is therefore possible that the fatty acids affect the structure and properties of the membranes modifying their activities and binding sites.

REFERENCES

Addicott FT (1982) Abscission. Univ of California Press, Berkley

Arditti J, Flick E (1976) Post-pollination phenomena in orchid flowers VI Excised floral segments of *Cymbidium*. Amer J Bot 63: 201-211

Ando T, Tsukamoto Y (1971) Inhibitory action of saturated fatty acids and their derivatives. Phytochem 10: 2143-2144

Ando T, Tsukamoto Y (1974) Capric acid a growth inhibiting substance from dormant *Iris holandica* bulbs. Phytochem 13: 1031-1032

Babiano MJ, Aldasoro JJ, Hernandez-Nistal J, Rodriguez D, Matilla A, Nicolas G (1984) Effect of nonanoic acid and other short chain fatty acids on exchange properties in embryonic axes of *Cicer arientinum* during germination. Physiol Plant 61: 391-395

Berrie AMM, Don R, Buller, D, Alam M, Parker W (1975) The occurrence and function of short chain fatty acids in plants. Plant Sci Lett 6: 163-173

Berrie AMM, Buller D, Don R, Parker W (1979) Possible role of volatile fatty acids and abscisic acid in the dormancy of oats. Plant Physiol 63: 758-764

Buller D, Parker W, Grant, Reid JS (1976) Short chain fatty acids as inhibitors of gibberellin-induced amylolysis in barley endosperm. Nature 260: 169-170.

Burg SP, Dijkman MJ (1967) Ethylene and auxin participation in pollen induced fading of *Vanda* orchard blossoms. Plant Physiol 42: 1648-1650

Gilissen LJW (1976) The role of the style as a sense-organ in relation to wilting of the flowers. Planta 131:201-202

Gilissen LJW (1977) Style-controlled wilting of the flower. Planta 133: 375-280

Gilissen LJW, Hoekstra FA (1984) Pollination-induced corolla wilting in *Petunia hybrida*. Rapid transfer through the style of a wilting-induced substance. Plant Physiol 75:496-498

Goh CJ, Halevy AH, Engel R, Kofranek AN (1985) Ethylene evolution and sensitivity in cut orchid flowers. Scientia Nortic 25: 57-67

Halevy AH, Mayak S (1979) Senescence and postharvest physiology of cut flowers, part I. Hort Rev 1: 204-236

Halevy AH, Mayak S (1981) Senescence and postharvest physiology of cut flowers, part II. Hort Rev 3: 59-143

Halevy AH, Whitehead CS, Kofranek AN (1984) Does pollination induce corolla abscission of cyclamen flowers by promoting ethylene production? Plant Physiol 75: 1090-1093

Hall IV, Forsyth FR (1967) Production of ethylene by flowers following pollination and treatments with water and auxin. Can J Bot 45: 1163-1166.

Hoekstra FA, Weges R (1986) Lack of control by early pistillate ethylene of the accelerated wilting of *Petunia hybrida* flowers. Plant Physiol 80: 403-408

Jackson PC, Taylor JM (1970) Effects of organic acids on ion uptake and retention in barley roots. Plant Physiol 46: 538-542

Kao CH, Yang SF (1983) Role of ethylene in senescence of detached rice leaves. Plant Physiol 73: 881-885

Kendall JN (1918) Abscission of flowers and fruits of the Solanaceae with special reference to *Nicotiana*. Univ Calif Pub in Botany 5: 347-428

Lovell PJ, Lovell PH, Nichols R (1987) The control of flower senescence in petunia (*Petunia hybrida*). Ann Bot 60: 49-59

Metzger JD, Sebasta DK (1982) Role of endogenous growth regulators in seed dormancy of *Avena fatua* 1. Short chain fatty acids. Plant Physiol 70: 1480-1485

Mohan Ram HY, Mathur G (1984) Flower changes in *Lantana camara*. J Exp Bot 35: 1656-1662

Mor Y, Halevy AH, Spiegelstein H, Mayak S (1985) The site of 1-aminocyclopropane-1-carboxylic acid in senescence carnation plants. Physiol Plant 65: 196-202

Nichols R (1977) Sites of ethylene production in the pollinated and unpollinated carnation (*Dianthus caryophyllus*) inflorescence. Planta 135: 155-159

Nichols R, Bufler, G, Nor Y, Fujino DW, Reid MS (1983) Changes in ethylene production and 1-aminocyclopropane-1-carboxylic acid content of pollinated carnation flowers. J Plant Growth Regulation 2: 1-8

Nichols R, Frost CE (1985) Wound-induced production of 1-aminocyclopropane -1-carboxylic acid and accelerated senescence of *Petunia* corollas. Scientia Hort 26: 47-55

Pêch JC, Latché A, Larrigaudiére C, Reid MS (1987) Control of early ethylene synthesis in pollinated petunia flowers. Plant Physiol Biochem 25: 431-437

Reid MS, Fujino DW, Hoffman NE, Whitehead CS (1984) 1-aminocyclopropane-1-carboxylic acid - the transmitted stimulus in pollinated flowers? Plant Growth Regul 3: 189-196

Stead AD, Moore KG (1979) Studies on flower longevity in *Digitalis*. Pollination induced corolla abscission in *Digitalis* flowers. Planta 146: 409-414

Stead AD, Moore KG (1983) Studies in flower longevity in *Digitalis*. The role of ethylene in corolla abscission. Planta 157: 15-21

Stewart RRC, Berrie AMM (1979) Effect of temperature on the short chain fatty acid induced inhibition of lettuce seed germination. Plant Physiol 63: 61-62

Strauss M, Arditti J (1982) Postpollination phenomena in orchid flowers X. Transport and fate of auxin. Bot Gaz 143: 286-293

Ulbright CE, Pickard RG, Varner JE (1982) Effects of short chain fatty acids on radicle emergence and root growth in lettuce. Plant Cell Environ 5: 293-301

Ulbright CE, Pickard RG, Varner JE (1982) Effects of short chain fatty acids on seedlings. Plant Cell Environ 5: 303-307

Wallner S, Kassalen R, Burgood J, Craig R (1979) Pollination, ethylene production and shattering in geraniums. Hort Science 14: 446

Whitehead CS, Halevy AH (1989a) The role of octanoic and decanoic acid in the ethylene sensitivity during pollination-induced senescence of *Petunia hybrida* flowers. Acta Hort (in press)

Whitehead CS, Halevy AH (1989b) Ethylene sensitivity: The role of short-chain saturated fatty acids in pollination-induced senescence of *Petunia hybrida* flowers. Plant Growth Regulation (in press)

Whitehead CS, Fujino DW, Reid MS (1983) Identificstion of the ethylene precursor, 1-aminocyclopropane-1-carboxylic acid (ACC) in pollen. Scientia Hortic 21: 291-297

Whitehead CS, Halevy AH, Reid MS (1984) Roles of ethylene and ACC in pollination and wound-induced senescence of *Petunia hybrida* L flowers. Physiol Plant 61: 643-648

Willmer CM, Don R, Parker W (1978) Levels of short chain fatty acid and of absisic acid in water-stress and non-stressed leaves and their effects on stomata in epidermal strips and excised leaves. Planta 139: 281-287

Yang SE, Hoffman NE (1984) Ethylene biosynthesis and its regulation in higher plants. Ann Rev Plant Physiol 34: 155-189

THE ABSCISSION PROCESS IN PEACH: STRUCTURAL, BIOCHEMICAL AND HORMONAL ASPECTS

A. Ramina[1], N. Rascio[2] and A. Masia[1]
Institute of Pomology[1]
Department of Biology[2]
University of Padova
35100 Padova, Italy

INTRODUCTION

The study of the abscission process is rather difficult in fruit because the sensitivity of the target cells is restricted in time (Ramina, 1981). Since leaf and fruit are homologous organs, it would be much more convenient experimentally to substitute fruit with leaves, where abscission can be induced readily by deblading at almost any time. This chapter describes our attempts to validate the use of leaf abscission as a model for fruit abscission by ascertaining if, in peach, structural, enzymatic and hormonal aspects of abscission in the two organs are closely similar.

STRUCTURAL ASPECTS OF FRUIT AND LEAF ABSCISSION

In fruit, the abscission zone responsible for the June drop (AZ3), located between the fruit and the receptacle, is pre-differentiated and consists of two different bands each made up of a few layers of cells with dense cytoplasm (Rascio *et al.*, 1985). The proximal band shows intercellular spaces, the distal one, on the contrary, is very densely packed. Both bands are made up of small groups of several cells only, each group being surrounded by a thick wall. This structure provides evidence of secondary cell division activity in AZ3. Embryoctomy activates the whole AZ3 system, leading to the same ultrastructural responses in cells of both the proximal and distal bands. However, because of its initial compactness (Fig. 1) changes are particularly evident in the distal band. Embryoctomy leads to a further enrichment of the AZ cell cytoplasm, as well as to an increase in secretory activity (Fig. 2), which is probably related to the production and extrusion of lytic enzymes attacking the cell walls. This cell wall digestion involves only the middle lamella of the thick outer wall of the cell packets (Fig. 3). As a consequence, intercellular spaces greatly increase (Fig. 4) and shedding occurs because of separation between the packets of cells (Fig. 5).

NATO ASI Series, Vol. H 35
Cell Separation in Plants
Edited by D. J. Osborne and M. B. Jackson

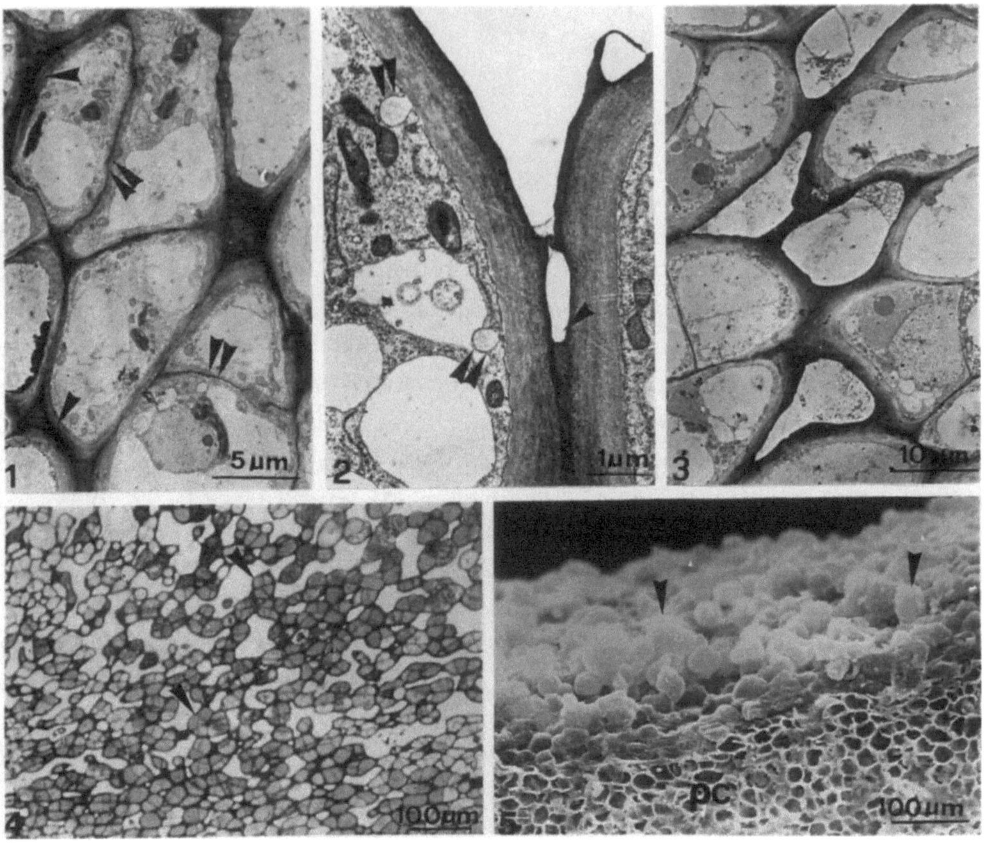

FIGURES 1-5. The abscission pattern in the distal band of peach fruit AZ3.
Fig. 1: Compact tissue before AZ activation. Small cells are surrounded by thick cell walls (arrows) and crossed by thin walls (double arrows). Cytoplasm is rather dense and no intercellular spaces are present. Fig. 2: Middle lamella digestion (arrow) in activated AZ. Note the secretory activity in the cell on the left (double arrows). Fig. 3: Large intercellular spaces are formed by digestion of the thick cell walls surrounding the small cell- clusters. Fig. 4: Note the loosening of the AZ tissue following the formation of ever larger intercellular spaces. Several cell packets (arrows) are still visible. Fig. 5: Intact roundish cell surfaces (arrows) on the abscission side of the fruit, above the sectioned cell of the adjoining parenchyma (pc).

Unlike the abscission zones of the fruit, the leaf AZ at the petiole/ shoot junction is neither pre-differentiated nor distinguishable before induction (Rascio *et al.*, 1987). Tissues surrounding the vascular system are made up of single parenchymatous cells (Fig. 6). After deblading, differentiation of the AZ occurs and about ten layers of cells become

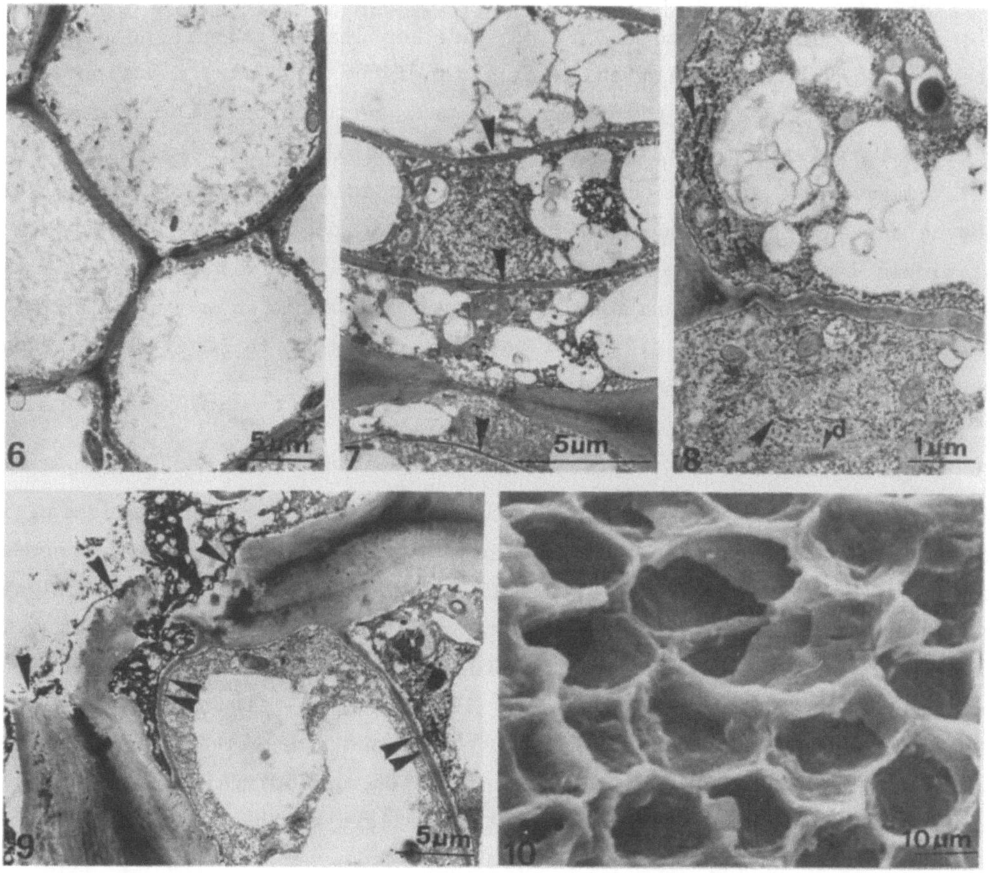

FIGURES 6-10. The abscission pattern of the AZ of peach leaf.
Fig. 6: Parenchymatous cells with a very large central vacuole and little cytoplasm in AZ before activation. Fig. 7: AZ cells after activation. Note the thin cell walls crossing them (arrows). Fig. 8: In activated AZ cells cytoplasm is enriched, with dictyosomes (d) and numerous rER (arrows) profiles. Fig.9: Swollen and fractured cell walls (arrows) in AZ. The thin walls formed by secondary cell division (double arrows) are not affected. Fig. 10: Broken cells on the abscission side of the leaf petiole.

crossed by several thin walls (Fig. 7), demonstrating recent secondary division of AZ cells. Vacuoles are smaller and cytoplasm more dense than before deblading (Fig. 8), with an increase in polysomes, dictyosomes and rER. Later, AZ cell walls swell and show areas of digestion and lysis. The cell wall digestion involves the entire parietal region, *without* preferential middle lamella dissolution and cell separation. The cell walls fracture (Fig. 9) and broken cells are visible on the abscinding surfaces of the petiole (Fig. 10) and shoot. In neither leaf nor fruit do digestion phenomena involve the thin cell walls formed by secondary division of AZ cells. They only affect the outer thick wall surrounding cell packets. The different resistance to digestion by thin and thick cell walls might be related to differences in chemical composition and/or insensitivity to lytic enzyme action. Several authors have described abscission as a consequence of middle lamella digestion and cell detachment, while others noticed cell wall fracture and cell breakdown (reviewed in Rascio *et al.*, 1985, 1987). Interestingly, these two different patterns can occur in AZs from different organs of the same species.

BIOCHEMICAL ASPECTS OF ABSCISSION

In the AZ systems of both fruit and leaves, induction of abscission is related to an increase in soluble protein synthesis which always precedes the rise in lytic enzyme activity. In agreement with the different ultra-structural patterns of dissolution and separation in fruit compared with leaves, different enzymes have been found to be active in fruit and leaf abscission of peach (Rascio *et al.*, 1985, 1987). In fruit, the main activity is ascribed to exopolygalacturonases and endocellulases. In leaves the enzymes primarily involved are cellulases and polygalacturonases with endoactivity. This different biochemical pattern can account for the lysis of the middle lamella in fruit but a more comprehensive digestion of the entire parietal region in leaves.

HORMONAL ASPECTS OF FRUIT ABSCISSION

Since plant hormones present in the fruit and in the leaf lamina are generally believed to be the chemical messengers responsible for abscission induction, our first approach was to investigate their levels in persisting and abscinding fruits. Attention was focused on auxins, growth inhibitors and ethylene (Ramina and Masia, 1980, 1982; Vizzotto *et al.*, 1988). The IAA concentration relates to the rate of fruit growth, being higher at the beginning of stage I (SI) of fruit development, and decreasing towards stage II (SII). Fruits with the greatest abscission potential contained the least IAA. As far as inhibitors are concerned, bioassays following TLC of the acidic fraction from pericarp and seed tissues, revealed two sharp regions of inhibitory activity in wheat coleoptile elongation- and cress seed germination-bioassays. Abscisic acid (ABA) and *p*-coumaric acid (PCA), identified by GC-MS analysis, were responsible for this inhibitory activity. The ABA content appeared to be positively related to the fruit growth rate, and negatively to fruit abscission potential. The concentration of free PCA, a cofactor of IAA oxidase, was highest in abscission-prone fruit and was increased by exogenous ethylene (Giulivo *et al.*, 1981). A significant rise in ethylene evolution has also been observed in whole fruits with high abscission potential, before they drop. Seeds from the same population of fruit showed a sharp peak of ethylene evolution in the middle of SI, concurrently with a drop in IAA. Thus no association appears to exist between fruit ABA levels and physiological drop. In contrast, the interaction between auxin and ethylene may well play a crucial role in the induction of abscission. Research on the interaction between exogenous ethylene, embryoctomy and auxin transport and metabolism through AZ3 indicates that abscission is preceded by (a) a reduction of $(1\text{-}^{14}C)$IAA transport through AZ3 cells (b) an increase in its decarboxylation and (c) changes in partitioning of the radioactivity among different solvents (Ramina *et al.*, 1986). In the EtOH soluble fraction, higher levels of free $(1\text{-}^{14}C)$IAA and lower amounts of $1\text{-}^{14}C$ IAASp with total radioactivity declining have been found. However, this was not confirmed by enzyme-linked immunoassay (ELISA) of endogenous free IAA in AZ3, although an increase of IAA-oxidase activity was found. One of the main features characterizing the induction phase of abscission is the change in polarity of the AZ tissues, with a reduction of IAA basipetal transport, followed by decreased acropetal translocation of (^{3}H) sucrose and a stimulation of callose deposition on sieve plates. We have evidence that exogenous ethylene, at least in peach, promotes fruitlet abscission indirectly, by reducing fruit growth potential. By applying AVG or $AgNO_3$ to the AZ3 of fruits in which the shedding process had been induced respectively by embryoctomy, or by Ethephon, it has proved possible to delay or prevent abscission, although the effects of the Ethephon on IAA transport and metabolism, and sucrose translocation, were maintained. In embryoctomized fruits, the rise in exopolygalacturonases and endocellulases at AZ3 level was prevented by AVG, while $AgNO_3$ did not affect the activity of these enzymes.

CONCLUSIONS

In peach, the homology between leaf and fruit does not imply an analogy in the development of morphological and biochemical events leading to abscission. It appears that different mechanisms operate in the two organs.

In fruit, the process is an activation of pre-differentiated regions while in leaf, differentiation of the abscission layer follows the induction of the shedding. Nevertheless, these findings support the concept of target cells: abscission occurs always in predetermined but not necessarily pre-differentiated regions. The process of differentiation may be associated with secondary cell divisions while activation always implies rejuventation of the cytoplasm necessary to produce responsive conditions in the cell.

As far as the regulation of the process is concerned, at least three aspects should be considered: (1) genetic determination of cellular competence for the shedding response; (2) differentiation of the AZ system; (3) activation of the cell separation process.

Most information from research in peach fruit concerns the AZ3 activation phase and is summarized as follows: (a) the signal activating the AZ3 system comes from the developing seed; (b) interaction between ethylene and auxin may play a crucial role in determining the signal; (c) signal transduction might be coupled to the mechanism of auxin polar transport - indeed, inhibition of the IAA polar transport itself may cause activation of AZ3; (d) ethylene is essential since it regulates, *in loco,* lytic enzyme synthesis and extrusion within AZ3.

REFERENCES

Giulivo C, Ramina A, Masia A (1981) Effect of (2-chloroethyl)phosphonic acid on abscisic acid and para-coumaric acid levels in peach fruit. Riv Ortoflorofrutt It 65: 381-387

Ramina A (1981) La dinamica della cascola ed alcuni aspetti fisiologici della abscissione nel diradamento chimico dei frutti di pesco (*Prunus persica* L. Batsch.). Proceedings Congress on I fitoregolatori nel controllo della produzione degli alberi da frutto. Ferrara, Italy: 9-32

Ramina A, Masia A (1980) Levels of extractable abscisic acid in the mesocarp and seed of persisting and abscising peach fruit. J Am Soc Hort Sci 105: 465-468

Ramina A, Masia A (1982) Levels of extractable para-coumaric acid in the exomesocarp and seed of persisting and abscising peach fruit. Sci Hort 16: 375-383

Ramina A, Masia A, Vizzotto G (1986) Ethylene and auxin transport and metabolism in peach fruit abscission. J Am Hort Sci 111: 760-764

Rascio N, Casadoro G, Ramina A, Masia A (1985) Structural and biochemical aspects of peach fruit abscission (*Prunus persica* L. Batsch). Planta 164: 1-11

Rascio N, Ramina A, Masia A, Carlotti C (1987) Leaf abscission in peach, *Prunus persica* (L.) Batsch: Ultrastructural and biochemical aspects. Bot Gaz 148: 433-442

Vizzotto G, Masia A, Bonghi C, Tonutti P, Ramina A (1988) IAA levels in *Prunus persica* L. Batsch in relation to fruit growth and development. Acta Hort (in press).

SECTION V

Cell Separation in Development

CELL SEPARATION: A DEVELOPMENTAL FEATURE OF ROOT CAPS WHICH MAY BE OF FUNDAMENTAL FUNCTIONAL SIGNIFICANCE

M.E. McCully
Department of Biology
Carleton University
Ottawa K1S 5B6
Canada

INTRODUCTION

The separation of cells and the synthesis and release of mucilage at the periphery of root caps are related developmental events which occur throughout the life span of a vascular plant. These processes are widely recognized, and almost inevitably referred to in accounts of root-soil interactions. Of the two, only mucilage production has been studied in detail, but almost always in seedling roots, axenically-grown, with high moisture. The mucilage is synthesized by cells lying at or within 2 or 3 cells from the surface of the intact root-cap tissue. Secreted mucilage may accumulate in deposits between the plasma membrane and the wall on the outer side of the secreting cells (see review by Rougier and Chaboud, 1985). It is still commonly stated that subsequent disintegration of the cell walls allows release of this mucilage (see, for example, Sticher and Jones, 1988). The chemical composition of the mucilage is well characterized (see Bacic *et al.*, 1986) and some of its physical properties have been studied (Guinel and McCully, 1986).

The almost total neglect of the structural and biochemical events which lead to abscission of the surface cells in the cap may be the result of the common assumption that these cells are programmed to die and slough off (see Guinel and McCully, 1987). It is not even known how extensive this abscission is during the lifetime of a plant, although it is generally assumed that the number of cells lost is large. The best data available are the conflicting results for the times of turnover of the approximately 10,000 cells in the cap of the seminal root of a young maize seedling -- 1 day (Clowes, 1971) or at least 4-7 days (Barlow, 1978). The problem is compounded because, in maize, for example, the approximately 70 other main root tips of a mature plant vary in size among themselves and are generally much larger than those of the seminal root, while branch root tips are of course much smaller.

The abscission of root-cap cells is unusual in several respects when compared with all other situations where plant organs are shed. For example, in a growing root there is a continuous replacement of lost cells by the cap meristem and, as they are separating, the

NATO ASI Series, Vol. H 35
Cell Separation in Plants
Edited by D. J. Osborne and M. B. Jackson

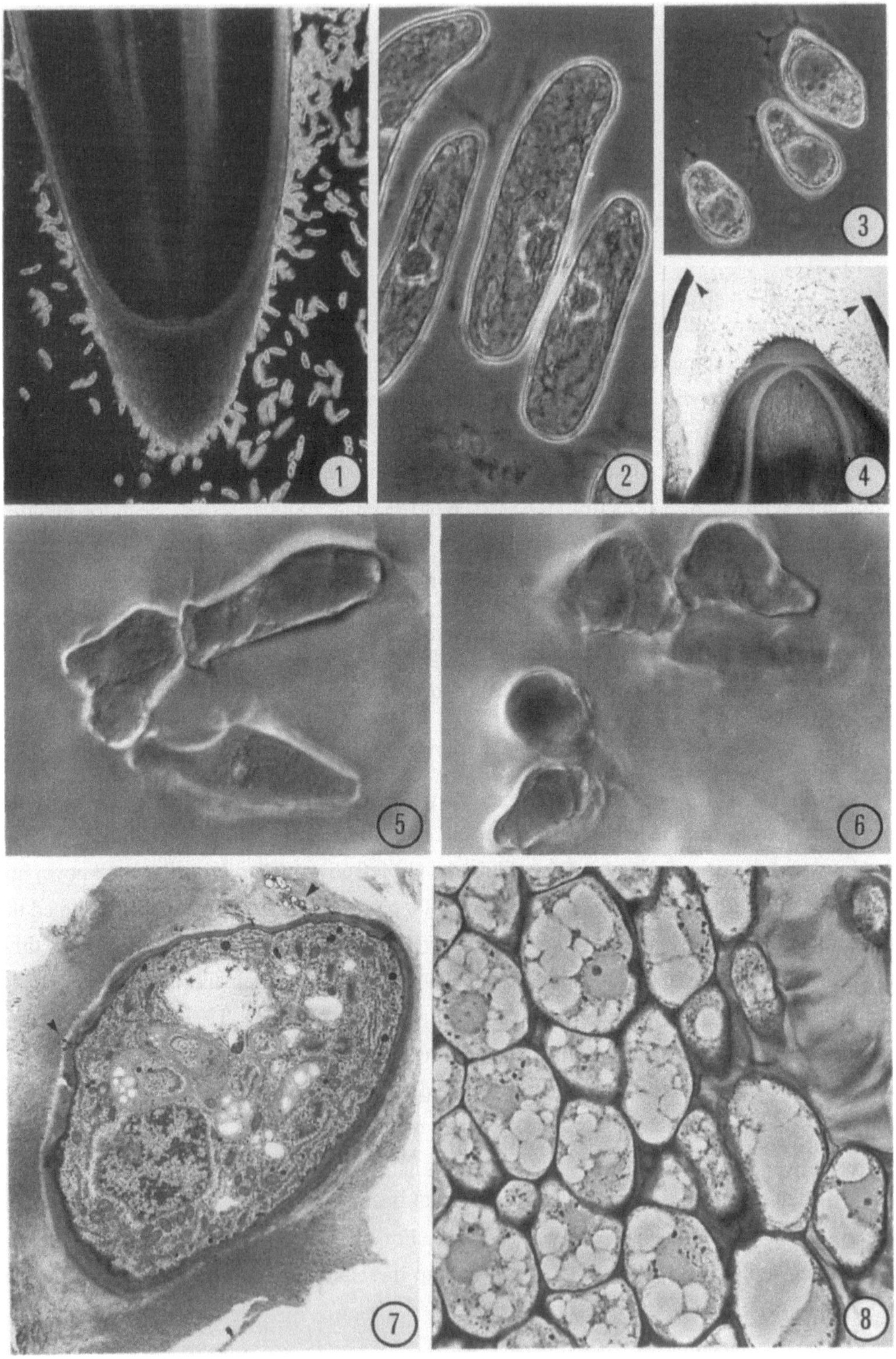
1
2
3
4
5
6
7
8

peripheral cells still lie very close to that meristem (as near as 2 to 3 cells in the upper flanks to somewhat fewer than 20 at the tip). The rates of cell abscission and cell replacement do not necessarily coincide: in maize there appears to be a relationship between the overall growth rate of the root and the size of the cap (slow growing roots have larger caps). A quite different, non-renewed abscission occurs in old branch roots of maize; the tips become determinate and either break off, or, as the apical meristem proper disappears by differentiation, the cap initials also fail and the cap is gradually completely abscised, leaving a smooth, rounded-off end (Fig. 16, McCully, 1987).

The following sections deal with some recent findings that are putting the process of root-cap cell abscission into a new perspective. Unless otherwise noted, all results refer to maize roots. The micrographs show tissues from the cultivar Seneca Chief

SEPARATION OF THE CAP CELLS

Spectacular numbers of detached cap cells remain trapped within the swollen mucilage which surrounds the tip of seedling roots growing in moist conditions (Fig. 1). Even at the low magnification of this micrograph, the heterogeneity of these cells, first noted by Guinel and McCully (1987), is apparent; those few cells which detach right at the tip of the cap are small and roughly spherical while those detaching further back are more sausage-shaped and larger. In apices of older roots, the difference between the two cell types can be very marked (Figs. 2 and 3). The spherical cells are all spent statocytes which have moved directly through the columella, the flank cells may or may not (depending on their position) have been statocytes (Guinel and McCully, 1987). Somewhat similar size and shape differences occur between cells

FIGURE 1. Tip of a 4-day old seminal root from a seed germinating aseptically on 1% agar showing detached cap cells suspended in a drop of root-cap mucilage. Hand-cut section, unstained, semi-dark field optics. X40

FIGURES 2 AND 3. Whole mounts of cells detached from the flanks and the tip respectively of a nodal root from the first tier on a 3-week old plant growing in nutrient solution culture. Phase contrast optics. X500.

FIGURE 4. Hand-cut section through a nodal root primordium not yet emerged through the stem surface (arrows mark where broken during sectioning) at node 7 of 2-month old field-grown plant. X10.

FIGURES 5 and 6. Whole mounts of cells detached at the flanks and the apex respectively of the primordium shown in Fig. 4. X275.

FIGURE 7. Transmission electron micrograph of a median section through a cell detached at the apex of the root cap. Note the remains of plasmodesmata (arrows) in the wall and in the surrounding matrix. Tannic acid-glutaraldehyde-osmium tetroxide fixation. X1,600.

FIGURE 8. Transverse section through the edge of the root cap proximal to the tip by about one quarter of the length of the cap. The peripheral cells are abscising. Axenically-grown seedling root. Glutaraldehyde-osmium tetroxide fixation. Toluidine blue staining. X640.

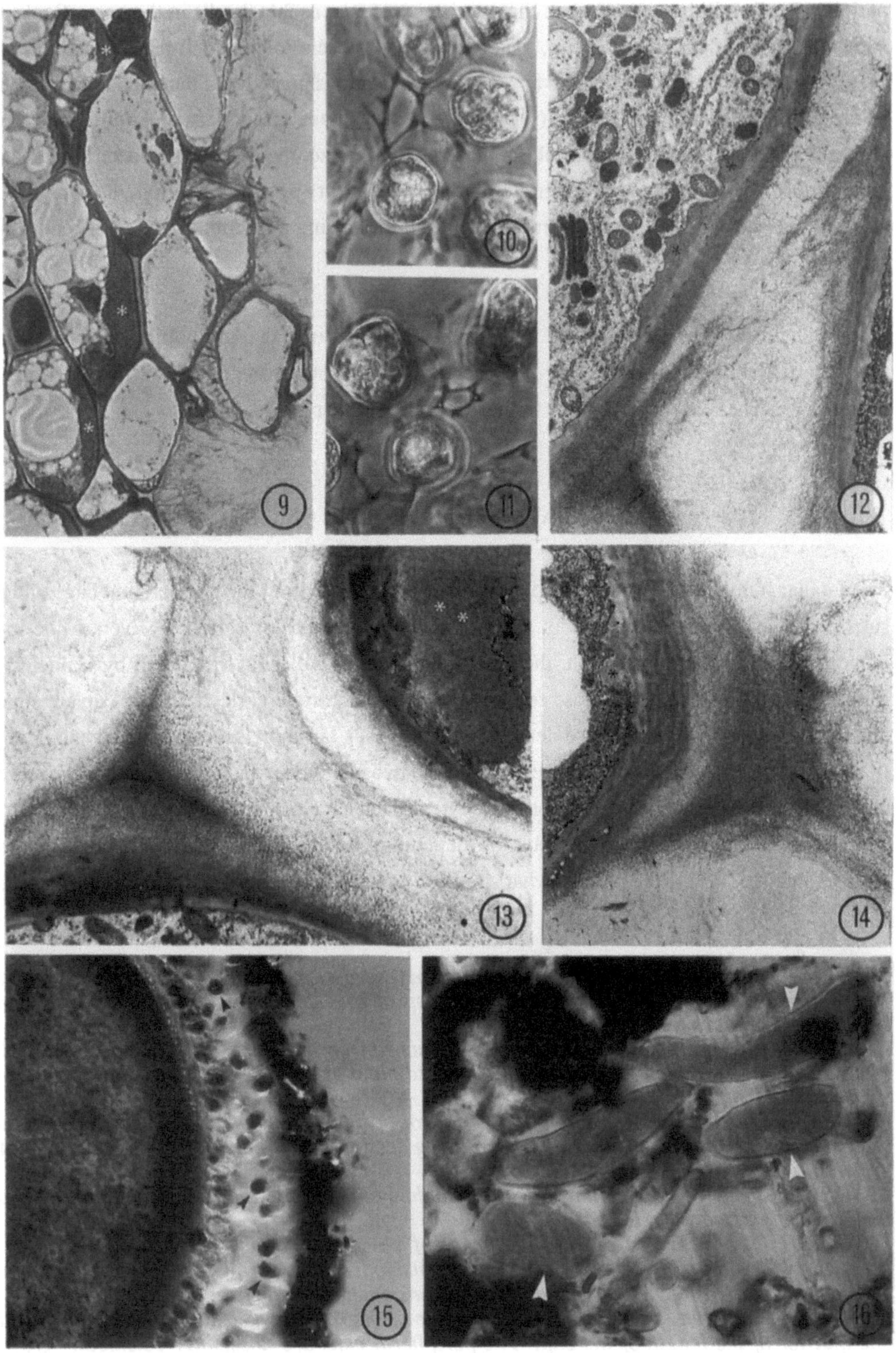
9
10
11
12
13
14
15
16

detached from the flanks or at the tip of very young nodal roots, still covered by intact stem and leaf-base tissues (Figs. 4-6), although the columella cells had not yet developed as statocytes. All the abscising cells have recently synthesized and secreted root-cap mucilage regardless of where they lie on the root periphery (Guinel and McCully, 1987). In all roots that have not been subjected to excessive drying, the predominant feature of the detached cells is that they remain alive; they show rapid cytoplasmic streaming, they can concentrate neutral red in their vacuoles (Figs. 5 and 6) and they are enclosed by a functional cell wall (Figs. 2, 3, 5, 6 and 7). The only obvious structural difference (apart from size and shape) between cells detaching at the tip and on the flanks of the cap is in the degree of vacuolation -- numerous small vacuoles in the former (Fig. 7) and large central vacuole in the latter (Fig. 8).

Nothing seems to be known of the signals which initiate the abscission of cells at root caps. The process is initiated even before the radical emerges from the seed (Fig. 33, Clarke *et al.*), or before nodal roots emerge from the stem (Figs. 4-6). The beginning of abscission seems closely tied to the initiation of mucilage synthesis. However, the beginning of cell separation (which appears as many as 4 cells in from the tissue periphery) precedes the deposition of the new mucilage between the plasma membrane and the wall (Fig. 9). It is only after the periplasmic deposits begin to move outside the wall that complete separation of the cells occurs. Remnants of these deposits remain for a time inside many of the detached cells (Figs. 12-14).

Clearly, the cell separation process involves some sort of loosening and swelling of the middle lamella (Figs. 12-14, 23, and 24). The wall proper remains intact, indeed these walls show heavy label when roots are incubated in radioactive sugars, and this incorporation of label

FIGURE 9. Longitudinal section through periphery of the cap of a pot-grown root at a position equivalent to that of the section in Fig. 8. Periplasmic deposits of mucilage are indicated (asterisks). Tannic acid-glutaraldehyde-osmium tetroxide fixation. Periodic acid-Schiff's reaction. Phase contrast optics. X640.
FIGURES 10 AND 11. Whole mounts of cells detached at the tip of a seminal root cap showing the polygonal net which remains from the middle lamella. Phase contrast optics. X370.
FIGURE 12. Separated cap cells. The extracellular material is a mixture of secreted mucilage and expanded middle lamellar material. Asterisks mark periplasmic mucilage. Tannic acid-glutaraldehyde-osmium tetroxide fixation. X5,800.
FIGURES 13 AND 14. As for Fig. 12 but including pieces of the polygonal net derived from corner regions of the middle lamella. X5,800.
FIGURE 15. Transverse hand-cut section of a field-grown nodal root near the proximal end of the root-cap flank which overlies the columnar epidermis. Section stained with neutral red at pH 7.5. The living detached cap cells (arrows) have accumulated the dye. X95
FIGURE 16. Epidermal strip from a soil-sheath bearing portion (15 cm from the tip) of a field-grown nodal root stained with neutral red at pH 7.5. Detached root-cap cells on the surface (arrows) have taken up the dye into their vacuoles. The dark material is soil. X350

coincides with the enlargement of these cells which accompanies their detachment (Guinel and McCully, unpublished). It is impossible in many places between detaching cells to distinguish expanded middle lamella from secreted mucilage. However, those regions of the middle lamella which lay at edges where 3 cell walls adjoined, or which lined small intercellular spaces at these locations do not disintegrate. These regions remain between detached cells as a polygonal net (Figs. 10-14). Phase contrast optics (Figs. 10 and 11), electron density after tannic acid-glutaraldehyde fixation (Figs. 12-14) and various histochemical tests (Guinel *et al.*, 1987) sharply distinguish this net from the surrounding mucilage.

There seems to be only a single instance where abscission of cap cells has been manipulated by the use of an exogenous substance. Bennet *et al.* (1987) found an increase in cap volume and delayed abscission in the presence of aluminium concentrations below those which affected cell production. The failure of abscission of the peripheral cells was accompanied by their failure to enlarge, disruption of their Golgi and reduced periplasmic deposition of mucilage.

THE FATE OF THE DETACHED CELLS

Living, detached cap cells which survive among root hairs in the soil sheath of field grown roots were first described in 1982 (Vermeer and McCully). This surprising observation has focussed attention on a neglected early study by Knudson (1919) which showed that cells detached from the cap of maize could live for many days in the non-sterile, nutrient solution in which the plants were growing. Guinel and McCully (1987) have shown that cells detached from axenically-grown roots will enlarge over at least 3 weeks when cultured on their own mucilage with no added nutrients; in appropriate culture media they continued to enlarge for as long as 3 months but did not divide. Hawes and Pueppke (1986) showed that detached cap cells from a wide range of monocotyledons and dicotyledons and from *Pinus* species were viable.

Neutral red staining is particularly useful for demonstrating the viability of detached cells in the rhizosphere (Figs. 15 and 16). Living cap cells are quite common as far as 15 to 20 cm from the tip in the soil sheaths of field-grown roots under moist conditions, though they are often hard to distinguish because of encrusting soil particles. Recent work has shown that sheathed regions of roots have a high water status (relative to older bare regions) because their late metaxylem is immature. This feature must be essential for the maintenance of the cap cells at the root surface (McCully and Canny, 1988). There are no data on the length of time cells can remain alive in the rhizosphere but it has to be at least several days.

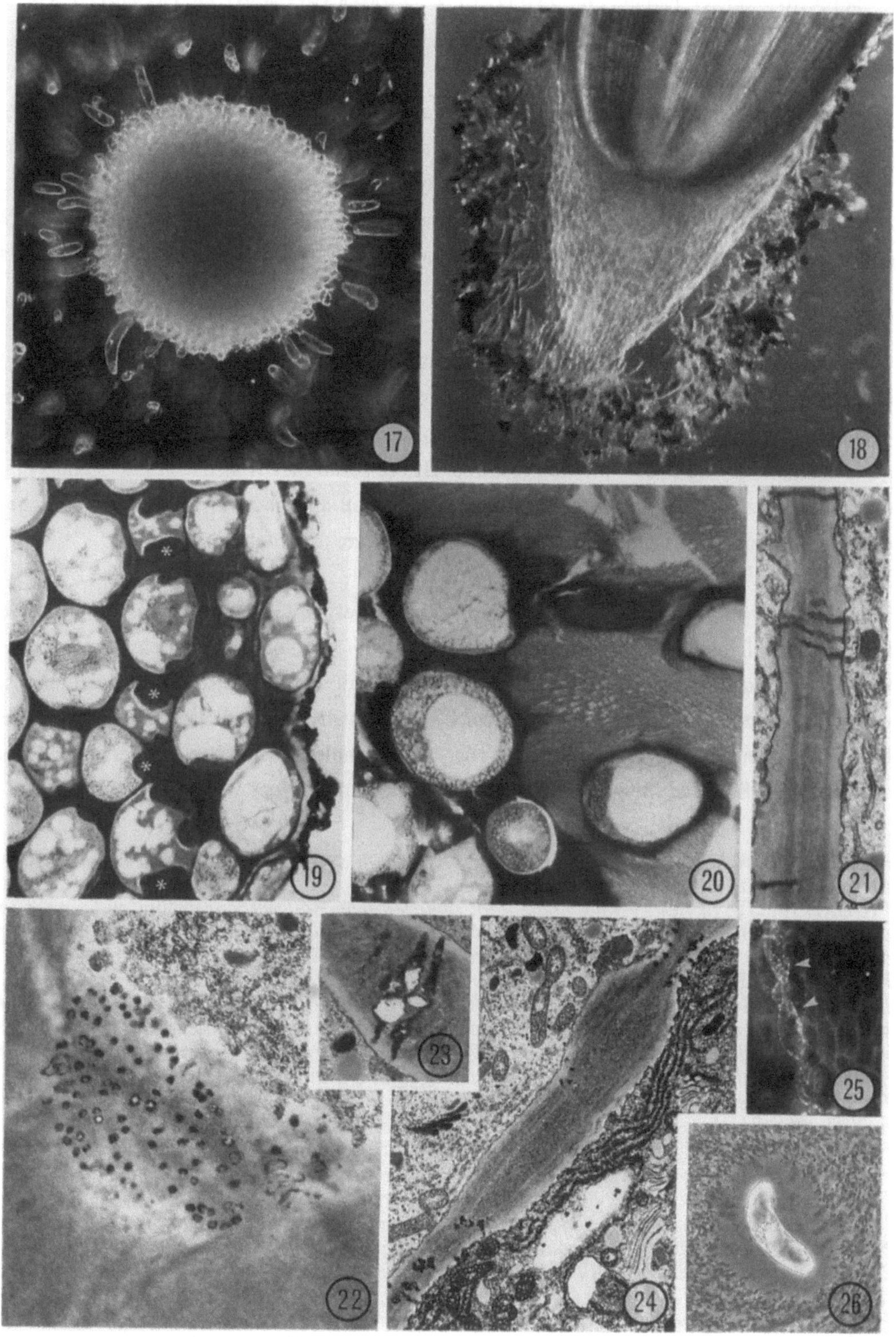
17
18
19
20
21
22
23
24
25
26

POSSIBLE FUNCTIONS OF PARTIALLY SEPARATED AND DETACHED CAP CELLS

Traditonally, the root cap has been thought of as a well-lubricated, continually-renewed boring device which is thrust ahead by the elongating root (e.g. Haberlandt, 1914). Certainly, this function is suggested by the appearance of a fully hydrated cap viewed as in Fig. 17, with its aspect of a firm core surrounded by a large volume of mucilage and numerous ball-bearing-like detaching and detached, turgid cells. There is evidence, however, that the root cap interacts with the soil milieu in more subtle ways.

Darwin in 1881 (see particularly the beautifully descriptive last paragraph of his book) was first to propose, on the basis of extensive experimental evidence, that the root cap is an exquisitely sensitive probe of the soil environment, detecting among other things moisture gradients and very light asymmetric pressures. These stimuli are transduced into signals that elict unequal elongation in the growth zone of the root, and hence bending toward higher moisture (hydrotropism) and away from the higher pressure (thigmotropism).

These two tropisms of roots have received relatively little attention in this century (see Ball, 1969) until quite recently. The report by Jaffe *et al.* (1985) of an ageotropic mutant of pea which is particularly amenable for the demonstration of root hydrotropism has revived interest in this phenomenon. Meantime, the importance of the sensitivity of root caps to touch has been beautifully demonstrated in several situations. For example Goss and co-workers (e.g. Goss and Russell, 1980) have shown that even very light pressure on a cap will quickly stop

FIGURE 17. A head-on view of an axenically-grown seedling root, the tip of which was surrounded by an expanded drop of mucilage. X80.
FIGURE 18. Tip of field-grown nodal root removed from soil without drying and placed in distilled water from 2 hours. The mucilage surrounding the tip is beginning to swell, detaching cells and displacing the adhering soil. Neutral red staining. X40.
FIGURES 19 AND 20. Transverse sections at the cap periphery mid-way along the length of the caps of field-grown nodal roots. The tissues were freeze-substituted into Spurr's resin and stained with toluidine blue at pH 11. Material is identical except that the tip shown in Fig. 20 was soaked in distilled water for 2 hours before freezing: periplasmic mucilage (asterisks) has moved through the walls and extracellular mucilage has expanded. X620.
FIGURE 21. Section through walls of adjoining cap cells just beginning to separate. Tannic acid-glutaraldehyde-osmium tetroxide fixation (also for Figs 23-24). X6,400.
FIGURE 22. Grazing section through primary-pit field between cells at separation stage equivalent to that shown in Fig. 21. X6,400.
FIGURE 23. Stretched plasmodesmata between separating cap cells. X6,400.
FIGURE 24. Relationship between plasmodesmata and ER in separating root cells. X4,000.
FIGURE 25. Hand-cut section through the periphery of a cap of field-grown nodal root soaked 2 hours in distilled water. Aniline-blue fluorescence of callose (arrows) prominent around detaching cells. X96.
FIGURE 26. Detached cap cell cultured in root-cap mucilage on nutrient-free agar, showing inhibition of bacteria isolated from maize rhizosphere. Nomarski optics. X190.

elongation. This elongation will resume in a few minutes, but with pressure as low as 20 to 50 kPa it is greatly reduced. Instead there is expansion in cell width in the elongation zone and this in turn effects passage through the soil by opening up a space ahead of the root into which the tip is thrust. This growth shift is induced by pressures on the cap which are only a fraction of the turgor pressure of the detaching cells (approximately 1.2 MPa, Guinel and McCully, 1987). Bandara and Fritton (1986) have demonstrated a further complexity of touch sensitivity in maize, which was described earlier in *Acer rubrum* (Wilson, 1967). When caps of elongating roots contact an angled barrier they are slid along the barrier by asymmetric extension of the elongation zone. When the tip passes over the end of the barrier, new asymmetry of elongation returns it to passage in the original direction.

Unlike gravity, which is detected deep within the cap, the transducers for moisture and touch, though presently unknown, are likely to be peripheral. Some structural features are suggestive of such a function. For example, root tips examined as they are collected from the field under normal moisture conditions are very different to the hydrated tip shown in Fig. 17. Soil particles interspersed with detached, living cells are tightly appressed to the cap surface. Only after about 2 hours soaking in water is there some expansion of mucilage and the surface accumulation is pushed out from the cap tissue (Fig. 18). After about 6 hours soaking such a root is surrounded by a fully expanded mucilage drop containing numerous detached cells. These observations are in accord with the finding that axenic root-cap mucilage has a water holding capacity less than that of any soil below field capacity (Guinel and McCully, 1986).

Sections of soil-grown roots made before and after periods of soaking, and prepared so as to avoid artifactual hydration of the mucilage (Figs. 19 and 20) indicate that during hydration most of the periplasmic deposits of mucilage leave the peripheral cells and these cells become separated by an expanded volume of intercellular mucilage. Electron micrographs of peripheral cells (Figs. 21-24) show that they are initially linked by numerous plasmodesmata. These plasmodesmata, in turn, connect directly to an extensive peripheral system of endoplasmic reticulum which is characteristic only of cells of the outer layers (Stevenson and Hawes, 1986). Stresses at the plasmodesmata where cells are separating from hydrating caps can be inferred from the ample callose deposits that form at the periphery of the separating cells (Fig. 25). Signals generated by these plasmodesmata could be passed to the ER network and hence, as in geotropism, by subsequent unknown steps to the root elongation zone. Asymmetric swelling of mucilage and cell abscission is seen in axenically-grown root tips which contact moist surfaces (Fig. 1). Whether such asymmetry of cell abscission occurs in soil under conditions where roots are responding to moisture gradients or light touch at their tips remains to be clarified, but a function of the detaching cells as sites of signal transduction is an attractive possibility.

Recent work (Gochnauer, Sealey, McCully unpublished) suggests a role for the detached cells. These cells thrive when cultured in their own mucilage in the presence of a variety of bacteria isolated from maize rhizospheres. The growth of some of these bacteria is clearly inhibited by the cap cells (Fig. 26) while that of the others is stimulated. Thus, in the rhizosphere, the detached cells may be involved in the development and maintenance of the distinctive bacterial flora that is associated with the sheathed regions of maize roots (Gochnauer *et al.*, 1989).

ACKNOWLEDGEMENTS

I thank Joan Mallett for technical assistance and Linda Sealey for Fig. 26.

REFERENCES

Bacic A, Moody S, Clarke AE (1986) Structural analysis of secreted root slime from maize (*Zea mays* L). Plant Physiol 80: 771-777

Ball NB (1969) Tropic, nastic and tactic responses. In: Steward FC (ed) Plant physiology a treatise, Vol 5A, Analysis of growth; behaviour of plants and their organs. Academic Press, New York

Bandara BW, Fritton DD (1986) Directional response of corn roots to physical barriers, Plant Soil 96: 359-368

Barlow PW (1978) Cell displacement through the columella of the root cap of *Zea mays* L. Ann Bot 42: 783-790.

Bennet RJ, Breen CM, Fey MU (1987) The effects of aluminium on root cap function and root development in *Zea mays* L. Envir Exp Bot 27: 91-104.

Clarke KJ, McCully MC, Miki NK (1979) A developmental study of the epidermis of young roots of *Zea mays* L. Protoplasma 98: 283-309.

Clowes FAL (1971) The proportion of cells that divide in root meristems of *Zea mays* L. Ann Bot 35: 249-261

Darwin C (1881) The power of movement in plants. Appleton, New York

Gochnauer MB, McCully ME, Labbe H (1989) Different populations of bacteria associated with sheathed and bare regions of roots of field-grown maize. Plant Soil 114: 107-120

Goss MJ, Russell RS (1980) Effects of mechanical impedance on root growth in barley (Hordeum vulgare L.). III. Observations on the mechanism of response. J. Exp Bot 31: 577-588

Guinel FC, McCully ME (1986) Some water-related physical properties of maize root-cap mucilage. Pl Cell Env 9: 657-666

Guinel FC, McCully ME (1987) The cells shed by the root cap of *Zea*: their origin and some structural and physiological properties. Pl Cell Env 10: 565-578

Guinel FC, Pickard BG, Varner JE, McCully ME (1987) Root-cap net in corn. In: Crosgrove DJ, Knieval DP (eds) Physiology of cell expansion during plant growth. Am Soc Pl Physiol, Rockland, Maryland, p 280

Haberlandt G (1914) Physiological plant anatomy 4th ed. Translated by M Drummond Macmillan, London

Hawes MC, Pueppke SG (1986) Sloughed peripheral root cap cells: yield from different species and callus formation from single cells. Am J Bot 73: 1466-1473

Jaffe MJ, Takahashi H, Biro, RL (1985) A pea mutant for the study of hydrotropism in roots. Science 230: 445-447

Knudson L (1919) Viability of detached root-cap cells. Amer J Bot 6: 309-310
McCully ME (1987) Selected aspects of the structure and development of field-grown roots with special reference to maize. In: Gregory PJ, Lake JV, Rose DA (eds) Root development and function. Cambridge University Press, London, p 53
McCully ME, Canny MJ (1988) Pathways and processes of water and nutrient movement in roots. Plant Soil 111: 159-170
Rougier M, Chaboud A (1985) Mucilages secreted by roots and their biological function. Israel J Bot 34: 129-146
Stephenson JLM, Hawes CR (1986) Stereology and stereometry of endoplasmic reticulum during differentiation in the maize root cap. Protoplasma 131: 32-46
Sticher L, Jones RL (1988) Monensin inhibits the secretion of α-amylase but not polysaccharide slime from seedling tissues of *Zea mays*. Protoplasma 142: 36-45
Vermeer J, McCully ME (1982) The rhizosphere in *Zea*: new insight into its structure and development. Planta 156: 45-61
Wilson BF (1967) Root growth around barriers. Bot Gaz 128: 79-82

CELL WALL CHANGES IN RIPENING PEACHES*

Ruth Ben-Arie, Lillian Sonego, Musia Zeidman and Susan Lurie
Department of Fruit and Vegetable Storage
Agricultural Research Organization, The Volcani Center
P.O.Box 6, Bet Dagan 50250
Israel

INTRODUCTION

Although cell wall hydrolysis can occur quite extensively in many species of fruit during the ripening process, this does not necessarily imply cell separation. Suprisingly, separation between cells in ripening fruits seems to have received little attention. The main focus of research has been on the softening process and the accompanying cell wall changes. There are, however, two clear examples of cell separation occurring in fruits.
The more obvious case is the gradual separation, which develops with the onset of ripening, between the peel and the flesh of certain fruits. In stone fruits, for example, this appears to be a separation between adjacent tissues by dissolution of part of the intervening tissue or cell wall, such as occurs in abscission. Although the two tissues differ structurally, they are cemented to each other in the immature fruit, but at a certain stage of maturity become easily separable. The final cleavage between them might conceivably be the result of cell wall degradation or cell disintegration.

The second instance of total cell separation occurring in fruits has been suggested in relation to the development of mealiness in over-ripe pome fruit. Mealiness, i.e., the apparent loss of juice in the fruit flesh, may occur also as a result of chilling injury, e.g. woolliness in peaches. As there is no excessive loss of water, which could explain the loss in juiciness in either case, it has been suggested that mealiness in pears results from both increased dissolution of the pectin in the middle lamella and increased rigidity in the cell wall due to restricted cellulose and hemi

* Contribution from the Agricultural Research Organization, The Volcani Center, Bet Dagan, Israel. No. 2513-E, 1988 series.

NATO ASI Series, Vol. H 35
Cell Separation in Plants
Edited by D. J. Osborne and M. B. Jackson

cellulose degradation (Yamaki *et al.*, 1983). Therefore, when the fruit is exposed to any kind of mechanical stress, the cells separate and slip easily past each other, instead of breaking and releasing their juicy contents. In the case of woolliness of peaches, resulting from over-extended cold storage, a different hypothesis has been suggested, based on the imbalance of pectin metabolism, incurred by differential low temperature effects on the activities of the enzymes involved (Ben-Arie and Sonego, 1980).

Although very little is known about complete cell separation in ripening fruits, the partial hydrolysis of certain cell wall components in fruit parenchymatic tissue has been widely studied and reviewed (Knee and Bartley, 1981; Dey and Brinson, 1984). A detailed structure of the primary cell walls of fruit parenchyma has not been formulated, but there is evidence that aspects of the model postulated by Albersheim's group (Keegstra *et al.*, 1973) for the primary wall structure of cultured sycamore cells, is applicable to apple fruit cell walls. The middle lamella, is apparently specifically metabolized early during ripening due to solubilization of the pectic polymers which later occurs also in the primary cell wall (Ben-Arie *et al.*, 1979; Huber, 1983). In fruits in which the softening process is gradual and not so extreme, such as the apple and clingstone peach, cell-wall turnover is probably dominant (Knee and Bartley, 1981; Labavitch, 1981) and exo-polygalacturonase (exo-PG) activity is found (Bartley, 1978; Pressey and Avants, 1978). However, many fruits show a rapid and extreme softening of the ripened pulp tissue and in these cases, endo-PG appears to be the key enzyme involved in wall solubilization (Wallner and Bloom, 1977; Pressey and Avants, 1978; Ahmed and Labavitch, 1980).

In a number of fruits, cellulase activity has been shown to increase with maturation and ripening (Pesis *et al.*, 1978; Awad and Young, 1979) although changes in cell wall cellulose content have generally not been assessed. Other polysaccharide hydrolases, such as glucanase, xylanase and a number of glycosidases (Bartley, 1974; Wallner and Walker, 1975; Ahmed and Labavitch, 1980; Paull and Chen, 1983), have also been found in various ripening fruits and might be involved in the degradation processes occurring in the cell wall. However, most investigators and reviewers are of the opinion that pectin solubilization is the predominant factor in cell-wall degradation in ripening fruits and that this is chiefly catalyzed by endo-PGs.

The role and fate of calcium in the cell-wall of ripening fruits have not received much attention. In unripe strawberries, more than half of the insoluble pectin is extractable in EDTA (Knee *et al.*, 1977), suggesting that it is probably stabilized in the wall by calcium. During ripening, there was an increase in water-soluble pectin, which was suggested to be due to a loss of the calcium-stabilized gel structure, as the strawberry has no endo-PG. In ripening tomatoes, too, solubilization of cell wall calcium has been indicated (Rigney and Wills, 1981).

In the following sections three degrees of cell separation in ripe peach fruits are described ultrastructurally and to some extent biochemically.

1. TOTAL CELL SEPARATION

Under the light microscope there is no observable change in soft ripe peaches in the region between the peel and the pulp. In mature ripe fruit, electron micrographs show the cell walls between the hypodermal layer and the parenchyma of the pulp to be fibrillar and organized with a clearly defined dense middle lamella. In soft ripe fruit this region is characterised by empty areas beginning in the region of the middle lamella and eventually encompassing a large portion of the cell wall area. Between the epidermal cells the walls remain intact and many plamodesmata can be observed transversing these walls.

Why should one wall of a cell be degraded and the other remain intact in the ripening process? One possibility could be differences in their chemical composition. Indeed, in the mature firm peach, major differences were found in the total amount of pectin in the peel and the pulp of the fruit and in the relative amounts of the pectic fractions (Table 1). The more abundant quantity of insoluble pectin in the fruit peel would necessitate more intensive pectolytic activity to degrade the cell walls of the peel.

TABLE 1 - *Pectic fractions in the peel and pulp of firm mature 'Hermosa' peaches.*

Fraction	Peel	Pulp
Total pectin (mg/100g f.wt.)	864.0	545.0
Water-soluble pectin (% of total)	2.7	44.0
Calcium pectate (% of total)	16.3	4.6
Insoluble pectin (% of total)	80.9	51.4

The second possibility to be considered is different enzymatic activity on either side of the separation layer. This is demonstrated in Fig. 1. Prior to softening and separation between peel and pulp, there is no PG activity in either tissue, but pectin-esterase (PE) activity in the peel is more than double its activity in the pulp. Moreover, it tends to increase in the peel with little change occurring in the pulp as the fruit ripens. On the other hand, PG activity increases much more rapidly in the pulp than in the peel. Therefore, both from the point of view of composition of pectic substances and the availability of pectolytic enzymes on either side of the cell wall intervening between the peel and pulp of the fruit, this region would appear to be the most susceptible to pectin degradation. In addition, the increased cellulolytic activity emerging from the direction of the peel might well contribute to the total separation of peel from pulp in the fully ripe fruit.

2. PARTIAL CELL SEPARATION IN SOFTENING FRUIT PULP

Electron micrographs show a loosening of the organized structure of the cell walls of the parenchyma as the fruit ripens and a decreased density of the fibrils, especially in the region of the middle lamella. This is accompanied by an increase in activities of both endo-PG and exo-PG and a rise in soluble pectin polysaccharides as the fruit softens (Pressey *et al.*, 1971). The rise in soluble pectin is chiefly at the expense of the insoluble pectin, but there is also an almost 50% reduction in the smallest pectic fraction, i.e., calcium pectate (Table 2). This is reflected in the changes

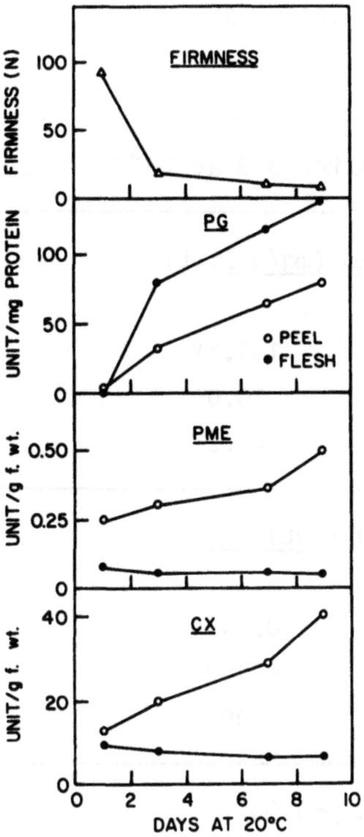

FIGURE 1 - *Polygalacturonase, pectin-esterase and cellulase (cx) activities in the peel and flesh of 'Swelling' peaches during softening at 20°C.*

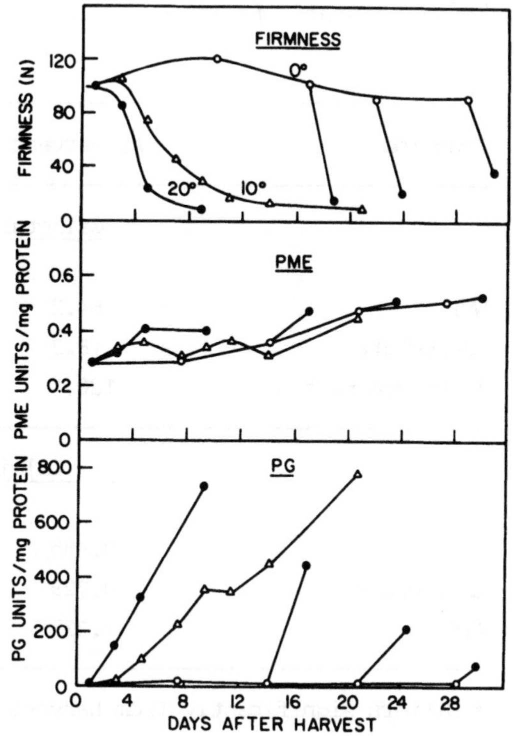

FIGURE 2 - *Firmness in relation to polygalacturonase and pectin-esterase activities in 'Hermosa' peaches during storage at different temperatures.*

in the calcium contents of the fractions, viz. a 22% increase in the calcium bound to the water-soluble fraction and a 33% decrease in the calcium content of the calcium pectate fraction. Additional evidence of the contribution of the calcium pectate fraction to the increased water-soluble pectin (WSP) in soft ripe fruit can be gained by examining the change in neutral sugar content of the WSP with ripening (Table 3). There was a more or less uniform decrease in the neutral sugar components of the WSP from ripe fruits, with an exceptionally high reduction in the rhamnose content. This can be interpreted as the contribution of the solubilized calcium

TABLE 2 - *Pectic fractions in ripening 'Summerset' peaches.*

Fraction	At harvest	After 3 d at 20°C
	Galacturonic acid (mg/g f.wt)	
WSP	64.2	151.1*
Ca-pectate	18.2	10.0*
Insoluble pectin	120.1	47.2*
	Calcium (mg/g f.wt)	
WSP	0.485	0.593*
Ca-pectate	0.729	0.490*
AIS	1.23	1.09*

* Differs significantly from harvest value at p=0.05

WSP, water-soluble pectin.

AIS, alcohol insoluble solids.

TABLE 3 - *Neutral sugar composition (mol %) of water-soluble pectin (WSP) and calcium pectate in ripening 'Summerset' peaches.*

Neutral	WSP			Ca-Pectate	
sugar	Hard	Soft	% Change	Hard	Soft
Galactose	11.0	7.0	-36.4	1.0	1.2
Arabinose	8.0	5.0	-37.5	0.6	0.8
Glucose	5.8	4.0	-31.0	1.2	1.0
Xylose	1.6	1.0	-37.5	0.5	0.4
Mannose	0.8	0.5	-37.5	0.15	0.2
Rhamnose	1.4	0.2	-85.7	N.D.	N.D.

N.D. = not detectable.

pectate fraction, which had a generally low neutral sugar content, and in which no rhamnose was detectable. It therefore seems that peach fruit softening is due mainly to the solubilisation of pectic substances in the middle lamella and, in part, to PG action following de-esterification of insoluble pectin by PE in the primary cell wall. However, without the increased activity of additional hydrolytic enzymes, such as cellulase which, as seen in Fig. 1, does not increase in activity in the flesh, dissolution does not occur and the cells do not truly separate.

3. CELL WALL CHANGES IN ABNORMAL SOFTENING OF PEACHES

Mealiness or woolly breakdown in peaches occurs as result of storage at 0°C for periods generally exceeding 2 weeks. During cold storage there is very little noticeable change in fruit firmness or ripening. The symptoms of the disorder become apparent upon transfer to higher temperatures for ripening. The fruit loses some of its ability to soften (Fig. 2) and it fails to become juicy. The development of the disorder has been attributed to an imbalance of pectin metabolism (Ben-Arie *et al.*, 1971). PE activity in the pulp was not affected by storage temperature, and there was a tendency for it to increase as the fruit senesced. PG activity increased in parallel with fruit softening and was markedly inhibited by reduced temperatures. After an extended period at 0°C the recovery of enzyme activity at higher temperatures was impaired, the softening of the fruit retarded, and woolly breakdown developed. Instead of cell separation, as has been shown in mealy apples (Ben-Arie *et al.*, 1979), the intercellular spaces in the fruit pulp became filled with a fairly dense matrix and the middle lamella appeared to widen and become less dense. Analysis of the pectic fractions indicated that some of the insoluble pectin was solubilized but the main change occurred in the calcium pectate fraction, which almost doubled in quantity during 3 days at 20°C after 1 month at 0°C (Fig. 3).

The distribution of calcium also showed a concomitant increase in the calcium pectate fraction (Fig. 4). During storage there was almost no change in the distribution of calcium between the pectin fractions. However, after 3 days post-storage at 20°C, there was significantly more calcium in the pectate fraction compared with fruit held under the same conditions after harvest. This increase appeared to be due to a reversed migration of calcium from both the soluble and the insoluble pectic fractions to the middle lamella.

The composition of the neutral sugars in the calcium pectate fraction (Table 4) supported the hypothesis that insoluble pectin from the primary cell wall had probably been de-esterified and then precipitated with available calcium as calcium pectate, causing an expansion of the middle lamella region and a filling of intercellular spaces with a calcium pectate gel. The appearance of rhamnose and the almost three fold increase in the percentage of galactose in the calcium pectate fraction are supporting indications of this process. The increase in the percentage of rhamnose in the WSP, in contrast to its decrease in the normal course of fruit softening, is additional evidence that the small but significant increase in WSP in the woolly peach did not derive from the calcium pectate fraction, but rather from the insoluble pectin, possibly due to limited PG activity.

TABLE 4 - *Neutral sugar composition (mol %) of WSP and Ca-P fractions in 'Summerset' peaches after 1 month at 0°C (A) plus 3 days at 20°C (B), in relation to galacturonic acid content.*

Neutral sugar	Neutral sugar/galacturonic acid x 100			
	WSP		Ca-P	
	A	B	A	B
Galactose	13	9	2.7	7.8
Arabinose	11	8	2.6	2.5
Glucose	6	2	2.2	7.6
Xylose	1	1	0.9	0.5
Mannose	1	0.4	0.2	0.5
Rhamnose	0.3	0.8	-	0.1

In conclusion, it appears that cell separation in fruits is a process which can be reversed under certain conditions due to an imbalance in pectolytic activity. However, as the chemical reactions are not truly inverted, the outcome is not a desirable reversal of fruit softening. Nevertheless, this finding has in some small way enhanced our understanding of the softening process.

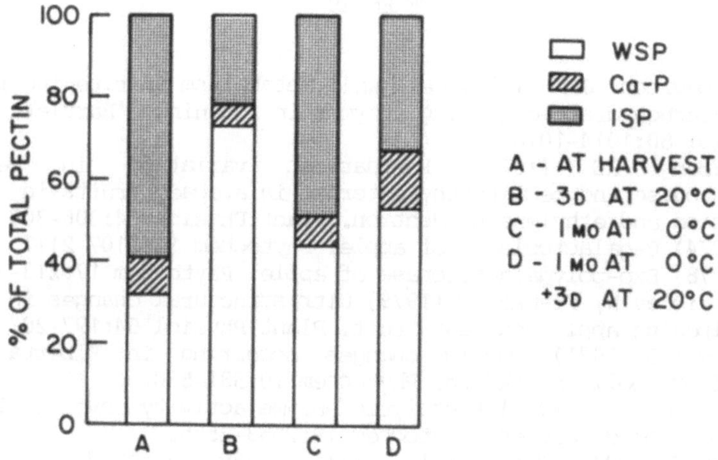

FIGURE 3 - *Changes in the pectic fractions from 'Summerset' peaches held under storage conditions conducive to woolly breakdown development.*

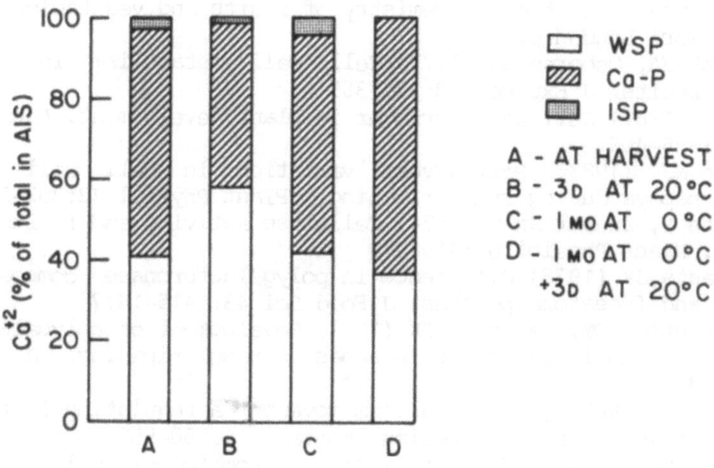

FIGURE 4 - *The distribution of cell wall bound calcium between the pectic fractions from 'Summerset' peaches held under storage conditions conducive to woolly breakdown development.*

REFERENCES

Ahmed AE, Labavitch JM (1980) Cell wall metabolism in ripening fruit. II Changes in carbohydrate-degrading enzymes in ripening 'Bartlett' pears. Plant Physiol 80:1014-1016

Awad M, Young RE (1979) Postharvest variation in cellulase, polygalacturonase and pectinmethylesterase in avocado fruits in relation to respiration and ethylene production. Plant Physiol 64:306-308

Bartley IM (1974) B-galactosidase of apple. Phytochem 13:2107-2111

Bartley IM (1978) Exo-polygalacturonase of apple. Phytochem 17:213-216

Ben-Arie R, Kislev N, Frenkel C (1979) Ultrastructural changes in the cell walls of ripening apple and pear fruit. Plant Physiol 64:197-202

Ben-Arie R, Lavee S (1971) Pectic changes occurring in Elberta peaches suffering from woolly breakdown. Phytochem 10:531-538.

Ben-Arie R, Sonego L (1980) Pectolytic enzyme activity involved in woolly breakdown of stored peaches. Phytochem 19:2553-2555.

Dey PM, Brinson K (1984) Plant cell walls. Adv Carbohyd Chem Biochem 42:265-382

Huber DJ (1983) The role of cell wall hydrolases in fruit softening. Hort Rev 5:169-219

Keegstra K, Talmadge KW, Bauer WD, Albersheim P (1973) The structure of cell walls. III. A model of the walls of suspension cultured sycamore cells based on the interconnections of the macromolecular components. Plant Physiol 51:188-196

Knee M, Bartley IM (1981) Composition and metabolism of cell wall polysaccharides in ripening fruits. In: Friend J, Rhodes MJC (eds) Recent advances in the biochemistry of fruits and vegetables. Academic Press, New York London p. 133

Knee M, Sargent JA, Osborne DJ (1977) Cell wall metabolism in developing strawberry fruits. J Exp Bot 28:377-369

Labavitch JM (1981) Cell wall turnover in plant development. Ann Rev Plant Physiol 32:385-406

Paull RE, Chen NJ (1983) Postharvest variation in cell wall degrading enzymes of papaya during fruit ripening. Plant Physiol 72:382-385

Pesis E, Fuchs Y, Zauberman G (1978) Cellulase activity and fruit softening in avocado. Plant Physiol 61:416-419

Pressey R, Avants JK (1978) Difference in polygalacturonase composition of clingstone and freestone peaches. J Food Sci 43:1415-1417

Pressey R, Hinton DM, Avants JK (1971) Development of polygalacturonase activity and solubilization of peaches during ripening. J Food Sci 36:1070-1073

Rigney CJ, Wills RBH (1981) Calcium movement a regulating factor in the initiation of tomato fruit ripening. HortSci 16:550-551

Wallner SJ, Bloom HL (1977) Characteristics of tomato cell wall degradation in vitro. Implications for the study of fruit softening enzymes. Plant Physiol 60:207-210

Wallner SJ, Walker JE (1975) Glycosidases in cell wall degrading extracts of ripening tomato fruits. Plant Physiol 55:94-98

Yamaki S, Sato Y, Machida Y (1983) Characteristics of cell wall polysaccharides and their degrading enzyme activities in mealy fruit and "Ishinashi" fruit of Japanese Pear. J Japan Soc Hort Sci 52:123-134

REGULATION OF AERENCHYMA FORMATION IN ROOTS AND SHOOTS BY OXYGEN AND ETHYLENE

Michael B. Jackson
Department of Agricultural Sciences
University of Bristol
AFRC Institute of Arable Crops Research
Long Ashton Research Station
Bristol, BS18 9AF
U.K.

INTRODUCTION

Excess water, especially around roots and the lower parts of the shoot, is a frequent cause of abnormally slow rates of gaseous exchange between plants and their environment (Armstrong, 1979) in a wide range of habitats that embrace tidal estuaries, river and lake margins, swamp and marshland, poorly drained agricultural land, rice paddies and irrigated land in arid areas. The challenge this presents to survival is compounded by activities of the soil microflora that initially competes with roots for oxygen and later, when the soil is rendered anaerobic, chemically reduces iron and sulphur, in sequence, to soluble phytotoxic forms (Ponnamperuma, 1972). Thus, it is not surprising that most species of land plants are severely injured by temporary soil flooding and are excluded completely from areas that experience prolonged waterlogging. However, some dryland species such as wheat, maize, tomato and sunflower can survive soil flooding by adaptation (Jackson and Drew, 1984). Their survival does *not* usually reflect a metabolic tolerance to anoxia *per se*. Instead, new roots and existing shoot bases *avoid* anoxia by forming highly porous (aerenchymatous) interiors that permit gas exchange along internal pathways of small resistance linking root cells with the air or with photosynthetically generated oxygen in the leaves. These pathways constitute a linearized form of cell separation.

For many wetland species, the formation of extensive aerenchyma especially in rhizomes, is an integral part of ordinary growth and development that may be considered constitutive and pre-adaptive. In some land plants, such as maize (McPherson, 1939; Norris, 1913), wheat (Trought and Drew, 1980), tomato, sunflower (Kawase and Whitmoyer, 1980), various forage crops (Arikado, 1955) and some trees (Coutts and Philipson, 1987; Topa and McLeod, 1986) aerenchyma is promoted by impeded aeration. How this may be brought about is the subject of the present chapter.

NATO ASI Series, Vol. H 35
Cell Separation in Plants
Edited by D. J. Osborne and M. B. Jackson

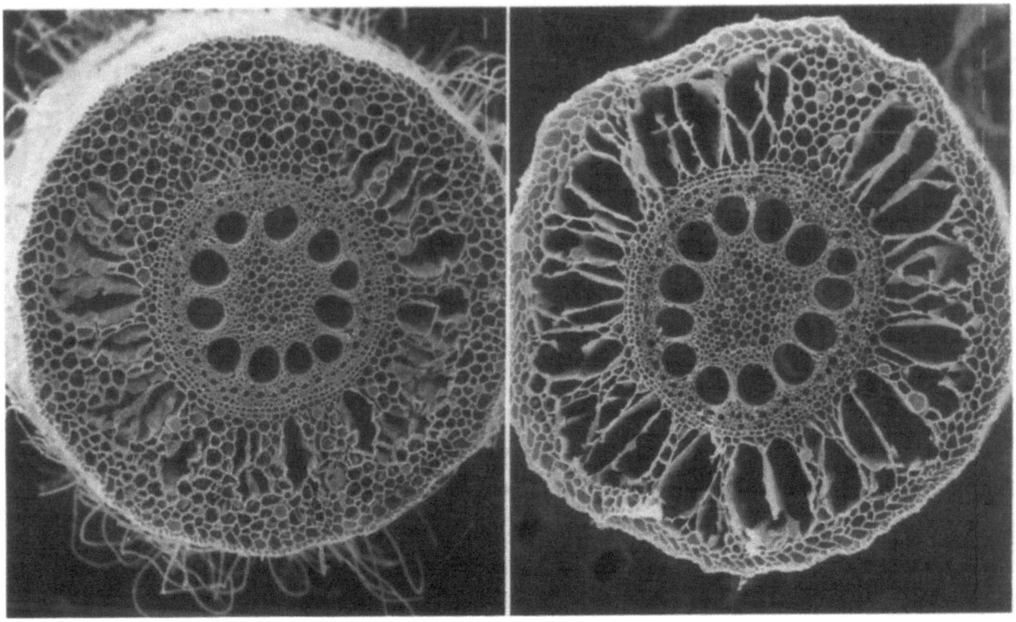

FIGURE 1. Scanning electron micrographs showing the degeneration of cortical cells to form aerenchyma in poorly aerated, adventitious roots of Zea mays. Left: 2-3 days old; Right: 6-7 days old.

AERENCHYMA DEVELOPMENT IN MAIZE ROOTS

a) *Regulation by oxygen and ethylene.* Aerenchyma results from the collapse of cortical cells to form large gas-filled lacunae (Fig. 1). An old idea that it was caused by death from anoxia (McPherson, 1939) was dispelled by experiments showing a complete *suppression* of aerenchyma by anaerobic conditions (Drew *et al.*, 1979). The extensive aerenchyma of adventitious roots that emerge and grow near the surface of flooded soil or into non-aerated nutrient solution must, therefore, have some other cause. In other morphological responses to flooding or 'submergence' such as fast underwater extension or epinastic leaf curvature, ethylene gas has been established as the principal mediating promoter (Jackson, 1982). The case for positive mediation by ethylene in aerenchyma development is equally strong. The results of several different kinds of experiments to be described, together satisfy strict criteria proposed by Jacobs (1959) and Jackson (1987) for causally implicating hormones in a developmental process.

Roots that form aerenchyma in stagnant conditions are enriched with ethylene, as judged by measurements of the gas extracted under partial vacuum (Drew *et al.*, 1979). This enrichment probably arises from faster ethylene biosynthesis induced by oxygen partial pressures of 3.0-12.5 kPa [the partial pressure of oxygen in air is 21 kPa, approximately] and entrapment of endogenously produced gas within roots by their static water covering (Jackson, 1982; Jackson *et al.*, 1985a). The increase in ethylene formation is associated with higher concentrations of 1-aminocyclopropane-1-carboxylic acid (ACC), the principal precursor of ethylene, especially in the apical 10 mm of the root, where ethylene production is the most intense (Atwell *et al.*, 1988). The efficacy of the conversion of ACC into ethylene does not appear to be increased by a partial shortage of oxygen. The biochemical basis of the promotion of ACC accumulation and its location within the root (cortex, stele, meristem) remain to be resolved. One testable possibility is that external oxygen partial pressures of 3.0-12.5 kPa induce localized anoxia (0 kPa oxygen) either in the extreme root tip, where respiration is fast and gas diffusion is slow because cells are densely packed, or in stelar tissue that is separated from external oxygen by the cortex and epidermis. If Bradford and Yang (1980) were correct in deducing that anoxia can promote ACC synthesis, then an anoxic tip or stele might export increased amounts of ACC to adjacent, better oxygenated cells, where its conversion to ethylene would be unhindered by the absence of oxygen. A likely outcome would be to raise the overall rate of ethylene biosynthesis.

Control of ethylene production may also reside in pH changes in the cytosol arising from fermentation. A lowering of pH can occur within a few minutes of withdrawing oxygen (Roberts *et al.*, 1984) as the Kreb's cycle fails in the face of accumulated NADH. Initially, lactic acid is formed (Smith and ap Rees, 1979) resulting in a fall in pH from 7.4 to approximately 5.9; it is maintained there for some time by an inhibition of further lactic acid production caused by acid-inhibition of lactate dehydrogenase (Davies, 1973). There is increasing evidence that other treatments which increase ethylene formation, such as with auxin or fusicoccin, also acidify the cytosol (Parish *et al.*, 1985). This may be a unifying mechanism by which several different treatments can stimulate ethylene synthesis. Low pH may have its effect by promoting transcription of the ACC-synthase gene (Bradford, 1988).

Several lines of evidence suggest that the extra ethylene in partially oxygen deficient roots promotes aerenchyma. For example, applying inhibitors of ethylene action such as aminoethoxyvinylglycine (AVG) (Jackson *et al.*, 1985a), cobalt chloride or amino-oxyacetic acid (Konings and de Wolf, 1984) inhibit aerenchyma development; the effect of AVG being demonstrably specific for ethylene biosynthesis since aerenchyma formation is almost totally retained when ACC is administered with the AVG (Jackson *et al.*, 1985a). Additionally, silver ions supplied to hypoxic roots at concentrations too small to be toxic but large enough to inhibit ethylene action also prevent aerenchyma formation in hypoxic roots (Drew *et al.*, 1981). Finally, supplying small concentrations of ethylene (>0.01 Pa) to well-aerated roots promotes an aerenchymatous structure that is indistinguishable from that formed under hypoxic conditions (Drew *et al.*, 1981; Konings, 1982).

In maize, adventitious roots emerge from the shoot base in whorls of 4-6 from pre-formed initials. Emergence is hastened by soil waterlogging or non-aeration of nutrient solution. Treatment of well-aerated plants with ethylene also has this effect, suggesting that the gas mediates in this response to waterlogging (Jackson *et al.*, 1981). A related finding is that the height of at least one of the stem-nodes from which these new roots emerge is increased when aeration of the original root system is impaired. This is a consequence of stimulated internode extension and is likely to bring newly emerging adventitious roots nearer to the soil or water surface and thus closer to atmospheric oxygen (Plate A). A causal role for ethylene was suspected, in view of the known growth promoting effects of the gas on aquatic (Musgrave *et al.*, 1972) and semi-aquatic species (Jackson, 1985) and even some dryland cereals (Pooviah and Leopold, 1973). However, our attempts to induce internode extension with applications of ethylene or ACC have failed, both on intact plants and on excised shoot bases similar to those used by Raskin and Kende (1984) to study ethylene-promoted stem extension in deep water rice. Alternative growth promoting signals might be partial oxygen shortage or an increase in carbon dioxide that inevitably accompany flooding, at least in the dark. Such effects of carbon dioxide and oxygen shortage are known in rice coleoptiles (Ku *et al.*, 1969) and some dicotyledonous aquatics (Suge and Kusanagi, 1971) but, in each of these cases, ethylene also stimulated elongation.

(b) *Influence of the root tip.* The principal role of the root tip is to supply files of cells that form the structure of the root. Some of these files, which produce aerenchyma, are presumably highly sensitive to ethylene or to degenerative enzymes that may be formed in some or all files in response to ethylene. Tonoplast breakdown in cells of the cortex, approximately 10 mm from the tip of hypoxic maize roots, is visible under the electron microscope and extensive cellular degeneration takes place in more mature cortical cells 30-40 mm from the tip (Campbell and Drew, 1983). The age at which cells first become responsive to ethylene is not known but is probably less than 1 day since cells 10 mm from the tip will have entered the zone of rapid expansion and be approximately 0.5-d old. It is also possible that in the presence of increased ethylene, cells of certain files are rendered defective during their initial structural assembly rather than selectively destroyed by increased degradative activity that is induced by ethylene at some later stage.

In addition to generating the cells involved in forming aerenchymatous tissues, the tip of the root can also influence the development of aerenchyma some considerable distance back from the tip. This is shown by the effect of removing the apical 5 or 10 mm of the root. Such tip-excision suppresses the lysis of cells 20 or 40 mm proximal to the root apex (Atwell *et al.*, 1988). Furthermore, osmotically stressing the apical 10 mm (0.25 MPa) with raffinose slows extension by two thirds and suppresses breakdown in cells exposed to 3 kPa oxygen, 20 mm from the tip. This suggests that for maximum aerenchyma development, a growing apex is required. One possible role for the apex is to supply the more proximal tissues with ACC. However, this was rendered less likely by finding *suppression* of aerenchyma rather than promotion, when the apical 10 mm of an otherwise hypoxic root was exposed to anaerobic conditions that favour ACC accumulation (Atwell *et al.*, 1988). The promoting influence from the apex could involve basipetally transported cytokinins, since Konings and de Wolf (1984) found more aerenchyma in roots that were given kinetin. Root tips are also rich in abscisic acid and auxin (Pilet, 1983), but, according to Konings and de Wolf (1984), these hormones suppress aerenchyma formation. Much more work on this aspect is required.

Although the apex clearly has a promoting role, close proximity is not an absolute requirement for aerenchyma to form. Konings (personal communication) has shown that impeding the aeration of a restricted length of root several days old and several centimetres from the apex promotes aerenchyma within the treated zone. Thus, it seems that mature cortical cells retain their ability to lyse in response to environmental conditions that favour accumulation of ethylene. However, it is clear that in roots extending into poorly aerated surroundings, cortical tissue close to the meristem quickly develops aerenchyma, the vigour of the response being enhanced by a growing root tip.

REGULATION OF AERENCHYMA FORMATION IN ROOTS OF SPECIES OTHER THAN MAIZE

Extensive aerenchyma develops in a many species, most notably in wetland plants (Justin and Armstrong, 1987). In some species, aerenchyma forms by cell separation without associated lysis, while in others, files of cells collapse to create the porous structure. Roots of barley, rice (Das and Jat, 1977), various forage crops (Arikado, 1961), wheat (Trought and Drew, 1980), lodgepole pine (Coutts and Philipson, 1987), and *Senecio jacobaea* (Smirnoff and Crawford, 1983) have been reported to increase in porosity by lysigeny when flooded, suggesting that the mechanisms of control established for maize may also apply to these species. However, detailed studies with rice roots suggest this supposition may not always be correct. Electron microscopy reveals that the sequence of structural degeneration that leads to cell collapse is different from that in maize. Cell wall changes *precede* visible loss of tonoplast integrity in rice (Webb and Jackson, 1986), but the reverse is true for maize (Campbell and Drew, 1983). Furthermore, these changes take place in younger cells in rice than in maize and cannot readily be accelerated by low oxygen or ethylene treatment or retarded by inhibitors of ethylene action [silver ions or carbon dioxide] or ethylene biosynthesis [AVG and cobalt chloride] (Jackson *et al.*, 1985b). Thus, in roots of rice, the rapid development of aerenchyma is a constitutive feature that does not require environmental cues mediated by ethylene for maximal or near maximal expression. Thus, maize seems an unsuitable model for rice and inevitably there must be doubt concerning its applicability to other true wetland species. The findings with maize will probably apply more closely to dryland species such as wheat and barley.

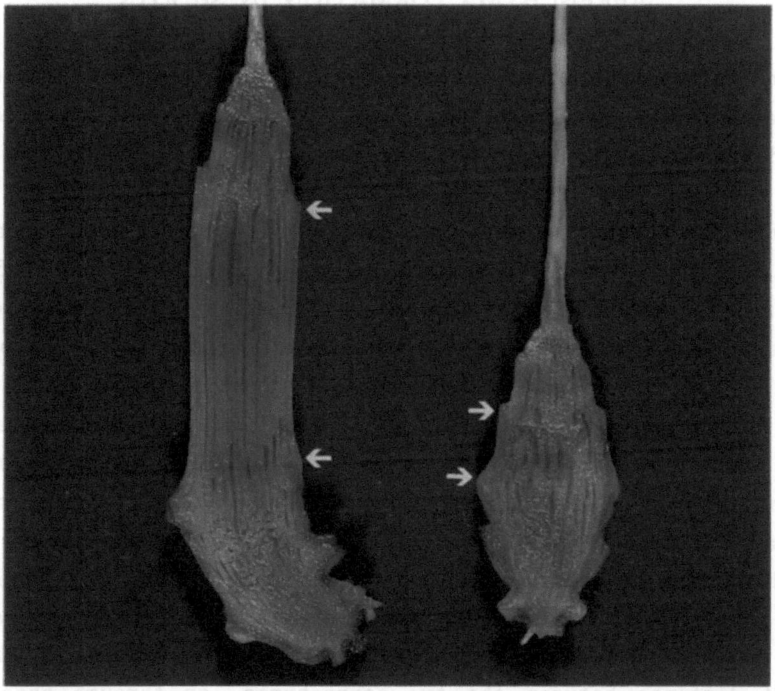

PLATE A. *Stems dissected from 3-4-week-old Zea mays. The stem on the left was taken from a plant that was submerged in non-aerated nutrient solution to a depth of 30 cm above the roots for 14 days. The increase in length of the longest internode (arrows) was 1.5 cm. (Jackson and Young, unpublished)*

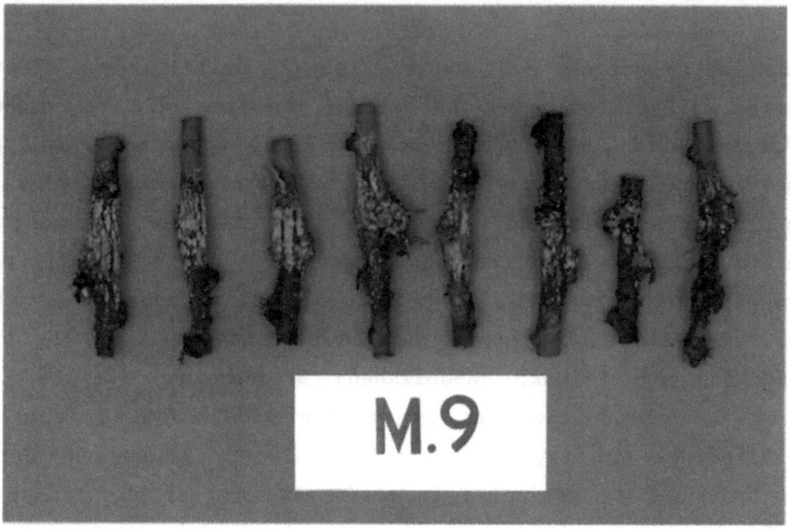

PLATE B. *Hypertrophic swelling at the stem base of rooted cuttings of apple (cv. M9) after flooding the soil for 4 weeks. Original root system removed for clarity. (Husainy and Jackson, unpublished)*

AERENCHYMA FORMATION IN SHOOTS

Aerenchyma is not restricted to roots. Rhizomes, stems and leaves can all possess aerenchyma and in wetland species, an extensive system of interconnected gas-filled lacunae ramifying through roots and shoots is a common feature (Johnson and Chrysler, 1938; Sifton, 1945) that is probably constitutive. However, stems and leaves of some dryland species that normally contain little aerenchyma can also develop this feature if aeration is impaired (Arikado, 1961). For example, if the shoot bases of intact maize plants are submerged in water for more than 3-5 days, the outer leaf sheaths develop large gas-filled cavities by lysigenous cell breakdown between the vascular strands (Fig. 2). Even aerated water is effective, although less so than non-aerated water. In the latter conditions, induced aerenchyma extends at least 5 cm above the water line. Serial sectioning shows that the aerenchyma in *roots* of maize is continuous, up to the point of attachment to the stem and separated from that of the leaf bases by only a few millimetres of non-aerenchymatous stem and the outer layers of root and leaf tissue. Thus, oxygen and other gases may be able to diffuse across the short distance between the two aerenchymas, or through space between overlapping leaf bases (Erdmann *et al.*, 1988). Leaf-base aerenchyma in maize can be stimulated by supplying 20 μM ACC daily for 7 days in aerated solution cultures (Jackson and Young, unpublished). This suggests that endogenous ethylene may also promote the process in the base of the leaf, as it does in roots.

In many woody species e.g. apple (Plate 2) or herbaceous dicotyledons, flooding the soil or enclosing the base of the stem in a sleeve containing water (Kawase, 1979; Jackson and Palmer, 1981) promotes stem hypertrophy (swelling) and an opening-up of lenticels and more internal tissues that may facilitate gaseous diffusion. Some cell separation and lysigeny occur in addition to cell swelling and enlargement. The resulting more open structure may, with some justification be classified as aerenchymatous. In sunflower, the effect of flooding can be simulated by applying sufficient auxin (Jackson and Palmer, unpublished) or ethephon (Kawase, 1974) to promote ethylene production, or by gassing with ethylene (Kawase, 1979, 1981). Flooding, or ethylene treatment, also increases the activity of carboxymethylcellulase estimated in crude extracts of *stem* tissue (Kawase, 1979). Furthermore, infiltrating stem sections with commercial cellulase preparations causes cell lysis and perhaps some cell swelling reminiscent of that seen in flooded plants (Kawase, 1981). Increased endogenous cellulase production, or activity, resulting in a softening or degradation of cell walls may, thus, be an essential part of the mechanism that brings about stem hypertrophy.

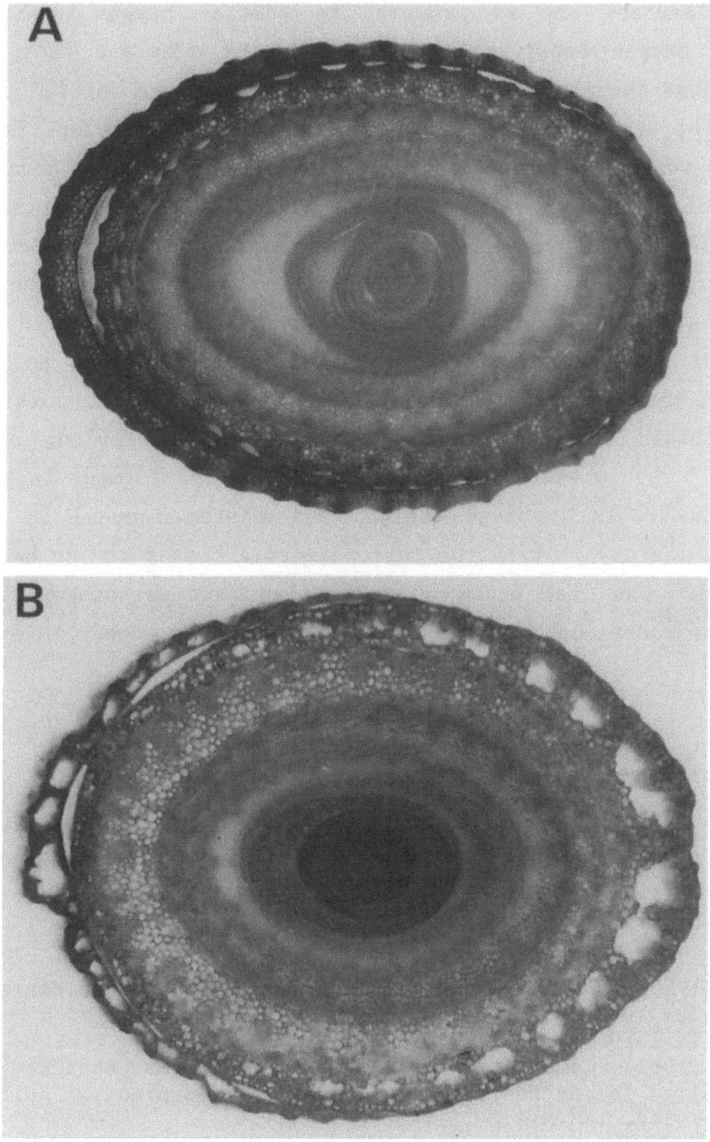

FIGURE 2. Aerenchyma in the outer-leaf bases in Zea mays. Transverse sections taken 3 cm above the base of shoots from plants that were (A) grown throughout with roots in aerated solution culture with the shoot base above the water surface, (B) grown for 10 days with the roots and lower 10 cm of shoot submerged in non-aerated solution culture (Jackson & Young, unpublished)

CONCLUSIONS

The formation of aerenchyma enables a wide range of species to survive and grow in poorly aerated environments. In maize roots and leaves, it results from cell lysigeny that separates files of cortical cells by longitudinal gas-filled lacunae, and is promoted by partial oxygen shortage. This response to oxygen deprivation is mediated by ethylene that is made in unusually large amounts in response to external oxygen partial pressures of 3-12 kPa. The evidence supporting a causal role for ethylene is varied and satisfies several different criteria for implicating a hormone in the regulation of a developmental process. Growing root tips also exert a promoting influence on aerenchyma development, which commences with tonoplast rupture that leads to cell wall degeneration as the cells age. Whether cells fail as a result of an inherent weakness induced by ethylene or as a result of activation of cell wall degrading enzymes, perhaps by ions leaking from the vacuole, remains to be established. The possibility that *de novo* synthesis of such enzymes is promoted by ethylene also remains unexplored. By whatever means aerenchyma is formed in maize, the aerenchymatous roots involved emerge from a stem induced to elongate by submergence, thereby raising the root primordia closer to the better aerated soil surface. The leaf bases of maize also form aerenchyma in a similar manner to the roots and this too may help to improve access to oxygen.

ACKNOWLEDGEMENTS

The substantial contribution of M.C. Drew, B.J. Atwell, S.C. Giffard, S.F. Young and T.M. Fenning to much of the work described is acknowledged.

REFERENCES

Arikado A (1955) Anatomical and ecological responses of barley and some forage crops to the flooding treatment. Bull Fac Agric, Mie University, Tsu Mie 11: 1-29

Arikado H (1961) Comparative studies on the gas content and oxygen concentration in the roots of lowland and upland plants. Bull Fac Agric, Mie University, Tsu Mie 24: 17-22

Armstrong W (1979) Aeration in higher plants. In: Woolhouse HW (ed) Advances in botanical research. Academic Press, London, pp 226-232

Atwell BJ, Drew MC, Jackson MB (1988) The influence of oxygen deficiency on ethylene synthesis, 1-aminocyclopropane-1-carboxylic acid levels and aerenchyma formation in roots of *Zea mays*. Physiol Plant 72: 15-22

Bradford KJ (1988) Is ethylene biosynthesis regulated by uncontrolled pH? Abstracts 13th International Conference on Plant Growth Substances, Calgary, Canada [abstract 729]

Bradford KJ, Yang SF (1980) Xylem transport of 1-aminocyclopropane-1-carboxylic acid, an ethylene precursor, in waterlogged tomato plants. Plant Physiol 65: 322-326

Campbell R, Drew MC (1983) Electron microscopy of gas space (aerenchyma) formation in adventitious roots of *Zea mays* L. subjected to oxygen shortage. Planta 157: 350-357

Coutts MP, Philipson JJ (1987) Structure and physiology of sitka spruce roots. Proc Royal Soc Edinburgh 93B, 131-144

Das DK, Jat RL (1977) Influence of three soil-water regimes on root porosity and growth of four rice varieties. Agron J 69: 197-200

Davies D.D. (1973) Control of and by pH. Symp Soc Exp Biol 27: 513-529

Drew MC, Jackson MB, Giffard S (1979) Ethylene-promoted adventitious rooting and development of cortical air spaces (aerenchyma) in roots may be adaptive responses to flooding in *Zea mays* L. Planta 147: 83-88

Drew MC, Jackson MB, Giffard SC, Campbell R (1981) Inhibition by silver ions of gas space (aerenchyma) formation in adventitious roots of *Zea mays* L. subjected to exogenous ethylene or to oxygen deficiency. Planta 153: 217-224

Erdmann B, Wiedenroth EM, Ostareck D (1988) Anatomy of the root-shoot junction in wheat seedlings with respect to internal oxygen transport and shoot growth retardation by external oxygen shortage. Ann Bot 62: 521-529

Jackson MB (1982) Ethylene as a growth promoting hormone under flooded conditions. In: Wareing PF (ed) Plant growth substances 1982. Academic Press, London, pp 291-301

Jackson MB (1985) Ethylene and the responses of plants to excess water in their environment - a review. In: Roberts JA, Tucker GA (eds) Ethylene and plant development. Butterworths, London, pp 241-265

Jackson MB (1987) A structured evaluation of the involvement of ethylene and abscisic acid in plant responses to aeration stress. In: Hoad GV, Lenton JR, Jackson MB, Atkin RK (eds) Hormone action in plant development - A critical appraisal. Butterworths, London, pp 189-199

Jackson MB, Drew MC (1984) Effects of flooding on growth and metabolism of herbaceous plants. In: Kozlowski TT (ed) Flooding and plant growth, Academic Press, Orlando, pp 47-128

Jackson MB, Palmer JH (1981) Responses of sunflowers to soil flooding. Plant Physiol 67: 58 (abstract)

Jackson MB, Drew MC, Giffard SC (1981) Effects of applying ethylene to the root system of *Zea mays* on growth and nutrient concentration in relation to flooding tolerance. Physiol Plant 52: 23-28

Jackson MB, Fenning TM, Drew MC, Saker LR (1985a) Stimulation of ethylene production and gas-space (aerenchyma) formation in adventitious roots of *Zea mays* L. by small partial pressures of oxygen. Planta 165: 486-492.

Jackson MB, Fenning TM, Jenkins W (1985b) Aerenchyma (gas-space) formation in adventitious roots of rice (*Oryza sativa* L.) is not controlled by ethylene or small partial pressures of oxygen. J Exp Bot 36: 1566-1572

Jacobs WP (1959) What substance normally controls a given biological process? 1. Formulation of some rules. Developmental Biol I: 522-533

Johnson DS, Chrysler MA (1938) Structure and development of *Regnellidium diphyllum*. Am J Bot 25: 141-156

Justin SHFW, Armstrong W (1987) The anatomical characteristics of roots and plant response to soil flooding. New Phytol 106: 465-495

Kawase M (1974) Role of ethylene in induction of flooding damage in sunflower. Physiol Plant 31: 29-38

Kawase M (1979) Role of cellulase in aerenchyma development in sunflower. Am J Bot 66: 183-190

Kawase M (1981) Effect of ethylene on aerenchyma development. Am J Bot 68: 651-658

Kawase M, Whitmoyer RE (1980) Aerenchyma development in waterlogged plants. Am J Bot 67: 18-22

Kelly MO, Bradford KJ (1986) Insensitivity of the diageotropica tomato mutant to auxin. Plant Physiol 82: 713-717

Konings H (1982) Ethylene-promoted formation of aerenchyma in seedling roots of *Zea mays* L. under aerated and non-aerated conditions. Physiol Plant 54: 119-124

Konings H, de Wolf A (1984) Promotion and inhibition by plant growth regulators of aerenchyma formation in seedling roots of *Zea mays*. Physiol Plant 60: 309-14

Ku HS, Suge H, Rappaport L, Pratt HK (1969) Stimulation of rice coleoptile growth by ethylene. Planta 90: 333-339

McPherson DC (1939) Cortical airspaces in the roots of *Zea mays* L. New Phytol 38: 190-202

Musgrave A, Jackson MB, Ling E (1972) *Callitriche* stem elongation is controlled by ethylene and gibberellin. Nature, New Biology 238:93-96

Norris F de la M (1913) Production of air passages in the roots of *Zea mays* by variation of the culture media. Proc Bristol Naturalists Soc 4: 134-138

Parish RW, Felle H, Brummer B (1985) Evidence for a mechanism by which auxins and fusicoccin may induce elongation growth. In: Trewavas AJ (ed) Molecular and cellular aspects of calcium in plant development. Plenum/NATO, New York London, pp 301-308

Pilet P-E (1983) Control of root growth by endogenous IAA and ABA. In: Jackson MB, Stead AD (eds) Growth regulators in root development, Monograph No 10. British Plant Growth Regulator Group, Wantage, pp 15-24

Ponnamperuma FN (1972) The chemistry of submerged soil. Advances in Agronomy 24: 29-96

Pooviah BW, Leopold AC (1973) Effects of ethephon on growth of grasses. Crop Sci 13: 755-758

Raskin I, Kende H (1984) Regulation of growth in stem segments of deep water rice. Planta 160: 66-72

Roberts JKM, Callis J, Wemmer, D, Walbot V, Jardetzky (1984) Mechanism of cytoplasmic pH regulation in hypoxic maize root tips and its role in survival under hypoxia. Proc Natl Acad Sci 81: 3379-3383

Sifton HB (1945) Air space tissues in plants. Bot Rev 11: 108-143

Smirnoff N, Crawford RMM (1983) Variation in the structure and response to flooding of root aerenchyma in some wetland plants. Ann Bot 51: 237-249

Smith AM, ap Rees T (1979) Effects of anaerobiosis on carbohydrate oxidation by roots of *Pisum sativum*. Phytochem 18: 1453-1458

Suge HM, Kusanagi T (1971) Ethylene and carbon dioxide: Regulation of growth in two perennial aquatic plants, arrowhead and pondweed. Plant and Cell Physiol 16: 65-72

Topa MA, McLeod KW (1986) Aerenchyma and lenticel formation in pine seedlings: A possible avoidance mechanism to anaerobic growth conditions. Physiol Plant 68: 540-550

Trought MCT, Drew MC (1980) The development of waterlogging damage in young wheat plants in anaerobic solution cultures. J Exp Bot 31: 1573-1585

Webb J, Jackson MB (1986) A transmission and cryo-scanning microscopy study of the formation of aerenchyma (cortical gas-filled space) in adventitious roots of rice. J Exp Bot 37: 832-841

CELL ISOLATION, RECOGNITION AND FUSION DURING SEXUAL REPRODUCTION

C.J. Keijzer and M.T.M. Willemse
Department of Plant Cytology and Morphology
Agricultural University
Arboretumlaan 4
6703 BD Wageningen
The Netherlands

INTRODUCTION

The process of sexual reproduction in higher plants can be subdivided into a number of developmental stages. Most of them include one or more examples of cell separation, recognition and fusion. Therefore, sexual reproduction can be considered a model system for studies of such cell interactions that also provides an opportunity to compare homologous processes in male and female development. In this chapter the main interactions during sexual reproduction which lead to cell separation and fusion are reviewed.

ISOLATION DURING SPOROGENESIS

The first cell separations in both the developing anther and ovule start before meiosis. In many plant species, the walls between the pollen mother cells (PMCs) lack plasmodesmata; this is in contrast to the other anther tissues (Keijzer and Willemse, 1988a). This pre-meiotic phenomenon, which may be considered as a first sign of mutual PMC isolation, can also be seen in female development, where plasmodesmata, initially connecting the enlarged megaspore mother cell (MMC) with the surrounding tissues, may disappear before meiosis (De Boer-de Jeu, 1979; Kapil and Bhatnagar, 1981). Subsequently, in meiotic prophase, both the PMCs and MMCs are isolated by callose deposition between the plasma membrane and the original cell wall. In anthers, this may occur asynchronously among the four locules, indicating that its direct induction is probably located inside the locule itself and not in other plant parts. The callose wall acts as a selective barrier (Heslop-Harrison and McKenzie, 1967), indicating changing development, and has a shaping function for the future pollen grains (Fig. 1) (Keijzer and Willemse, 1988a). After meiosis, each individual microspore is surrounded by an additional thin layer of callose and accordingly becomes (selectively) isolated. This final callose may function as a template for the pattern of sporopollenine deposition in the exine (Waterkeyn and Bienfait, 1971) which starts immediately upon callose formation, synchronously with the deposition

NATO ASI Series, Vol. H 35
Cell Separation in Plants
Edited by D.J. Osborne and M.B. Jackson

of sporopollenine on the orbicule-bearing membranes around the secretory tapetal cells.

During female meiosis the callose deposition is generally incomplete, since it does not occur around the non-degenerating functional megaspore (Fig. 2) (Kapil and Bhatnagar, 1981). The callose wall expresses the polar selection of the megaspore and coincides with its degeneration, probably preventing nutrient flow.

A final main process in male sporogenesis is the synchronous dissolution of the callose and the original walls of both the tetrads and the tapetal cells, initiated by the tapetum (Keijzer and Willemse, 1988a), which accordingly is responsible for the separation of the future pollen grains. Due to this dissolution, secretory-type tapetal cells become protoplasts, merely surrounded by the sporopollenine-containing tapetal membranes. In contrast, intruding-type tapetal protoplasts lack such membranes along their tangential inside. They intrude into the locule and may mutually fuse to form a syncytium (Pacini *et al.*, 1985). Their plasma membranes directly contact all the pollen grains in the locule (Pacini and Keijzer, 1989), in contrast with the situation in most species with secretory tapeta. Here the locular fluid acts as an intermediate for the contact of most of the pollen grains with the tapetum. From this point of view the fluid may be considered as part of the tapetal domain, filled with the microspore population which in its turn can be seen as a (polygenic) chimeric tissue.

The tapetum-pollen interaction is a one-way system (Keijzer and Cresti, 1987) transferring nutrients via the germ pores (Rowley and Flynn, 1971) and recognition substances, including those effecting sporophytic incompatibility, and enzymes into the exine cavities (Knox, 1984). In female development, the integumentary tapetum also transfers nutrients, often directly contacting the developing embryo sac after absorption of the nucellus and interruption of the bordering cuticles (Tilton and Lersten, 1981). Whether this interaction also is a one-way system seems unlikely, since the changing of the integumentary tapetum into a protective structure of the seed coat (Kapil and Tiwari, 1978) will depend upon embryogenesis, i.e. events inside the embryo sac.

ISOLATION DURING GAMETOGENESIS

Inside the developing pollen grains, a sequence of specialized cell divisions and separations takes place. The division of the microspore into a vegetative and a generative cell is immediately followed by transitory callose deposition between the two cells. Since the initially wall-bound generative cell never borders a germ pore, it is isolated from both the vegetative cell and the locular fluid for some time, indicating an independent course (Keijzer and Willemse, 1988b). After the subsequent callose dissolution, the generative protoplast is pinched off from the intine and finally floats freely in the vegetative cell, packed by a sphere of vegetative cell plasma membrane. Initially the two membranes may be separated by a wall, but plasmodesmata are absent, indicating a minimal interaction. After the subsequent division of the generative cell (species-dependent either before or after pollen germination), the two sperm cells may remain mutually connected by plasmodesmata (Fig. 3), both ensheathed by the same vegetative plasma membrane-envelope (Wilms, 1980), but they lack a wall. They may be associated with the vegetative nucleus to form the male germ unit (Matthys-Rochon and Dumas, 1988). The nature and function of these associations is still unclear, no interactions have been demonstrated up to now. A relationship between the spatial positioning inside the pollen tube and a predestination for either the egg cell or the central cell is presumed (Russell, 1985; Matthys-Rochon and Dumas, 1988). In line with this, an interesting recognition mechanism may exist between one particular sperm cell and the egg cell on one hand and between the other sperm cell and the central cell on the other. Russell (1985) found organelle differences between the two sib sperm cells, due to polarity in the generative cell (Russell *et al.*, 1988) and preferential fertilization of each of the two distinguishable sperm cells with either the egg cell or the central cell. Since plasma membrane fusion precedes fertilization *sensu stricto*, membrane differences between these two cell pairs of fusion partners may exist.

After female meiosis, the single surviving haploid spore generally undergoes three subsequent mitotic cycles, turning it into an 8-nucleate coenocyte. Pre-existing polarity, including nuclear migration (Willemse, 1981) precedes the formation of cell walls, resulting in a mature embryo sac. In the egg apparatus, parts of the cell walls are specialized (Tilton and Lersten, 1981) to allow interaction and fusion with the pollen tube and its contents. The synergids may be considered homologous with the pollen vegetative cell. The thick part of their walls bordering the micropyle (filiform apparatus) is probably homologous with the germ pores of the pollen grains. Both, wall types (Fig. 4) could be pre-conditioned to allow passage by the pollen tube. Moreover, the filiform apparatus may be the site through which recognition substances for the pollen tube are transferred from the secretory synergids into the micropyle. In contrast, the chalazal side of the synergids, like the bordering parts of the egg and the central cell, may lack a cell wall, presumably in order to allow easy sperm cell passage and fusion, respectively.

RECOGNITION DURING POLLEN GERMINATION AND TUBE GROWTH

A highly specialized mechanism of cell dispersal is effected during pollen shedding. In the dehiscing anther, the pollen exines border the orbicule-bearing tapetal membranes. In entomophylous species they are stuck together by lipid-rich pollenkitt, ready to be picked up by pollinators (Fig. 5). Due to the non-wettability of both sporopollenine and pollenkitt, this system can hardly be influenced by the hydration state of the cells and the environmental humidity, which might cause either undesired extra sticking, or, in case of rain, undirected pollen shedding (Keijzer, 1987). In general, the pollination events are based upon an increasing stickiness when shifting from pollen-anther, via pollen-pollinator towards pollen-stigma (Woittiez and Willemse, 1979; Willemse, 1985). The stigma papillae are covered either by exudate or a proteinaceous pellicula. The latter, together with the shape, mass and exine pattern help to keep the pollen stuck to the stigma.

The interaction between pollen grain and stigma starts with recognition. The first observable process following pollination is the species-restricted hydration of the pollen grain, which is considered as a positive reaction on recognition. Subsequently, it becomes activated and forms a pollen tube, whereas the other germ pores are closed with callose. The pollen tube penetrates the stigma and grows apoplastically through either a canal or transmitting tissue to the ovary (Wilms, 1980; Willemse and Franssen-Verheijen, 1986; Van Roggen *et al.*, 1988). Passing these receptive tissues, it forms a callose wall (Fig. 6), which keeps the gametophyte and sporophyte partners separated, acting as a molecular sieve. However, the extreme tube tip lacks callose and acts as a place for uptake of nutrients and signals from the environment. Recognition may also turn into rejection, which may result either from interspecificity, leading to different reactions, based upon a set of penetration genes (Hogenboom, 1984) or from intraspecific incompatibility, based upon special multiple allelic S-genes, which is set out below in more detail.

In the process of recognition, S-locus specific glycoproteins are involved, which are not homologous between both partners (Nasrallah *et al.*, 1985; Takayama *et al.*, 1987). During hydratation, proteins are synthesized in the papillar cells. In line with this, Sarker *et al.* (1988) demonstrated that glycoproteins facilitate hydration. Whether glycoproteins mediate recognition with other molecules on the pollen grain is still uncertain. In case of the sporophytically encoded incompatibility, molecular complexes formed by glycoproteins and pollen exine proteins should prevent recognition symptoms. Based on such molecular couplings different models are proposed (Gaude *et al.*, 1985; Shivanna and Johri, 1985) which occur and function in the interface between pollen and papillar cells, either intra- or extracellular or both. In addition, Cornish *et al.* (1987) demonstrated that cells producing the protein component are present not only in the stigma, but along the entire pathway of pollen tube growth down the solid style. Also, in hollow styles and liquid transmitting tissue one can imagine diffusion of such substances along the pathway. This means that the continuous presence of the S-glycoprotein affects the pollen tubes and a gametophytic incompatibility reaction is gradually built up as the pollen tube proceeds.

The diversity of incompatibility models results partly from the different inhibition reactions which occur. The inhibition on the stigma, expressed in a lack of hydration, is a reaction of sporophytic incompatibility. Both the germ pores of the pollen grain and the stigma papillae may develop callose (Fig. 7). Gametophytic incompatibility shows different reactions: after pollen grain hydration, germination takes place, but the pollen tube is short, shows a thick callosic wall or cannot enter the stigma. In that case, the papillar cells of the stigma can develop a callose wall. In other cases, the pollen tube penetrates the style but shows retarded growth and finally stops. Such tubes show many callose plugs and finally a swollen callose tip, probably due to faulty mechanisms of nutrient uptake. A final gametophytic incompatibility reaction may be found near the ovary, where either pollen tubes fail to penetrate the micropyle or (after fertilization) the zygote aborts, sometimes covered by a callosic wall (Seavey and Bawa, 1986).

FGURE 1. During male meiosis, the orginally irregularly shaped pollen mother cells (walls indicated by an arrow) turn into spherical shaped cells due to the deposition of (selectively isolating) callose (between arrowheads). This enables them to divide into four equally shaped and sized microspores, excluding shape and size to be selective during pollination.

FIGURE 2. During female meiosis, the opposite can be observed. Here the three degenerating embryo sac mother cells are covered with callose, in contrast with the persistent one (arrow), which accordingly is not selectively isolated from the surrounding tissues and is not forced into a callose- dependent shape.

FIGURE 3. The interconnection of two sib sperm cells can be easily visualised (arrow) using fluorescence microscopy after staining their nuclei with DAPI.

FIGURE 4. The massive lobes of the synergid cell walls facing the micropyle, the so-called filiform apparatus, are specialised to let the pollen tube penetrate one of these cells. The part of this wall next to the micropyle fluoresces after aniline blue staining, due to the presence of callose.

FIGURE 5. The non-wettable sites (yellow-white) in the locule of a mature anther become visible after extended staining an epoxy-section with basic fuchsin. In this insect-pollinating species the exines of the pollen grains are stuck to each other and to the orbicules of the tapetal cells with pollenkitt (arrows). They can only be removed by pollinators, hardly by rain.

FIGURE 6. Thanks to the presence of callose in the pollen tube wall, the (gametophytic) pollen tube content remains separated from the (sporophytic) stylar tissue.

FIGURE 7. Due to sporophytic incompatibility, callose is deposited in papillae of the stigma.

FIGURE 8. In interspecific crosses, pollen tubes find the micropyles of the ovules, but do not penetrate them, indicating different signals for the pathways toward and into the micropyle, respectively.

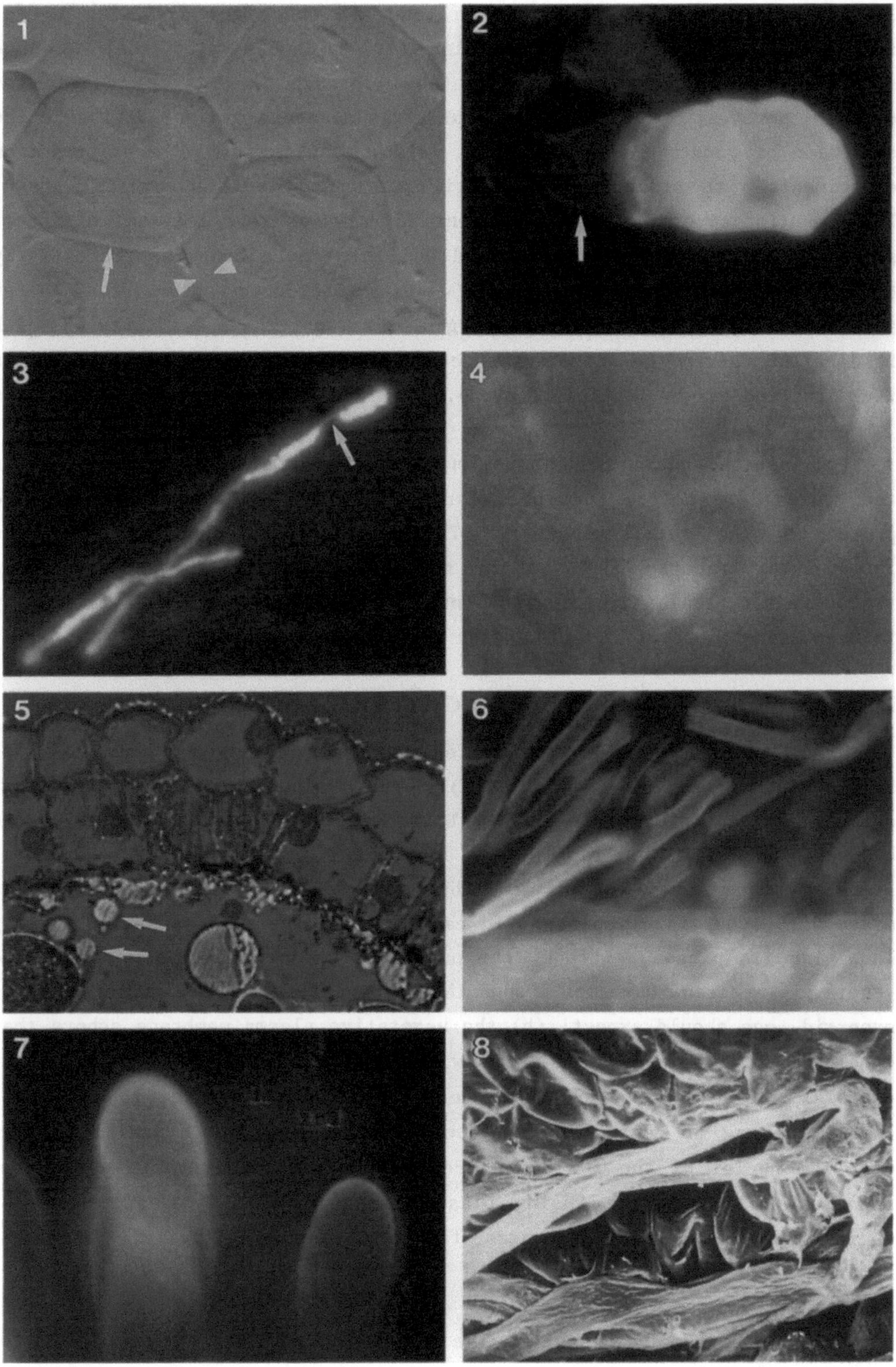
1
2
3
4
5
6
7
8

The mechanism that guides pollen tubes in the right direction is far from clear. Mascarenhas and Machlis (1964) demonstrated that in some species, a Ca-gradient directs the tubes through the stigma towards the ovules. However, inverted grafting of style segments did not influence the direction of growth (Iwanami, 1959). Also, results from *in vitro* attraction experiments with pollen tubes, pistil parts and isolated ovules (Rosen, 1968) contradict this finding. Recent SEM studies revealed that different stimuli lead the pollen tubes toward and into the micropyle, respectively (Fig. 8) (Van Roggen *et al.*, 1988). As mentioned above, the latter attraction might be a result of synergid excretion, the specificity of which is doubtful, given the different successful interspecific cross experiments *in vitro* carried out by Zenkteler (1986).

Pollination and subsequent pollen tube growth may induce reactions in the ovules long before the tubes enter the ovary. These reactions, like protein synthesis (Deurenberg, 1976) and callose deposition in the integumentary tapetum (Esser, 1963), may have to do with the preparation of the ovule for final attraction and fertilization. They can be artificially induced using gibberellins and auxins (Beasly and Jensen, 1985).

FUSION DURING FERTILIZATION

Ultrastructural studies of the fusion processes during fertilization are scarce (Wilms, 1980; Mogensen, 1988). The fusions probably occur rather quickly and accordingly are difficult to detect. Problems are also caused by the hidden position of the egg apparatus in most species. The most probable sequence of fusions will be: (a) the vegetative cell cytoplasm mixes with synergid cytoplasm. In case of a degenerated synergid, already loosening its plasma membrane, the pollen tube plasma decays, as does the vegetative membrane envelope around the sperm cells. In the case of a vital synergid, one should expect: (b) the vegetative plasma membrane sphere that envelops the sperm cell to fuse with synergid plasma membrane at the chalazal pole; (c) in each sperm cell, the plasma membrane to fuse with either the egg or central cell plasma membrane; (d) sperm nuclei to fuse with the nuclei of the two latter cells. During these events a directed recognition between the fusing plasma membranes is supposed, which may include the 'sperm cell towards female target specificity', mentioned before.

Many induction pathways of the processes during and shortly after fertilization remain to be elucidated. For example, which of the fusion steps mentioned above induces embryo and endosperm development? An answer to this question will help us to develop ways to artificially induce gynogenesis or apomixis. Pollination, pollen tube growth or (single) fertilization of only the central cell might induce these processes (Asker, 1980), but the exact localization of the stimuli is still unclear.

CONCLUSIONS

Many of the signals, stimuli and pathways of cell interactions during sexual reproduction are still unknown, especially in sporo- and gametogenesis. Incompatibility mechanisms have received the most attention and research is now focused on their molecular basis (Linskens, 1985). The discovery of association and predestination of the sperm cells (Russell, 1985) recently focused attention on their interaction with their sib sperm cell, the vegetative nucleus and the female target cells (Keijzer *et al.*, 1988).

REFERENCES

Asker S (1980) Gametophytic apomixis: elements and genetic regulation. Hereditas 93: 277-293

Beasly CA, Jensen WA (1985) Cotton ovule culture: hormonal effects and role of synergids in fertilization. In: Chapman GP, Mantell SH and Daniels RW (eds) Experimental manipulation of ovule tissues. Longman, New York, pp 15-23

De Boer-de Jeu MJ (1979) Megasporogenesis. Wageningen Univ. Papers 78-16: 1-122

Cornish EC, Pettitt JM, Bonig I, Clarke AE (1987) Developmentally controlled expression of a gene associated with self-incompatibility in *Nicotiana alata*. Nature 326: 99-102

Deurenberg JJM (1976) *In vitro* protein synthesis with polysomes from unpollinated, cross and self pollinated *Petunia* ovaries. Planta 128: 29-33

Esser K (1963) Bildung und Abbau von Callose in den Samenanlagen der *Petunia hybrida*. Z Bot 51: 32-51

Gaude T, Palloix A, Herve Y, Dumas C (1985) Molecular interpretation of overcoming self-incompatibility in *Brassica*. In: Willemse MTM van Went JL (eds) Sexual reproduction in seed plants, ferns and mosses. Pudoc, Wageningen, The Netherlands, pp 102-104

Heslop-Harrison J, McKenzie A (1967) Autoradiography of soluble (2-^{14}C)thymidine derivatives during meiosis and microsporogenesis in *Lilium* anthers. J Cell Sci 2: 368-400

Hogenboom NG (1984) Incongruity: non-functioning of intercellular and intracellular partner relationship through non-matching information. Encycl Plant Physiol 17: 640-654

Iwanami Y (1959) Physiological studies of pollen. J Yokohama Munic Univ 116: 1-137

Kapil RN, Bhatnagar AK (1981) Ultrastructure and biology of female gametophyte in flowering plants. Int Rev Cytol 70: 291-341

Kapil RN, Tiwari SC (1978) The integumentary tapetum. Bot Rev 44: 457-490

Keijzer CJ (1987) The processes of anther dehiscence and pollen dispersal II. The formation and the transfer mechanism of pollenkitt, cell wall changes in the loculus tissues and a function of the orbicules in pollen dispersal. New Phytol 105: 499-507

Keijzer CJ, Cresti M (1987) A comparison of anther development in male sterile Aloe vera and male fertile *Aloe ciliaris*. Ann Bot 59: 533-542

Keijzer CJ, Willemse MTM (1988a) Tissue interactions in the developing locule of *Gasteria verrucosa* during microsporogenesis. Acta Bot Neerl 37: 493-508

Keijzer CJ, Willemse MTM (1988b). Tissue interactions in the developing locule of *Gasteria verrucosa* during microgametogenesis. Acta Bot Neerl 37: 475-492

Keijzer CJ, Wilms HJ, Mogensen HL (1988) Sperm cell research: the current status and applications for plant breeding. In: Wilms HJ Keijzer CJ (eds) Plant sperm cells as tools for biotechnology. Pudoc, Wageningen, The Netherlands, pp 3-8

Knox RB (1984) The pollen grain. In: Johri BM (ed.) Embryology of angiosperms. Springer, Berlin, New York, pp 197-271

Linskens HF (1985) Recognition during the progamic phase. In: Cresti M, Dallai R (eds) Biology of reproduction and cell motility in plants and animals. University of Siena, Italy, pp 21-31

Mascarenhas JP and Machlis L (1964) Chemotropic response of the pollen of Antirrhinum to calcium. Plant Physiol. 39: 70-77

Matthys-Rochon E and Dumas C. (1988) The male germ unit: retrospect and prospects. In: HJ Wilms and CJ Keijzer (eds) Plant sperm cells as tools for biotechnology. Pudoc, Wageningen, The Netherlands. pp. 51-60

Mogensen, H.L. (1988). Exclusion of male mitochondria and plastids during syngamy in barley as a basis for maternal inheritance. Proc Natl Acad Sci USA 85: 2594-2597

Nasrallah JB, Kao TH, Goldberg ML, Nasrallah ME (1985) A cDNA clone encoding an S-locus-specific glycoprotein from *Brassica oleracea*. Nature 318: 263-267

Pacini E, Franchi GG, Hesse M (1985) The tapetum: its form, function and possible phylogeny in embryophyta. Pl Syst Evol 149: 155-185

Pacini E, Keijzer CJ (1989) Ontogeny of invasive, non-syncytial tapetum of *Cychorium intybus*. Pl Syst Evol, in press

Van Roggen PM, Keijzer CJ, Wilms HJ, van Tuyl JM, Stals AWDT (1988) An SEM study of pollen tube growth in intra- and interspecific crosses of *Lilium* species. Bot Gaz (in press)

Rosen WG (1968) Ultrastructure and physiology of pollen. Ann Rev Plant Physiol 19: 435-462

Rowley JR, Flynn JJ (1971) Migration of lanthanium through the pollen wall. Cytobiol 3: 1-12

Russell SD (1985) Preferential fertilization in *Plumbago*. Proc Natl Acad Sci USA 82: 6129-6132

Russell SD, Strout GW, Thompson RA, Mislan TW, Schoemann LM (1988) Generative cell polarization in *Plumbago zeylanica*. In: HJ Wilms and CJ Keijzer (eds) Plant sperm cells as tools for biotechnology. Pudoc, Wageningen, The Netherlands, pp 17-26

Sarker RH, Elleman CJ, Dickinson HG (1988) Control of pollen hydration in *Brassica* requires continued protein synthesis, and glycosylation is necessary for intraspecific incompatibility. Proc Natl Acad Sci USA 85: 4340-4344

Seavey SR, Bawa KS (1986) Late-acting self-incompatibility in angiosperms. Bot Rev 52: 195-219

Shivanna KR, Johri BM (1985) The angiosperm pollen. Wiley Eastern Limited. New Delhi.

Takayama S, Isogai A, Suzuki A (1987) Sequences of S-glycoproteins, products of the *Brassica campestris* self-incompatibility locus. Nature 326: 102-105

Tilton VR, Lersten NR (1981) Ovule development in *Ornithogalum caudatum* with a review of selected papers on angiosperm reproduction III. New Phytol 88: 477-504

Tilton VR (1981) ibid. IV. New Phytol 88: 505-531

Waterkeyn L, Bienfait A (1971) On a possible function of the callosic special wall in *Ipomoea purpurea*. Grana 10: 13-20

Wilms HJ (1980) Pollen tube penetration and fertilization in spinach. Acta Bot Neerl 30: 101-122

Willemse MTM (1981) Polarity during megasporogenesis and megagametogenesis. Phytomorph 31: 124-134

Willemse MTM (1985) Pollenkitt, a glue? Acta Bot Neerl 34: 132

Willemse MTM, Franssen-Verheijen MAW (1986) Stylar development in the open flower of *Gasteria verrucosa*. Acta Bot Neerl 35: 297-309.

Woittiez RD, Willemse MTM (1979) Sticking of pollen on stigmas: the factors and a model. Phytomorph 29: 57-63

Zenkteler M (1986) Sexual reproduction in plants by applying the method of test tube fertilization of ovules. In: W Horn, CJ Jensen, W Odenbach and O Schieder (eds) Genetic manipulation in plant breeding. W de Gruyter, Berlin, New York, pp. 415-423

PECTINACEOUS BEADS AND PECTINASE ON CALLUS CELL SURFACES IN GRAFT UNIONS AND IN CULTURE

Christopher E. Jeffree, Fiona Gordon and Michael M. Yeoman
Department of Botany
University of Edinburgh
King's Buildings
Mayfield Road
Edinburgh, EH9 3JH, U.K

INTRODUCTION

It has been known for more than a century that the surfaces of cells bordering tissue air spaces are often ornamented with beads or elongated protuberances of material apparently containing pectin, as judged by their reactivity with ruthenium red (reviews by Kisser, 1928; Carlquist, 1956; Carr and Carr, 1975; Butterfield, Meylan and Exley, 1981). Pectic beads occur in all major groups of vascular plants and on cells from a wide variety of tissue types (Kisser, 1928). They have been observed by SEM in fern mesophyll by Carr and Carr, (1975) and in the ground parenchyma of *Cocos nucifera* stems by Butterfield *et al.*, (1981) who reported that the beads stained with ruthenium red, although other tests for pectins were inconclusive. Davies and Lewis (1981) reported pectic projections on callus cells of *Daucus carota* consisting of more than one compartment, the outer one often containing oil droplets. They proposed that this oil may spread over the cell surface to provide a waterproof barrier, while Lipetz (1970) suggested it may be a prerequisite for wound healing. Nevertheless the function of pectic beads is still almost entirely obscure.

Pectinaceous beads occur on cells at the surfaces of opposing stock and scion in the graft union (Jeffree and Yeoman, 1983). We have shown that they may coalesce to form a 'middle lamella', and since they sometimes appear to contain myelin figures have speculated that they represent caches of prepackaged enzymes, and/or recognition molecules with a role in cellular recognition during graft formation (Yeoman, 1984). In this study, simple cytochemical tests have been used to obtain basic information about the properties of the surfaces of Solanaceous callus cells. This information will be used as a framework for further investigation of the chemical basis of cell-cell recognition at graft unions.

MATERIALS AND METHODS

Plant material. Grafted tomato (*Lycopersicon esculentum* cv. Ailsa Craig) internodes were produced in 'whole' plants or in culture by methods previously described (Jeffree and Yeoman, 1983; Parkinson and Yeoman, 1982). Callus cultures of *L. esculentum,*

NATO ASI Series, Vol. H 35
Cell Separation in Plants
Edited by D. J. Osborne and M. B. Jackson

Nicandra physaloides and *Datura stramonium* tissue were initiated from internode explants on Murashige and Skoog medium containing 1.2% agar, 2% sucrose, and 0.2mg/l of kinetin and indole-3-acetic acid (IAA).

Low temperature scanning electron microscopy. Specimens which had been submerged in aqueous media were rinsed in distilled water and blotted to remove excess surface water before cryofixation. Tissue specimens were attached to a specimen stub, using Tissue-Tek II OCT (Miles Laboratories) as adhesive, and were then cryofixed by immersion in melting nitrogen at -210°C in an Emscope SP2000 Cryopreparation system. Cryofixed specimens were transferred under vacuum to the precooled (-160°C) specimen stage of a Cambridge Instruments S250 Mk1 microscope. Any ice contamination of the specimen was removed by warming to -75°C (etching) while under continuous observation in the SEM (Jeffree, *et al.*, 1987a). Etched specimens were recooled to below -130°C, shrouded, transferred to the SP2000 workchamber and sputter-coated with gold (20 mA, 1.5 - 2 minutes, 13 Pa) before examination in the SEM at accelerating voltages below 10kV.

Chemical fixation of cells and cell walls for SEM. Specimens were fixed for 4 hours in 2.5% glutaraldehyde plus 2% depolymerized paraformaldehyde in 0.1 M cacodylate buffer, pH 7.2, at 20°C, oxidised in 1% aqueous sodium *m*-periodate, and then incubated in half-saturated aqueous thiocarbohydrazide at 20°C for 2 h, washed thoroughly in water at 30°C and postfixed in 2% aqueous osmium tetroxide. Specimens were dehydrated in acetone and critical point dried via carbon dioxide. This procedure (GAPATCO) provided optimum preservation of cell morphology and cell wall fine structure, and improved the electrical conductivity of specimens.

Histochemical staining and light microscopy. Acid polysaccharides were detected in fresh and pectinase-extracted hand sections of graft and callus tissue using the following reagents: Ruthenium red (RR) made up as a 0.1% (w/v) solution in distilled water or in dilute ammonia, pH8.5. (Sterling, 1970): Alcian blue 8GX as a 1mg/ml solution in acetate buffer at pH 4.0: Hydroxylamine/Ferric chloride as described by Reeve (1959): Coriphosphine O (Pfaltz and Bayer Inc., Waterbury, CT 06708, USA) brief immersion in a 0.03% w:v aqueous solution. Specimens were viewed by brightfield microscopy (yellow-orange stain) or by fluorescence at 630nm (red) under 546nm (green) excitation (Ueda and Yoshioka, 1976; Weis, Polito and Labavitch, 1988): β-glucosyl Yariv antigen (Yariv, 1967) as a stock solution (1mg in 100μl DMSO) diluted 1:10 with 1% NaCl immediately before use.

β-1,4 glucans were detected using 0.01% aqueous Calcofluor White M2R (Hughes and McCully, 1975), or 0.1% aqueous Congo red. Pectin was extracted from cells using commercial preparations of pectinase (Sigma P2401 from *Rhizopus;* Sigma P5146 from *Aspergillus*) and an endopectinase (gift of G. Plaistow, Dalgety Foods Ltd.) produced by expression of the *Erwinia carotovora* gene in *Escherichia coli* (Zinc and Chatterjee, 1985). Pectinase was dissolved in 0.1M pH 4.0 acetate buffer to a concentration of 45 Units/ml.

Detection of Lipids. Fresh hand sections were stained in a saturated solution of Sudan Black B in 70% ethanol for 15 minutes, briefly washed in a large volume of 70% ethanol and mounted in water. Alternatively, 5 μl of saturated Nile red (Kodak, 52445) in acetone was dispersed in 2ml of water. Sections were briefly immersed in the dye and examined by fluorescence microscopy using green excitation and red emission (Reichert G1 filter module) (Greenspan, Mayer and Fowler, 1985).

Immunofluorescence microscopy. Polyclonal rabbit antibodies against Tomato polygalacturonase and tomato pectinesterase were the gift of Dr. G. A. Tucker, University of Nottingham, Dept. of Applied Biochemistry and Food Science. In double-diffusion against the purified antigens and tomato internode proteins both preparations produced single precipitin bands. Hand sections of tissue were fixed for 1 hour in ice-cold 2% (para)formaldehyde in 0.1M pH 7.2 phosphate-buffered saline (PBS) containing 0.02% w/v Tween 20 and 0.02% sodium azide (PBS-Tween). After washing in PBS-Tween, non-specific protein binding was blocked by immersion in 1% low-fat dried milk in PBS-Tween (PBS-Marvel-Tween) for 30 minutes at 20°C. Sections were then incubated for 90 minutes at 37°C in the primary antibody diluted 1:100 in PBS-bovine serum albumin (BSA)-Tween. After extensive washing in PBS-Tween sections were incubated for 90 minutes at 37°C in fluorescein isothiocyanate (FITC)-conjugated goat anti-rabbit immunoglobulin G (IgG), washed in PBS-Tween and mounted in 90% glycerol in 0.1M PBS, pH 7.2, containing 1mg/ml of *p*-phenylenediamine. Epifluorescence microscopy was performed on a Reichert Polyvar microscope using a B1 filter module designed for use with FITC.

Immune blotting. Cell surface antigens were directly blotted by tissue contact with nitrocellulose membranes. Blotting membranes were blocked with PBS-Marvel-Tween and probed with 1:100 dilutions of a specific antibody (anti-tomato polygalacturonase or anti-tomato pectinase) in PBS-BSA-Tween. Antibody binding was visualized by incubation in gold-conjugated goat anti-rabbit IgG antibody.

Detection of pectinase and cellulase activity. Pectinase activity at callus cell surfaces was detected by placing the undamaged surface of the tissue in contact with 1mm-thick substrate-gels containing 0.1% pectin, 0.1M sodium acetate buffer pH 4.0, and 1% agarose (Reid and Collmer, 1985). After incubation for 3-8 hours the gels were stained in 0.02% aqueous ruthenium red. Similarly, cellulase activity was detected with agarose substrate-gels containing 0.5% low-viscosity carboxymethyl cellulose (Sigma C-8758) in pH 5.0, 0.1 M sodium acetate buffer. Gels were stained with 0.1% aqueous Congo red and destained in 1M NaCl.

Detection of free polysaccharides with water-agarose gels. Callus tissue was placed for 3 to 8 hours on aqueous 1% agarose gels, to detect the presence of soluble polysaccharides on the cell surfaces. The gels were stained with ruthenium red or Congo red, as for substrate-gels, to detect pectins or β-1,4 glucans, or with β-glucosyl Yariv antigen to detect arabinogalactans.

RESULTS

Surface structure of callus cells. Cells adjacent to the cut surface of stock and scion of tomato internode grafts proliferate to give rise to callus by 48 hours after grafting. At 24 hours after grafting, the cell surfaces are smooth, but subsequently beads form on the outer surfaces of the cell walls (Fig.1; Jeffree and Yeoman, 1983). The appearance of these beads is suggestive of liquid droplets with a wide range of size (1 to 4 μm) and shape. Some are elongated normal to the cell surface by as much as twice their diameter. Similar beads occur on the surfaces of callus cells growing in culture (Fig. 17). Beads on cultured callus cells of *L. esculentum* and *D. stramonium* also ranged in size from ca. 1 μm to ca. 4 μm, and showed considerable variation in morphology, ranging from turgidly hemispherical or elongated, to wrinkled, low profile mounds. Beaded callus cells have been observed in this laboratory in grafted and cultured cells of a wide variety of species, e.g. *Pelargonium zonale* hybrid (wounded petiole tissue in culture, Schofield, 1987), *Cuscuta campestris* (wounded internodes in culture), *Picea sitchensis* (graft union, Barnett and Weatherhead, 1988), *Hyoscyamus niger* (callus in culture), graft unions and callus of *N. physaloides, L. esculentum and D. stramonium,* and callus of *Daucus carota* (Jeffree and Yeoman, 1983).

Distribution of beads on the cell surface. On callus cells in the graft union of Solanaceous grafts beads may be evenly distributed over the surface, (Fig. 17) or show a polar distribution, over the end of the cell distal from the parent tissue, (Fig. 16). In culture, beads are observed on the youngest daughter cells, on expanding cells, and on older cells. Their distribution is highly variable, but they may occur on any part of the cell surface, and at densities up to 2.0 x 10^6 mm^{-2}. During cell separation, sheets or veils of remnant wall tissue stretch between daughter cells (Fig. 18), finally separating into strands reminiscent of those seen between palisade mesophyll cells of leaves. (Davies and Lewis, 1981; Jeffree, Dale and Fry, 1986). As the cells separate fully, the veil breaks, and its remnants are visible on the cell surface (Fig. 18, arrows). These remnants appear to be distinct in respect both of their morphology and distribution from the beads.

Some tissue cultures of *L. esculentum* contained nodules of highly-differentiated tissue embedded in rapidly-proliferating, essentially undifferentiated, callus cells bearing numerous beads on their surfaces. Beads were absent from the cell surfaces of the more highly-differentiated tissue masses (Fig. 19).

Cytochemical reactivity of cell surfaces and beads in Datura stramonium callus. Beads on cryofixed cells were subjected to freeze-drying at -70°C in the SEM specimen chamber (see Fig. 6 in Jeffree and Yeoman, 1983) in order to distinguish them from droplets of water or aqueous solutions which might be condensed onto or forced from the cell during cryofixation (Jeffree, *et al.*, 1987). The beads survived complete freeze-drying of the tissue without major reduction in size. Beads were also observable by light microscopy in hand sections of tissue immersed in aqueous media (see e.g. Figs. 1-6), and could be prepared for SEM and TEM by a variety of chemical fixation and embedding protocols (Figs. 16-24, this paper; Jeffree and Yeoman, 1983). Treatment of the tissue with 1% sodium *m*-periodate followed by Schiff reagent resulted in a general intense magenta staining of the whole cell surface, including the beads (Fig. 1). Both the general cell surface and the beads reacted positively with ruthenium red (Fig. 2), with alcian blue at pH 4.0, and also with coriphosphine O (Fig. 13). Both beads and the cell surface also stained vigorously with Calcofluor white (Fig. 4), and the beads were also strongly stained with Congo red. The majority of beads accumulated Sudan black B, and showed intense red fluorescence with Nile red (Fig. 3), but some of them showed no affinity for either lipid stain. Nile red did not stain the general cell surface. Beads, and the callus cells at the tissue-air interface showed no or feeble affinity for β-glucosyl Yariv antigen. However, extracellular polysaccharides at the interface between callus cells and the agar medium were strongly Yariv-positive. Beads were intensely stained by Evans blue, which did not stain the cell surface or the cytoplasm of living cells.

Effects of pectinase on the cell surface and beads. Treatment of callus cells from Tomato homografts with a 2% solution of commercial *Rhizopus* pectinase (Sigma P2401) produced cell surface erosion-pits which were interpreted as being due to removal of pectinaceous beads (Fig. 20; Jeffree and Yeoman, 1983). Subsequently, experiments using pectinase from *Aspergillus niger* (Sigma P5146) and pure *Erwinia carotovora* endopectinase have shown that beads on *D. stramonium* callus cells survive prolonged exposure to pectinases, although their reactivity with RR (Fig. 5) and alcian blue is abolished. Following 7 hours in either pectinase at 30°C, the beads could still be observed by LTSEM (Fig. 21) or following chemical fixation using GAPATCO (Figs. 22–24). Specimens incubated for 7 hours in Sigma P5146 pectinase showed evidence of cell surface erosion (Fig. 24), the smooth surface layer being stripped to reveal microfibrils beneath. Beads in the immediate vicinity are apparently intact. Sigma P5146 pectinase was also shown to contain cellulase activity detectable at dilutions of up to 10^5-fold from the commercial solution when tested in CMC-agarose substrate-gels (Fig. 10). The same test revealed no cellulase activity in the pure *Erwinia* pectinase preparation, which caused no visible erosion of the cell surface or of the beads (Fig. 22).

Tests for mobile enzymes and polysaccharides at the cell surface. Pectin-agarose substrate-gels placed in contact with undamaged *D. stramonium* and *N. physaloides* callus cells for 8 hours at 20°C showed regions close to the tissue which were depleted of pectin (Fig. 7, arrowheads). At the same time, material which stained with RR at greater intensity than the background pectin-gel was also secreted into the gel (Fig. 7). Similarly, RR-positive material diffused into aqueous 1% agarose gels (containing no pectin) placed in contact with the callus surfaces (Fig. 9). Comparable tests using CMC-agarose gels showed regions of CMC-depleted gel beneath the tissue after 8 hours incubation at 20°C, following staining of the gel with Congo Red (Fig. 8). Water-agar gels showed negative reactions both with Congo red and β-glucosyl Yariv antigen after incubation overnight in contact with *D. stramonium* callus. Vertical sections of the callus, placed flat on pectin and CMC substrate-gels showed vigorous pectinase and cellulase activity in positions corresponding with the surfaces of the callus mass (Fig. 11, arrowheads). Similarly, secretion of RR-staining material from slices of callus tissue into aqueous agarose gels was predominantly from the original external surfaces of the callus. Callus placed on substrate-gels containing 0.1% pectin, 10mM $CaCl_2$, pH8.5 produced regions of turbid appearance (Fig. 12), suggesting modification of the gel structure. Staining with RR revealed regions of reaction product which could not be extracted by washing in water, which removed some of the soluble pectin from the gel.

Pieces of *D. stramonium, L. esculentum and N. physaloides* callus which were directly blotted on strips of nitrocellulose blotting membrane transferred nitrocellulose-binding material (presumably proteins) which was recognized by polyclonal anti-tomato poly-galacturonase antiserum but not by a polyclonal anti-tomato pectinesterase antiserum (Figs. 14,15).

Immunofluorescence microscopy. Sections of *Datura stramonium* callus stained with anti-tomato pectinase (rabbit) and FITC-anti-rabbit (goat) showed localisation of the antigen in extracellular matrix material at the base of the callus, and at the surfaces of cells at the tissue-air interface. Some beads showed intense reaction (Fig. 6), though the response was patchy. The sections showed no fluorescence above background when probed with anti-tomato pectinesterase antibody.

DISCUSSION

Davies and Lewis (1981) reported that beads on *D. carota* callus cells are associated with multivesicular bodies (lomasomes) and occur on cells with extensive rough endoplasmic reticulum (RER) and golgi, usually taken to indicate vigorous secretory activity. Our observations would support this view (see e.g Jeffree and Yeoman, Fig. 7e). Neither the distribution of the beads, nor their morphology are consistent with the possibility that they originate as remnants of the cell wall during cell separation. It seems most probable that the beads arise during active secretion of cell wall polysaccharides and other materials onto the cell surface.

The observations reported here are consistent with the view, expressed by other workers, that cell-surface beads contain pectin-like substances, but are not exclusively composed of pectin. Other authors have encountered apparently contradictory results of tests for pectins when applied to cell surface protuberances. Thus Davies and Lewis (1981) noted that pectic beads on *D. carota* cells gave a positive reaction to ruthenium red, but produced a negative reaction to the hydroxylamine-ferric chloride test for pectin (Reeve, 1959). In the material we have examined, beads generally react positively with RR, hydroxylamine-ferric chloride, alcian blue at low pH and coriphosphine O. Since arabinogalactans were not detected at the callus surface, these reactions all indicate the presence of pectin. The strong positive reaction of the beads with Calcofluor and Congo red also indicates that they contain β-1,4 linked glucans, in apparent contrast with the observations of Davies and Lewis, who reported the beads on carrot callus cells to give negative responses to tests for cellulose (zinc-chloro-iodine and IKI-H_2SO_4). Although beads were observed to react both with Sudan black B and Nile red, indicating that they contain lipids, no evidence was obtained in either *D. stramonium or L. esculentum* tissue of separate compartmentation of the lipids, such as observed by Davies and Lewis (1981), nor of any lipid core, as was indicated by our earlier observations (Jeffree and Yeoman, 1983).

Immunofluorescence microscopy and immunogold dot-blotting provided evidence for the existence of free (not cell wall-bound) pectinase at the cell surfaces in callus from all three Solanaceous species examined. That pectin in the vicinity of the undamaged tissue is digested in pectin-agarose substrate-gels appears to confirm this observation. Pectinesterase activity (which might consolidate pectin to form a middle lamella between opposing cells) could not be confirmed in any of the tests, although the modification of gel structure by the tissue in 0.1% pH 8.5 pectin agarose gels requires an explanation.

These observations demonstrate that enzymes (pectinase, cellulase) capable of releasing fragments of the cell wall polysaccharide which might function as recognition molecules in the graft union are present at the surfaces of callus cells in significant amounts. From the point of view of our hypothesis that pectic polysaccharides may act as recognition molecules in grafts (Yeoman, 1984; Jeffree *et al.*, 1987b), an extremely interesting observation is that acid pectic polysaccharides appear to be secreted into agarose gels from cell surfaces at the tissue air interface. This polysaccharide survives in gels in the presence of pectinase concentrations which are sufficient to degrade 0.1% citrus pectin. This implies that the material is not simply polygalacturonic acid, but may be a distinct pectic polysaccharide which is either resistant to, or protected from, pectinase activity, as for example, would be the rhamnogalacturonans, RG-I and RG-II (Fry, 1988). Since the most likely source of pectic fragments with recognition potential will be these relatively complex and pectinase-resistant side-chains or 'hairs', the identification of this polysaccharide is obviously of some importance, despite our experimental observation (Jeffree, *et al.*, 1987b) that incompatibility cannot be transferred to homografts by agarose gels containing exudate from an incompatible partner.

REFERENCES

Barnett JR, and Weatherhead I (1988) Graft formation in Sitka spruce: a scanning electron microscope study. Ann Bot **61**: 581-588

Butterfield BG, Meylan BA, Exley RR (1981) Intercellular protuberances in the ground tissue of *Cocos nucifera L.* Protoplasma **107**: 69-78

Carlquist S (1956). On the occurrence of intercellular pectic warts in Compositae. Am J Bot **43**: 425-429

Carr SGM, Carr DJ (1975) Intercellular pectic strands in parenchyma: studies of plant cell walls by scanning electron microscopy. Aust J Bot **23**: 95-105

Davies WP, Lewis BG (1981) Development of pectic projections on the surface of wound callus cells of *Daucus carota* L. Ann Bot **47**: 409-413

Fry SC (1988) The growing plant cell wall: chemical and metabolic analysis. Longman Scientific and Technical, London

Greenspan P, Mayer EP, Fowler SD (1985) Nile red: a selective fluorescent stain for intracellular lipid droplets. J Cell Biol **100**: 965-973

Hughes J, McCully ME (1975) The use of an optical brightener in the study of plant structure. Stain Technol **50**: 319-329

Jeffree CE, Dale JE, Fry SC (1986) The genesis of intercellular spaces in developing leaves of *Phaseolus vulgaris* L. Protoplasma **132**: 90-98

Jeffree CE, Yeoman MM (1983) Development of intercellular connections between opposing cells in a graft union. New Phytol **93**: 491-509

Jeffree CE, Read ND, Smith JAC, Dale JE (1987) Water droplets and ice deposits in leaf intercellular spaces: redistribution of water during cryofixation for scanning electron microscopy. Planta **172**: 20-37

Jeffree CE, Yeoman MM, Parkinson M, Holden, MA (1987) The chemical basis of cell to cell contact and its possible role in differentiation. **In:** Jackson, MB, Mantell, SH and Blake, J (Eds) Monograph 16 "Advances in the chemical manipulation of plant tissue culture." British Plant Regulator Group, Bristol, pp. 73-86

Kisser J (1928) Untersuchungen uber das Vorkommen und die Verbreitung von Pektinwarzen. Jahrb fur wiss Bot **68**: 206-232

Lipetz J (1970) Wound healing in higher plants. Internat Rev Cytol **27**: 1-28

Parkinson M, Yeoman MM (1982) Graft formation in cultured explanted internodes. New Phytol **91**: 711-719

Reeve RM (1959) A specific hydroxylamine-ferric chloride reaction for histochemical localisation of pectin. Stain Technol **34**: 209-211

Reid JL, Collmer A (1985) Activity stain for rapid characterisation of pectic enzymes in isoelectric focussing and sodium dodecyl sulfate-polyacrylamide gels. Appl Environ Microbiol

Schofield AD (1987) Cellular interactions between host and parasite. PhD Thesis, University of Edinburgh

Sterling C (1970) Crystal structure of ruthenium red and stereochemistry of its pectic stain. Am J Bot **57**: 172-175

Ueda K, Yoshioka S (1976) Cell wall development in *Micrasterias americana*, especially in isotonic and hypertonic solutions. J Cell Sci **21**: 617-631

Weis K G, Polito VS, Labavitch JM (1988) Microfluorometry of pectic materials in the dehiscence zone of almond *(Prunus dulcis* [Mill.] DA Webb) fruits. J Histochem and Cytochem **36**: 1037-1041

Yariv J, Lis H, Katchalski E (1967) Precipitation of arabic acid and some seed polysaccharides by glycosylphenylazo dyes. Biochem J **105**: 1-20

Yeoman MM (1984) Cellular recognition systems in grafting. **In:** Linskens HF, Heslop-Harrison J (eds) Cellular interactions. Encyclopedia of Plant Physiology. Springer Verlag, Berlin, Heidelberg, New York, pp 453-472

Zinc RT, Chatterjee AK (1985) Cloning and expression in *Escherichia coli* of pectinase genes from *Erwinia carotovora* ssp. *carotovora.* Appl Environ Microbiol **49**: 714-717

LEGENDS FOR FIGURES 1-12.

Figs. 1-6: Cytochemistry of D. stramonium cell surfaces: BF = Brightfield illumination; FL = Epifluorescence

Fig. 1: Positive reaction with Schiff reagent after sodium m-periodate oxidation. BF. Mag. x550.

Fig. 2a and b: Ruthenium red staining of beads, shown at two levels of focus. BF. Mag. x 550.

Fig. 3: Red fluorescence from Nile red-stained beads under green excitation. FL. Reichert G1 module. Mag. x 550.

Fig. 4: Calcofluor white: both the cell surface and beads are fluorescent under UV excitation. FL. Mag. x 220.

Fig. 5: Negative reaction to Ruthenium red in a cell extracted for 7 hours in E. carotovora pectinase. BF. Mag. x 550.

Fig. 6: Indirect immunofluorescence micrographs of D. stramonium callus cells labelled with anti-tomato pectinase and FITC-goat anti rabbit IgG. a) Mag. x 550, b) Mag. x 220.

Fig. 7: pH 4, 0.1% pectin-agarose gel after incubation for 8h in contact with D. stramonium callus. RR-stained. Regions of pectinase activity are shown (arrowheads). RR-staining material (magenta regions darker than background) is also secreted into the gel.

Fig. 9: 1% aqueous agarose gel after 8 hours in contact with D. stramonium callus, showing RR-staining secretions.

Fig. 8: 0.5% CMC-agarose gel, pH5, after 8 hours in contact with D. stramonium callus, showing regions of cellulase activity. Congo red stain.

Fig. 10: Assay for cellulase activity. Wells in a CMC-agarose substrate-gel, pH 5, contain two-fold dilutions of Sigma P5146 pectinase from 4-fold (well 1) to $6.5x10^4$ -fold (well 15), and water (well 16). Cellulase activity is still detectable in well 15.

Fig. 11: 0.5% CMC-agarose gel, pH5, after 8 hours in contact with 1-mm thick slices of D. stramonium callus, showing most intense cellulase activity associated with the surfaces of the callus mass (arrowheads). Congo red stain.

Fig. 12: Unstained substrate-gel containing 0.1% pectin + 10mM $CaCl_2$, pH 8.5, showing turbid regions visible in darkfield lighting after 8 hours incubation in contact with D. stramonium callus.

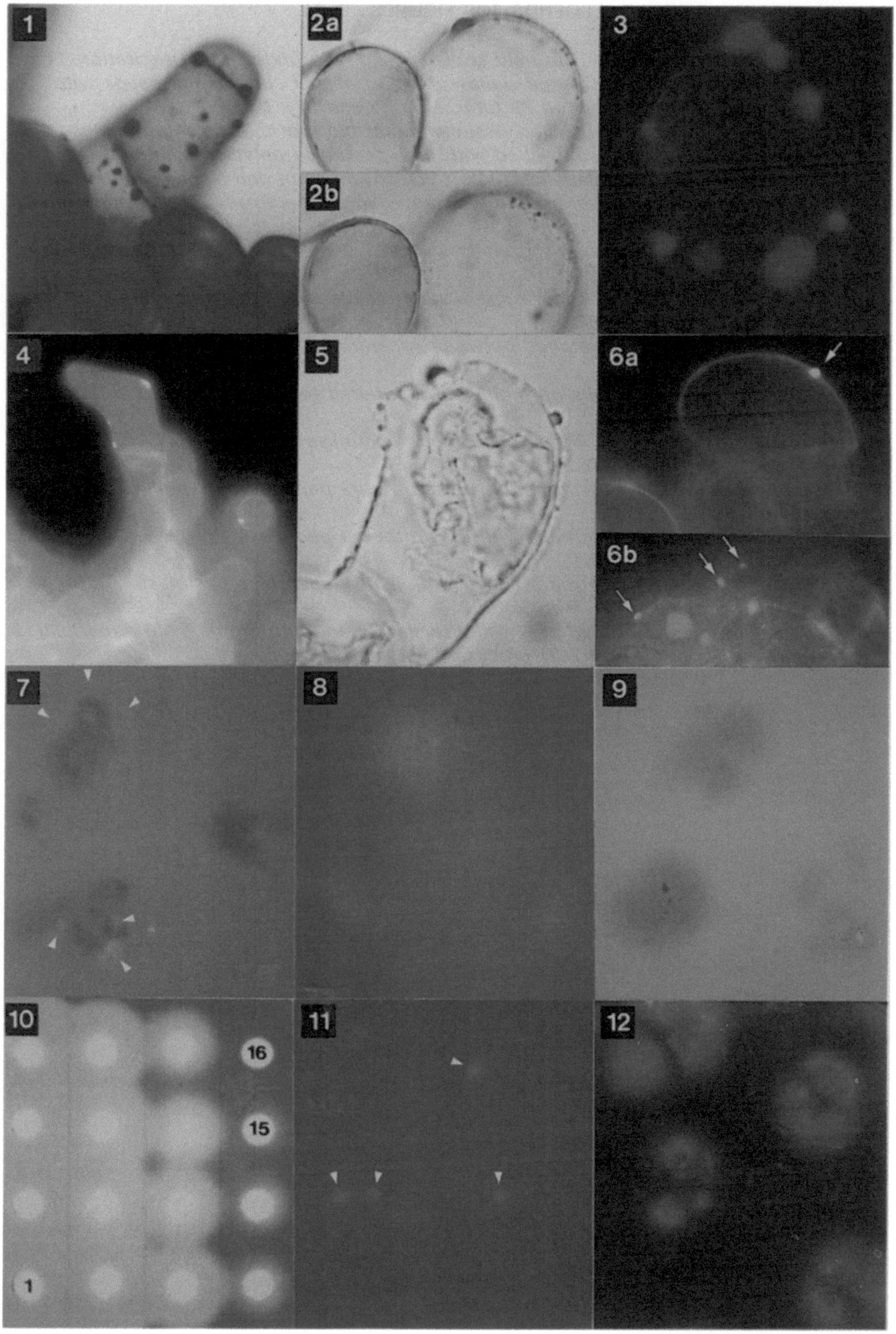
1
2a
2b
3
4
5
6a
6b
7
8
9
10
16
15
1
11
12

LEGENDS FOR FIGURES 13-24

Fig. 13: a) Red fluorescence of the field shown in Fig. 13b under green excitation. FL. Reichert G1 module. b: Orange-yellow staining of cell surface and beads with Coriphosphine O. BF, Wratten 75 filter. Mag. (a and b) x 550.

Fig. 14: Indirect immunogold-labelled tissue-contact blots of L. esculentum callus surfaces on nitrocellulose, probed with anti-pectinase polyclonal antibody.

Fig. 15: Indirect immunogold-labelled blots of D. stramonium and N. physaloides callus surfaces on nitrocellulose, probed with anti-tomato pectinase (a) and anti-tomato pectinesterase (b).

Fig. 16: Callus cell in a tomato autograft at 96 hours with a polar cap of beads 1-3um in diameter. LTSEM. Mag. x 450.

Fig. 17: Cell from a culture of Datura stramonium callus with a dense covering of beads. LTSEM. Mag. x 860.

Fig. 18: Cell wall remnants stretched between separating D. stramonium callus cells. LTSEM. Mag. x 615.

Fig. 19: Smooth, unbeaded cells on highly-differentiated tissue in a tomato culture. LTSEM. Mag. x 370.

Fig. 20: A cell surface treated with 2% Rhizopus pectinase pH4, 1.5h. GAPATCO process, SEM. Mag. x 11500.

Fig. 21: Intact beads after treatment with E. carotovora pectinase, 7h, pH4. LTSEM. Mag. x 2670

Fig. 22: Intact beads after treatment with E. carotovora pectinase, 7h, pH 4. GAPATCO process, SEM. Mag. x 1300.

Fig. 23: Intact beads after treatment with Sigma 5146 Aspergillus pectinase, 7h, pH 4. GAPATCO process, SEM. Mag. x 1900.

Fig. 24: Intact beads and eroded cell wall after treatment with Sigma 5146 Aspergillus pectinase, 7h, pH4. GAPATCO process, SEM. Mag. x 4500.

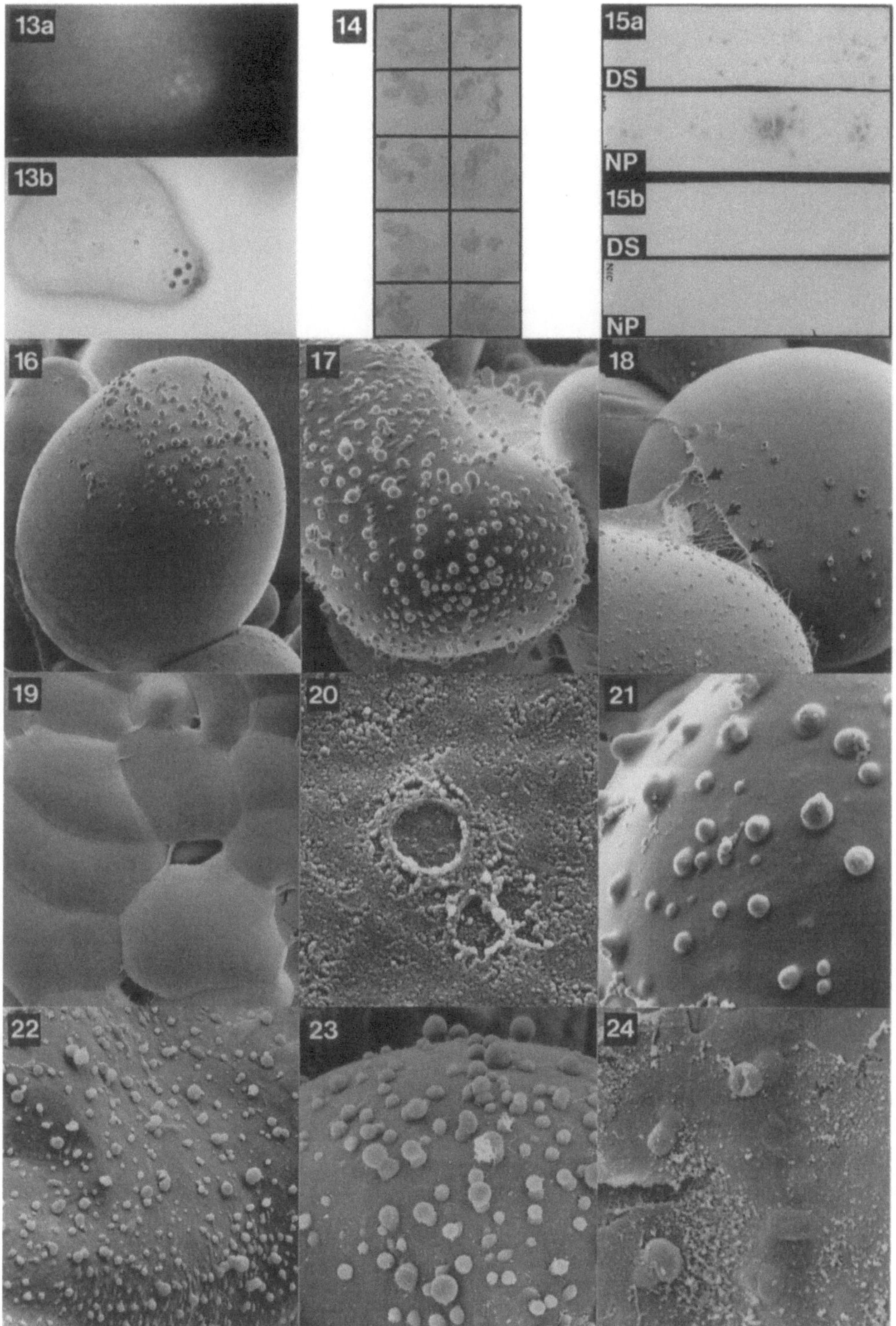
13a
13b
14
15a
DS
NP
15b
DS
NP
16
17
18
19
20
21
22
23
24

SIGNALS BETWEEN PLANT AND BACTERIAL CELLS: SPECIFIC DOMAINS IN CELL WALLS, A NEW INSIGHT

Edward C. Cocking
Plant Genetic Manipulation Group
University of Nottingham
Nottingham
UK.

The bacterium, Rhizobium, selectively infects legume roots and forms root nodules, which if effective, fix atmospheric nitrogen into ammonia that can be assimilated for plant growth. There are many cellular recognition phenomena which occur during the infection of legume roots by Rhizobium; attachment of rhizobia to legume host root hairs is considered an early step of cellular recognition (Dazzo 1985). There is no doubt that the location of the receptors involved in such cellular recognition has been hampered by the fact that plant cells are usually surrounded by a tough cell wall and this barrier must be penetrated to permit identification of receptors, and an understanding of signalling between plant and bacterial cells. One way of studying the plasma membrane of plant cells is to remove the cell wall and to isolate the living protoplast. Research with isolated protoplasts has provided considerable information regarding the structure, chemistry and function of plant plasma membranes (Fowke, 1985). Whilst the use of such protoplasts is highly attractive, the fact that protoplasts are isolated from the plant prevents any study of the signalling and interaction at the surface of the protoplast in situ within the plant itself.

Until recently, apart from the possibility of using laser microbeam cell surgery (Tsukakoshi *et al.*, 1984), the likelihood of the use of intact functional plants for such investigations seemed remote. However in 1985 it was found possible to achieve rapid enzymatic degradation of the cell wall of the apices of root hairs from a wide range of crop species. Thereby it was possible to expose plasma membranes at root hair apices whilst maintaining the functional integrity of the plant (Cocking, 1985). Under suitable conditions root hairs are present on the roots of most, if not all, angiosperms and gymnosperms, and an ability to remove readily their apical cell walls has provided an opportunity to study interaction of the plasma membrane with foreign DNA, viruses and micro-organisms.

NATO ASI Series, Vol. H 35
Cell Separation in Plants
Edited by D. J. Osborne and M. B. Jackson

There is good evidence for cellulose microfibrils arranged at random over the hemispherical tip of the root hair, with the extreme tip of the hair being the site of cellulose synthesis (Belford and Preston, 1961). The specific infection of legume root hairs by Rhizobium involves many steps of cellular recognition which culminate in the formation of a root nodule that reduces atmospheric nitrogen into ammonia for the host plant. As discussed by Dazzo and Gardiol (1984) these positive cellular recognitions are believed to arise from a specific union, reversible or irreversible, between chemical receptors on the surface of interacting cells. This hypothesis implies that communication occurs when cells that recognize one another come into contact, and therefore the complimentary components of the cell surfaces have naturally been the focus for most biochemical studies. The infection process in the Rhizobium-legume symbiosis is an elegant sequence of cellular recognitions. The "lectin recognition hypothesis" hypothesised that recognition between rhizobia and the root hair involved a binding of the plant lectins on the root hair to unique carbohydrates found exclusively on the bacterial symbiont as a prelude to infection. A unique feature of symbiotic nitrogen fixing associations, particularly between Rhizobium and legumes, is that bacterial invasion of the host root hair is mediated by a tubular structure known as the infection thread. The infection thread grows inwards reaching and spreading to the cells that become meristematic as a result of the infection, and restricting the invading microbe to an extra cytoplasmic compartment.

A new insight into the control of cellular recognition between Rhizobium and the root hair was provided by the finding that enzymatic treatment of clover root hairs removes a barrier to Rhizobium-host specificity. It was already well established that Rhizobium loti and Rhizobium 101/U nodulate Lotus species but not Trifolium repens. It was observed (Al Mallah *et al.* 1987) that a barrier to this specificity could be removed by enzymatically degrading the cell wall at the apices of root hairs of Trifolium repens using a cellulase-pectolyase enzyme mixture and inoculation with R. loti or R. 101/U, in the presence of polyethylene glycol. This treatment enabled R. loti to induce the formation of nitrogen fixing nodules on Trifolium repens, and identified the cell wall at root hair apices as a target site for detailed studies on signalling between plant and Rhizobium cells. The marked stimulation of nodule formation in white clover by polyethylene glycol (PEG), following this enzymatic treatment and inoculation with R. loti or R. 101/U, may be related to the known stimulatory effect of PEG on the uptake of micro-organisms into plant cells following enzymatic degradation of their cell walls (Davey and Power, 1975). It is noteworthy that a detailed study of the structural basis for infection of root hairs of Trifolium repens by Rhizobium species specific to T. repens, indicated that rhizobia directly penetrate the existing root hair wall by a process involving the alteration and degradation of cell wall polysaccharides at a very localized site. It was also indicated that the attack on the cell wall was probably enzymatic, emanating from the attached or enclosed bacterial cells. Thus the hair wall is completely degraded at the site of contact with the enclosed colony while adjacent wall areas appear modified but structurally intact (Callahan and Torrey, 1981). It has been suggested that localized activation of host root polygalacturonase might result in both localised degradation of the hair cell wall, and localized release of pectic elicitor substances that induce deposition of infection thread material (Bauer, 1981). It has also been hypothesized that there are host-secreted fragments bound with proteins or glycoproteins that function by signalling the initiation of the complex reactions leading to symbiosis (Keen and Holliday, 1982). There have been significant advances in our understanding of transmembrane signalling and the role of inositol lipids in animal cells (Berridge and Michell, 1988), but no comparable system has been, as yet, found in plant cells. What the present enzymatic treatment of root hairs provides is the opportunity to focus attention on the plant plasma membrane and its properties in this and other respects when interacting with bacteria such as rhizobia.

Is the plasma membrane at the tip of the root hairs of legumes different in its interaction and signalling responses from that at the tip of the root hairs of a non-legume such as oil seed rape, wheat or rice? Very interestingly, recent work suggests that legumes and non-legumes are comparable in this respect. Our recent finding that treatment of root hairs of clover seedlings with cell wall degrading enzymes removes a barrier to Rhizobium-host specificity led us to investigate whether a similar enzymatic treatment, already known to expose the plasma membrane at the tips of rice root hairs (Cocking, 1985), could enable such treated rice seedlings to be nodulated by rhizobia. We found that nodular structures were produced on the roots of rice seedlings when the roots were treated with a cellulase-pectolyase enzyme mixture, and inoculated with either Rhizobium or Bradyrhizobium in the presence of PEG (Al Mallah *et al*. 1988).

These studies now enable the interaction of rhizobia at the plasma membrane of the tip of root hairs of both legumes and non-legumes to be investigated without any direct involvement of components of the plant cell wall. This basic study could provide a new insight and ultimately a capability of producing effective nodules on non-legumes when enzyme-treated root systems are interacted with rhizobia.

REFERENCES

Al-Mallah MK, Davey MR, Cocking EC (1987) Enzymatic treatment of clover root hairs removes a barrier to rhizobium-host specificity. Bio/Technology 5:1319-1322

Al-Mallah MK, Davey MR, Cocking EC (1988) Formation of nodular structures on rice seedlings by rhizobia. J Exp Bot (submitted)

Bauer WD (1981) Infection of legumes by rhizobia. Ann Rev Plant Physiol 32:407-449

Belford DS, Preston RD (1961) The structure and growth of root hairs. J Exp Bot 12:157-168

Berridge MJ, Michell RH (1988) Inositol lipids and transmembrane signalling. Phil Trans Royal Society B 320:235-436

Callaham DA, Torrey JG (1981) The structural basis for infection of root hairs of Trifolium repens by Rhizobium. Can J Bot 59:1647-1664

Cocking EC (1985) Protoplasts from root hairs of crop plants. Bio/Technology 3:1104-1106

Davey MR, Power JB (1975) Polyethylene glycol-induced uptake of micro-organisms into higher plant protoplasts: an ultrastructural study. Plant Sci Lett 5:269-274

Dazzo FB, Gardiol AE (1984) Host specificity in Rhizobium-legume interactions In: Verma DPS, Hohn Th (eds) Genes involved in microbe-plant interactions. Springer-Verlag, New York, pp 3-31

Dazzo FB (1985) Receptors for attachment of Rhizobium to legume root hairs. In: Chadwick CM, Garrod, DR (eds) Hormones, receptors and cellular interactions in plants. Cambridge University Press, Cambridge, pp 319-331

Fowke C (1985) The plasma membrane of higher plant protoplasts. In: Chadwick CM, Garrod DR (eds) Hormones, receptors and cellular interactions in plants. Cambridge University Press, Cambridge, pp 217-235

Keen NT, Holliday MJ (1982) Recognition of bacterial pathogens by plants. In: Mount MS, Lacy GH (eds) Phytopathogenic prokaryotes. Academic Press, London, pp 179-217

Tsukakoshi M, Kurata S, Nomiya Y, Ikawa Y, Kasuya T (1984) A novel method of DNA transfection by laser microbeam cell surgery. Appl Phys B 35:1-6

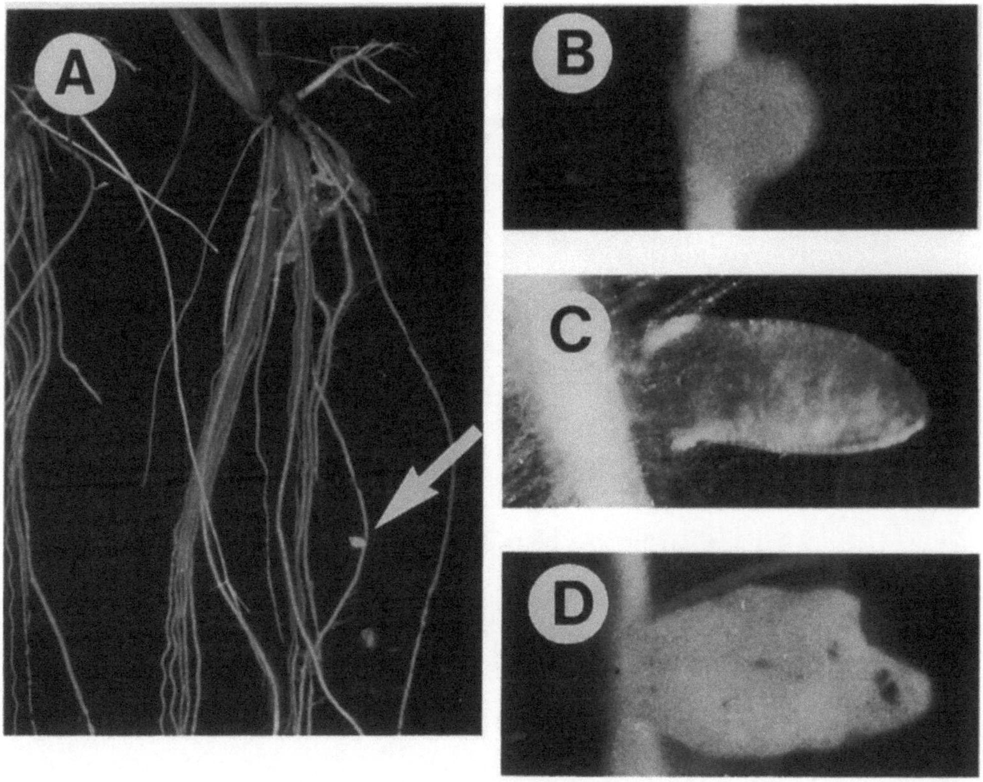

Figure 1. Nodular structure produced on rice roots by Rhizobia following treatment with a cellulase-pectolyase enzyme mixture and PEG.

SECTION VI

Crop Production and Harvesting

BUD, FLOWER AND FRUIT DROP IN CITRUS AND OTHER FRUIT TREES

N. Kaska
Department of Horticulture
Faculty of Agriculture
University of Çukurova
Adana, Turkey

INTRODUCTION

Fruit trees usually take many years before they produce a crop. This delay is perhaps longest in pistachio (about 15-20 years) grown for its nuts in some arid areas, and in seedling trees of walnut (10-15 years). In citrus, apple, pear and sweet cherries the delay is approximately 3-5 years but shorter in plums, peaches, apricots and figs. However, the time before first harvest can be shortened by certain cultural and technical measures e.g. girdling and ringing. It is quite obvious that the first requirement for fruiting is the formation of flower buds. But, between the initiation of flower buds and the differentiation to flowers, and from the flowering to fruit setting, many physiological processes must take place before a yield of fruit is achieved. One of the most important of these is the prevention of abscission of generative organs such as flower buds, flowers, small fruit and ripened fruit.

At flowering time, thousands of flowers are seen on a mature tree, but most of them separate from the tree and drop soon after flowering. Flowers may become pollinated and fertilized but during this time or just a little later, many more flowers abscind. When the surviving fruit start enlarging a new wave of abscission occurs and some of the partially developed fruit are lost. When the weather warms and relative humidity becomes low as the season progresses, a third wave of fruit loss (the June drop) takes place, when small fruit 0.5 to 1 cm in diameter are shed. After the June drop, the remaining fruit persist on the tree unless there is an adverse climatic event such as late frost or hail. Nearer maturation time, fruit which were injured for any reason and become infected via these wounds are also lost by abscission. After fruit maturation time is completed, some pre-harvest drop can also occur. These successive episodes of flower or fruit abscission are not accidental, they are regulated and take place interdependently at quite distinct times each year during the growing season. Each of these drops and their possible cause will be discussed in the following sections.

NATO ASI Series, Vol. H 35
Cell Separation in Plants
Edited by D. J. Osborne and M. B. Jackson

BUD DROP

Deciduous fruit. Bud drop is mostly encountered in apricots with long chilling requirements when they are grown in warm, subtropical areas. In such varieties and towards spring, flower buds with primordia already developed are shed with an abscission layer at the base, leaving only leaf buds surviving on the branches. In these cases, few, if any flowers open on the tree. Otherwise, the tree looks normal and healthy because of the full canopy of leaves. The main cause of shedding is the lack of chilling. In fact, varieties with a short chilling requirement bloom normally in warm winters and bear satisfactory fruits. From the practical point of view, it is important to establish orchards of short-chilling varieties in warm winter areas. In such places, precocious production may be possible and such production is obviously popular with the growers. To obtain economical fruiting from the long-chilling varieties, it is necessary to spray the trees with growth regulators or other chemicals such as DNOC, KNO_3, GA (gibberellin) and thiourea. Examples of such applications can be seen in South Africa, Israel and the Mediterranean Coast of Turkey. These chemicals break the dormancy of flower buds and as a result, no shedding occurs. In contrast, when the chilling requirements of apples and pears are not satisfied the flower buds are not shed. Instead, flower primordia within the buds remain undeveloped and in spring they emerge from the bud as degenerated organs. In peaches, flower-bud shedding is uncommon. The buds usually remain dried and undeveloped on the spurs.

The physiology of flower bud shedding in fruit trees grown in warm-winter zones has not been studied extensively. However, it has been attributed to increases in ABA and decreases in GA in the buds. In fact, GA applications in late winter may prevent flower bud shedding. There is no doubt that further studies on this phenomenon will be of considerable interest scientifically and be of benefit to commercial fruit growing.

Citrus. Although flower bud shedding in Citrus is not common, investigations have revealed that many unopened flower buds can be shed. For example, in one survey, Washington Navel, Valencia and Shamouti oranges lost 48.5%, 35.5% and 15.5% respectively of the total flower buds (Erickson, 1968). According to Desai *et al.* (1985) the percentage orange flower buds lost can be as high as 33.4%. Normally, citrus trees form very many flower buds. Even when large numbers of them are shed, the remainder will show quite a dense bloom overall. In the above mentioned experiments with

Washington Navel and Valencia, the total number of flower buds per tree were 198693 and 73378 respectively. Flower bud drop in Citrus seems closely related to nutrition. For example, Shavit (Erickson, 1968) has shown that the rate of flower bud drop is 35.8% in Zn-deficient trees, compared with 15.6% in normal trees.

FLOWER DROP

Large numbers of opened flowers drop soon after flowering. Some of these flowers are defective, but the majority are staminate. In Shamouti oranges, 77.0% of the flowers shed were found to have developmental abnormalities and in addition, 6.2% of the flowers were without pistils. The percentage of flowers abscinded in Washington Navel and Valencia was found by Erickson (1968) to be 16.7% and 3.0% respectively. In peaches, all the flowers which contain aborted pistils are lost. Obviously, in such flowers, pollination and fertilization could never take place.

A second wave of flower abscission, which overlaps the first, is caused by a lack of fertilization and possibly micronutrient deficiencies. In one experiment, it was found that the level of flower drop was 49.2% in Zn-deficient trees and only 34.4% in the non-deficient trees. In parallel to this experiment, Kogzi lime and oranges sprayed with 0.6% $ZnSO_4$ and 20 ppm 2,4-D have yielded more fruit than the untreated trees (Babu *et al.*, 1982). In the abscinded unfertilised flowers, the abscission layer usually forms at the distal end of the pedicel.

Manabe *et al.* (1982) demonstrated that fruit drop was much decreased when the flowers were thinned severely. On the other hand, Xu *et al.* (1982) effectively prevented the loss of many small fruit by applying 400 ppm benzyladenine (BA) to the flowers and to small fruits 0.5 cm in diameter. Gibberellin A_3 applied at 50 ppm at the same time was not effective.

According to Eti (1987) the number of dropped flowers in Clementine mandarins was increased when they were self-pollinated (Table 1 and 2). Although in 1983 the number of flowers lost was lower in the open pollinated and selfed trees than in cross-pollinated trees, the situation was reversed in 1984. In both 1983 and 1984, GA treatment decreased flower drop. These values in Table 1 and 2 show clearly that the cause of flower drop was a lack of fertilization. Growth regulators play an important role in this respect.

TABLE 1. Fruit set in Clementine mandarins after open and cross pollination and GA treatment in 1983 (Eti, 1987)

Treatment	No. of flowers	No. of small fruit	No. of fruit fruit after June drop	No. of fruit before harvest
Open pollination (Control)	1770	60.0	2.2	0.9
Clem. x Clem.(Selfing)	1498	51.9*	0.9	0.2
Clem. x Local Mandarin	1208	74.1***	10.1***	6.7***
Clem. x Local Orange	1075	63.2	4.2	3.0*
Clem. x Sour Orange	1399	77.3***	10.4***	8.9***
Clem. x Fremont	1262	70.4**	13.9***	10.9***
GA_3 12.5 ppm	1457	81.2***	17.5***	5.1***
GA_3 25 ppm	1882	77.2***	13.4***	5.4***
GA_3 50 ppm	1343	88.9***	20.1***	12.6***

Mean separation within columns by Students 't' test, at 5% (*), 1% (**) or 0.1% (***)

TABLE 2. Fruit set in Clementine mandarins after open and cross pollination and GA treatment in 1984 (Eti, 1987)

Treatment	No. of flowers	No. of small fruit	No. of fruit after June drop	No. of fruit before harvest
Open pollination (Control)	1703	88.5	2.0	1.8
Clem. x Clem. (Selfing)	1650	79.6	1.4	0.7
Clem. x Local Mandarin	1713	41.0***	9.8	6.9
Clem. x Local Orange	1052	55.7***	4.3	3.9
Clem. x Sour Orange	1527	33.3***	11.4	8.8
Clem. x Interdonato	1129	66.4**	5.1	3.8
GA_3 12.5 ppm	999	87.2	33.8***	24.1***
GA_3 25 ppm	1124	86.5	39.2***	31.5***
GA_3 50 ppm	922	90.2	36.2***	34.8***
GA_3 100 ppm	975	91.8	55.0***	46.4***
GA_3 200 ppm	986	84.5	56.1***	49.7***

Mean separation within columns by Students 't' test, at or 0.1% (***)

It is known that with the exception of a few varieties such as Tuono, Genco and Troito, almonds are self incompatible. Eti *et al.* (1988) have found that in this species the percentage of small fruit shed from the trees was more than in the June drop and much less severe in selfed trees than in cross-pollinated trees (Fig. 1).

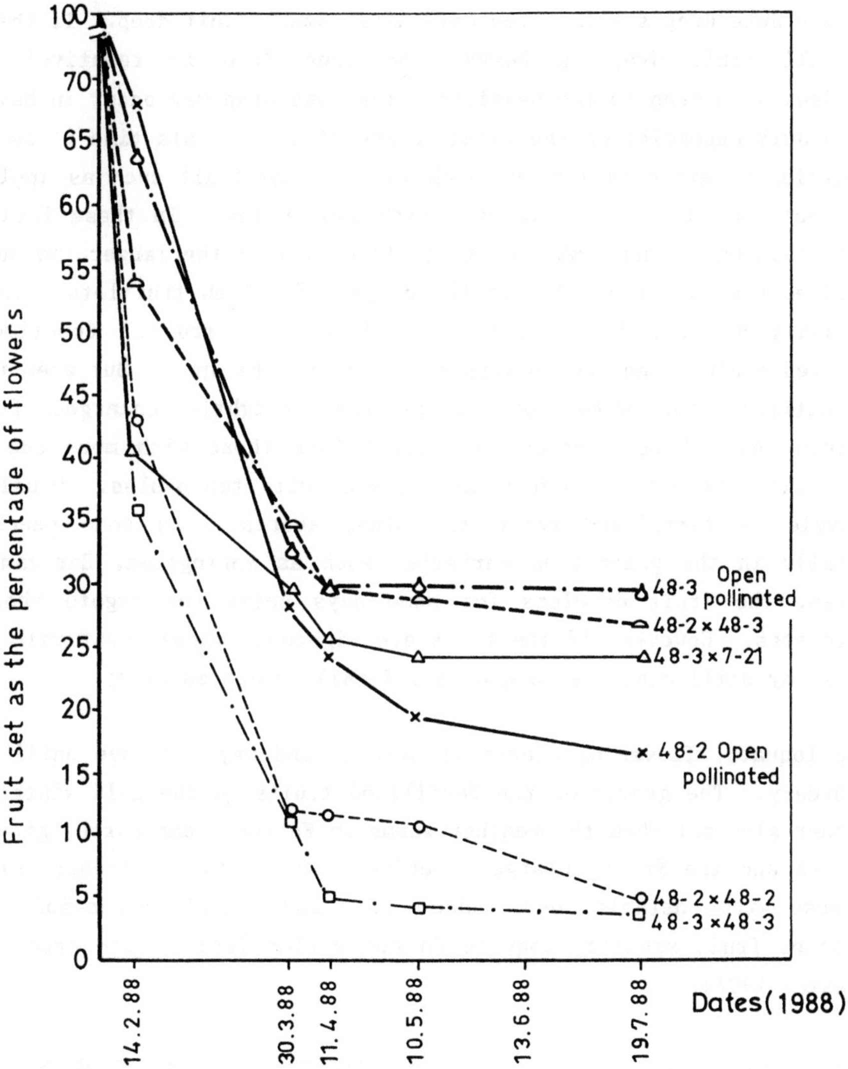

FIGURE 1. Effects of selfing, crossing and open pollination on the fruit set of some local almond varieties in Turkey (Eti, et al., 1988). '48.3', '48.2' and '7.21' are code numbers for the three varieties used in this experiment.

SMALL-FRUIT DROP

This drop usually occurs within a few days after petal fall. In apples and peaches it lasts 2 to 3 weeks. The main cause of this drop is a lack of fertilization. However, in stone fruits some very small but fertilized fruits may also abscind due to embryo and zygote abortion.

JUNE DROP

Generally June drop starts a few days after small fruit drop. If the flower and small fruit drop is heavy, the June drop is relatively light. Otherwise, June drop is the heaviest. The June drop may occur in May, June, even in July depending on the location and site. At this time it is usually the partially developed fruits which fall. Pome fruit such as apple, pear and loquat contain 5 carpels, each with two ovules. In these fruit, June drop is related to the number of fertilized ovules; the larger the number of fertilized ovules the smaller is the number of fallen fruitlets. Thus, the probability of drop is greatest in pathenocarpic fruit. Sometimes, one fertilized ovule is enough to keep the fruit on the spur. But when there is a competition for water or assimilates, slightly enlarged fruitlets containing only 1 to 3 seeds may drop before those with more seeds. In stone fruit, there is only one carpel, each with two ovules. Usually only one ovule is fertilized while the other aborts. In some peaches and especially in the precocious varieties such as Springtime, Earlyamber and Earlyred, the fruit develops for some days until the zygote aborts and deteriorates. However, if the trees are vigorous, these fruit with aborted zygotes may still continue to grow and finally reach maturity.

In loquats, flowering starts in October and may continue until January or February. The growth of the fertilised fruits in the cold winter months is rather slow but when the weather warms in February and March, growth rate increases and the fruit enlarge. During this period, parthenocarpic fruit and those with unstable seeds cannot grow and therefore abscind. Unlike many other fruit species, loquats do not suffer from a June drop (Erdogdu and Kaska, 1987).

In Citrus, the June drop can be rather heavy and results from the formation of a layer between the pedicel and the fruit (Xu *et al.*, 1982). Depending on the extent of the preceeding abscission of flowers and small fruit, the severity of June drop can be high e.g. 21.0% in lemons, 27.2% in Washington Navel, 16.1% in Valencia oranges, 23.8 to 45.1% in Shamouti oranges (Erickson, 1968) and 57.8 to 86.5% in Clementine mandarins (Eti, 1987). Consequently, in Citrus, only 3.5% (in Shamouti), 0.2% (in Washington Navel), 1.0% (in Valencia), 1.4% (in Clementine mandarins) of the flowers develop into harvestable fruit (Erickson, 1968; Eti, 1987). These rates are somewhat higher in loquat (4-9%) (Erdogdu and Kaska, 1987) and even in self-sterile almonds (approximately 11%).

In an experiment on the effect of irrigation frequency on fruit drop in Marsh Seedless grapefruits, Kamber *et al.* (1988) revealed that the June drop is followed by other less intense episodes of fruit drop until harvest, giving a total drop of pre-mature fruit of more than 40% in both experimental years (Fig. 2).

Causes of June Drop. Many investigations indicate that the June drop may result from: (1) abnormalities in fertilization, zygote abortion or degeneration, (2) competition for food and nutrients, (3) water stress and balance, (4) climatical factors.

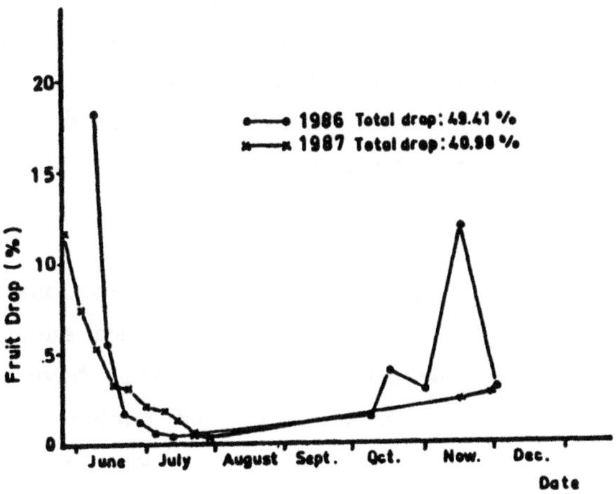

FIGURE 2. Periodic fruit drop following the June drop in Marsh seedless grapefruits in the Eastern Mediterranean region of Turkey (Kamber et al., 1988)

At the beginning of the June drop, small fruit either unfertilized or partially fertilized, are the first to fall. These fruit are of different sizes due to variation in the duration of the flowering period. It is known that in wind pollinated dioecious plants such as pistachio nuts, a lack of pollination is the main cause of the drop. This can be seen when hot, dry winds prevail at the time of pollination and where the number of pollinator or male trees in an orchard is less than one in ten. Pistachio fruit with aborted zygotes tend to be retained on the cluster until the harvest although it is, of course, of no economic value.

A similar type of empty shell can be seen in hazel which is monoecious and blooms in winter. The Turkish hazel nut plantations are on the Black Sea coast which is foggy most of the winter, resulting in poor pollination. Consequently, many female flowers shrivel and drop. As in the case of pistachio, some of the fruit can grow even with an aborted zygote and persist on the plants until they are harvested as empty nuts.

Normally, large fruit with increased seed number are the ones which persist on the tree. In parthenocarpic varieties of Citrus such as Washington Navel, Shamouti oranges, Eureka, Interdonato and "Demre Dikensiz" (a Turkish local variety) lemons, the rate of drop is much higher than in seeded varieties. However, in Seedless Satsuma Mandarin and Thompson Seedless and Marsh Seedless grapefruits, June drop is light (Özsan and Bahçecioglu, 1970). On the other hand in Clementine mandarins, a heavy June drop is seen every year. This is due to self sterility of this variety and to inefficient cross-pollination. In triploid pome fruit, abnormalities in fertilization are very well known. Most of the seeds with chromosome numbers less than 2n=34 tend to abort if there is any shortage of water, assimilates or inorganic nutrient.

When there is a very heavy fruit set, trees tend to lose most fruitlets from weak and old branches. In this case, there is not enough leaf surface to supply photosynthates for the fruit. Fruit which have the most seeds persist on the tree and those with smaller seed number are apt to fall. However, the vigour of the branches play an important role as well. Therefore, some fruit with a small number of seeds may persist on stronger spurs and fruit with a large number of seeds may fall from weaker shoots. In apple and pears generally, the king flowers grow more vigorously than the others since they develop before the pollination or fertilization of the others and thus have a competitive advantage. In contrast to raspberries, there is no flower or fruit drop in strawberries. Some of the small late opened flowers cannot set fruit but they do not abscind and shrivel attached to the cluster.

Bark-ringing on the stems or branches of Citrus can increase fruit sets, thus reducing flower and fruit abscission (Goren and Monselise, 1971; Chundavat and Randhava, 1972, 1973).

Clementine mandarins (Yesiloglu, 1988) increased fruit set and yields by single or double ringing of the stem or by applying GA (Table 3). In these experiments, a double ring 0.5 cm wide was found more effective than a single ring and yield was increased approximately two-fold in 1984 and 1985.

*TABLE 3. Effects of Girdling and GA applications on the number of fruit after June drop and before harvest (%) (Yesiloglu, 1988)**

Treatment	1984		1985		1986	
	After June drop	Preharvest drop	After June drop	Preharvest drop	After June drop	Preharvest drop
Control	4.00	3.20	6.00	5.00	7.50	7.00
GA	11.40	5.80	30.20	26.80	8.40	7.20
One ring	20.60	15.40	3.20	2.60	5.40	4.20
Two rings	11.00	8.20	29.60	27.40	8.00	7.80
One ring + GA	25.40	10.60	18.80	17.40	4.40	4.00

Treatment	Yield (kg/tree)		
	1984	1985	1986
Control	31.43	46.43	68.70
GA	36.30	103.45	75.13
One ring	86.57	17.16	65.44
Two rings	84.56	107.73	78.70
One ring + GA	69.49	44.60	72.28

* Girdlings were applied at the beginning of flowering. GA was applied at the end of flowering as 20 ppm. Application repeated every year.

According to the experiments of Kobel (Özbek, 1977), the osmotic potential of the leaves is about half that of young fruit. When evapotranspiration rate is high, absorption of water from the soil can be insufficient to meet this demand, resulting in plant water deficits. Because young fruit have a higher (less negative) osmotic potential than the leaves, it is these fruit which desiccate first. If the stress persists, the low osmotic potential of the leaves becomes insufficient to prevent desiccation and they too wilt and then abscind. In extreme cases, as in the walnuts, even thin terminal branches become dried and then separate with smooth abscission layers. The investigations of Birand (Özbek, 1977) in arid soils of Central Anatolia have shown clearly how the osmotic potential of the leaves of different drought resistant fruit species can change (Table 4). In this area, as the dry summer season progresses, the air temperature increases and relative humidity decreases gradually. As shown in the Table, the osmotic potentials decline gradually and in driest months like August and September they reach a minimum (lowest negative value). Under these extreme conditions even though their osmotic pressures are at their most negative the leaves drop since the roots are unable to absorb water from the soil. In Table 4, the value for *Crateagus* is not given for September because the leaves had already fallen due to the drought. The osmotic potentials of the leaves of the same species in humid areas were found much higher (Table 5).

In *Citrus*, the severity of the June drop is determined mostly by the very warm weather and the prevailing dry northerly winds which blow in summer. To growers of *Citrus* in Mediterranean coastal regions, this wind and its associated high temperatures is a major concern. The wind can blow for 3 to 4 days continuously and partially desiccate the plants by increasing transpiration rates considerably. Fruit and even leaf abscission take place as a consequence in all the deciduous fruit trees but especially in Washington Navel oranges. In this orange, the abscinded fruit are

TABLE 4. Changes in osmotic potential (atm.) of the leaves of some drought resistance fruit trees grown in arid soils around Ankara (Özbek, 1977)

Fruit species	Months			
	April	June	August	September
Pyrus elaeagrifolia (Wild pear)	-17.6	-24.2	-34.0	-34.9
Amygdalus orientalis (Almond)	-18.8	-25.5	-36.3	-37.8
Crateagus orientalis (Hawthorn)	-16.4	-21.4	-29.0	

TABLE 5. Changes in osmotic pressure (atm.) of the leaves of wild pear and almond grown in humid and arid soils (Özbek, 1977)

	Climatic Regions	
	Humid	Arid
Pyrus elaeagrifolia (Wild pear)	-14.0	-36.3
Amygdalus orientalis (Almond)	- 8.7	-34.0

usually 1 to 3 cm in diameter and their navel ends dried. Generally, some of the lost fruit are damaged by the carob moth (*Ectomyelois ceratoniae* Zell), but the full effect of this moth appears when the fruits are a little larger. In the summer of 1988, Turkish and Florida growers identified critical times when damage resulting from these winds and hot weather is most severe (Acuff, 1988). In such conditions, the grower should take all possible measures to prevent water deficits by irrigation.

It is known that severe hail storms can cause fruit drop. The injured fruits persist on the tree but although the wounds heal they are usually infected by fungi. Such fruit ripen on the tree earlier than normal and thus fall before commercial harvest time. In some regions, these hail storms occur in most years and since there is no remedy, these areas are not recommended for horticulture. However, in places such as Murre near Islamabad (Pakistan) growers have been encouraged by government to cover the apple trees with nets to prevent hail damage. Naturally, the economics of this measure depend on the price of the crop.

PRE-HARVEST DROP

Near to maturity, an abscission layer forms between the pedicel of the fruit and the spur. Fruit then separate from the plant. If the fruit are not harvested before the formation of this layer they are liable to drop. This is called the pre-harvest drop. On the other hand, fruit damaged by hail, agricultural equipment, codling moth etc., ripen prematurely and consequently drop before harvest. Pre-harvest drop is especially prevalent in William's pear amd Mackintosh apples. It is now common practice to prevent this by spraying 5 to 20 ppm naphthaleneacetic acid or naphthaleneacetamide on the trees (Chandler, 1951; Özbek, 1977; Stan *et al.*, 1984). SADH (daminozide) has also been used to extend the harvest season by preventing pre-harvest drop (Kücükaydin and Kaska, 1987; Brudel *et al.*, 1983; Edgerton, 1988). In contrast to the USA, most citrus fruit in Europe and Asia is consumed fresh rather than as juice. Therefore the growers wish to prevent pre-harvest drop, especially in grapefruit and oranges. For this purpose, plant growth regulators such as NAA, 2,4-D and GA have been used successfully (Ferguson *et al.*, 1984; Deidda and Dettori, 1982; Gill *et al.*, 1980; Hizal and Kaska, 1985) to prevent the processes of cell separation at the abscission zone.

Many experiments with Ethrel show this compound reduces the force required to detach the fruit. Ethrel causes early formation of an abscission layer between the pedicel and cluster base and at the same time promotes colouring and ripening. This chemical has given very successful results with Kütahya, the most popular cherry variety in Turkey (Kaska and Pekmezci, 1974; Paydas *et al.*, 1988). Unfortunate side effects include shedding of leaves and gummosis. However, in Kütahya sour cherry, even at 800 ppm, leaf shedding and gummosis was minimal. In some fruit such as olive and sour cherry, hand picking of fruit is tedious, expensive and time consuming. The growers wish to detach the fruit by shaking the tree or its branches. This has become possible after spraying the trees with Ethrel and related substances.

REFERENCES

Acuff G (1988) Drought Teaches Lessons. Fruit Grower 108: 22

Babu RSH, Rajput CBS, Rath S (1984) Effects of zinc, 2,4-D and GA_3 on fruiting of Kogzi lime and *Citrus aurantifolia* Swingle. Indian J Hort 41: 216-220

Brudel F, Sandke G, Scheffel HD (1983) Verlustsenkung durch Verminderung des Vorerntefruchtfalls bei Apfel. Under Anwendung von Mitteln zur Steuerung biologischer Prozesse. Gartenbau 30: 85-87

Chandler VH (1951) Deciduous orchards. Lea and Febiger, Philadelphia

Chundawat BS, Randhava GS (1972) Effect of ringing and root pruning on fruit set, fruit drop and quality of marsh seedless grapefruit (*Citrus paradisi* Macf.). Haryana J Hort Sci 1: 19-22

Chundawat BS, Randhava GS (1973) Effect of plant growth regulators on fruit set, fruit drop and quality of Foster and Duncan cultivars of grapefruit, (*Citrus paradisi* Macf.) Haryana. J Hort Sci 2: 6-13

Deidda P, Dettori S (1982) The efficiency of the isopropyl ester of 2,4-D in preventing fruit drop in Marsh Seedless grapefruit (in Italian). Rivista della ortoflorofruitticoltura Italiana 66: 239-246

Desai UT, Chaudhari SM (1985) Studies on flower bud and fruit drop in sweet orange cv. Current Research Reporter 1: 61-63
Edgerton LJ (1988) Don't forget Alar advantages. Amer. Fruit Grower 108: 9
Erdogdu H, Kaska N (1987) Investigations on the phenological and pomological characteristics of some local and foreign originated loquat varieties under Adana ecological conditions (in Turkish). J Sci Eng 1: 83-92
Erickson LC (1968) The general physiology of citrus. In: Reuther W, Batchelor LD, Webber HJ (eds) The citrus industry, Rev edn. University of California, p 97
Eti S (1987) Uber des Pollenschlauchwachstum und die Entwicklung der Samenanlagen in Beziehung zum Fruchtansatz under zur Fruchtqualitat bei der Mandarinens orte "Clementine" (*Citrus reticulata* Blando). Dissertation, University of Hohenheim
Eti S, Paydas S, Küden A, Kurnaz S, Kaska N (1988) Untersuchungen über Befruchtungsbiologie und Embryoentwicklung bei einiger Mandelsorten in Çukurova. Doga Bilim Dergisi (in press)
Ferguson L, Davies FS, Ismail MA, Wehaton TA (1984) Growth regulator and low-volume irrigation effects on grapefruit quality and fruit-drop. Scientia Hort 23: 35-40
Goren R, Monselise SP (1971) Effects of ringing on yields of low-bearing orange trees (*Citrus sinensis* L. Osbeck). J Hort Sci 46: 435-441
Gil SG, Cruz CE, Garcia EMG, Martino PA, Diaz ZF (1980) Induccion de color rojo en Manzanas can ethephon. Ciencia Investigacion Agraria 7: 77-88
Hizal AY, Kaska N (1985) Investigations on the flower and fruit drops in Citrus (in Turkish). Derim 2: 3-25
Kamber R, Oguzer V, Yazar A, Dincer D, Kaplankiran M, Sahin H (1988). A research on determination of the water consumption of grapefruit (Marsh Seedless) under Yesilkent Dörtyol light soil conditions (in Turkish). Doga Bilim Dergisi (in press)
Kaska N, Pekmezci M (1974) Effects of 2-chloroethylphosphonic acid (Ethrel) on reducing fruit removal force of Kütahya sour-cherry (in Turkish). UAFA Yearbook 1973, 23: 1-2 and 87-101
Kücükaydin HH, Kaska N (1987) Effect of Alar on the colouring and some quality factors of Amasya appple grown in Nigde and Pozanti district (in Turkish). J Sci Eng 1: 51-60
Manabe K, Humuro M, Ashizawa M (1982) Relationship between flower thinning and leaf number and mineral content and the incidence of fruit dropping in satsumas (in Japanese). Tech Bull Faculty of Agric Kagawa University 33: 165-173
Özbek S (1977) General fruit growing (in Turkish) ÇÜZF Publication No 11, Ankara
Özsan M, Bahçecioglu HR (1970) Investigations on the characteristics of Citrus species and varieties grown in Mediterranean Region under different ecological conditions (in Turkish). TÜBITAK-TOAG Pub No 10:108, Ankara
Paydas S (1988) Effects of different concentrations of Ethrel on the fruit removal force of Kütahya sour-cherry grown in Pozanti (in Turkish) (in preparation)
Stan S, Cotorobai M, Panea T (1984) The use of auxins in fruit thinning and preventing the pre-harvest fruit drop on Cure pear variety. Acta Hort 161: 171-176
Yesiloglu T (1988) Effects of girdling and GA applications on leaf carbohydrates and mineral nutrient composition in Clementine mandarins in relation to fruit yield and quality. Ph D thesis (in Turkish) Doga Bilim Dergisi
Xu E, Tang I, Cherg D, Zhang I, Xie D, Ziang R (1982) Effect of BA and GA on fruit set and quality of Navel orange (in Chinese). Acta Hort Sinico 9: 5-12

POST-HARVEST BERRY DROP AND ITS CONTROL IN CERTAIN GRAPE VARIETIES

F. Ergenoglu
Department of Horticulture
Faculty of Agriculture
University of Çukurova
Adana, 01330, Turkey

INTRODUCTION

The pre- and post-harvest drop of fruit has long been a serious problem. Depending on species, variety and the climate three broad periods of drop have been shown to occur in fruit trees. These have been named "flower drop", "small fruit drop" and "June drop" (see N. Kaska, this volume). The main reasons for these losses, which involve separation between the fruit stalk (pedicel) and branch, or between fruit stalk and fruit, are thought to arise mainly from poor pollination, non-fertilization of the ovules or inadequate nutrition of fruit (Ozbek, 1977). The abscission process itself is regulated by endogenous growth hormones and thus can also be influenced by exogenously applied plant growth regulators. Pull force measuring systems, used to test berry removal force quantitatively, have been used to measure the effect of plant growth regulators applied to control berry shatter (Cooper *et al.*, 1984; Craker and Abeles, 1969; Ağaoğlu and Çelik, 1978). Although fruit drop in apple, pear, peach and citrus causes loss of some yield even in the most heavily cropping years, natural thinning is beneficial economically because it permits the remaining fruits to attain greater size. But in contrast, berry drop of grape ("shatter") reduces quality by spoiling the shape of the cluster. Thus, it diminishes the marketable yield of top quality fruit clusters, leading to a loss of economic return.

One type of shatter, seen in the varieties Kandahor and Rish Baba, is a result of brittle cluster or weak stem. Thus, when they are packed in boxes carelessly or loosely or when the package of grapes is improperly or roughly handled, the clusters break up. The weak pedicels of Thompson Seedless and Monukka also cause shatter in the form detachment of individual berries (Winkler *et al.*, 1974; Fidan *et al.*, 1984). Nickell (1984) has stated that plant growth retarding chemicals can enhance the flexibility of the cluster framework leading to reduced shatter from mechanical stresses during handling.

NATO ASI Series, Vol. H 35
Cell Separation in Plants
Edited by D. J. Osborne and M. B. Jackson

The kind of shatter in which the "brush" is pulled-out from the berry is named "wet drop". The other type of berry drop, in which an abscission layer is formed which separates the berry from the pedicel, ("dry drop") has been shown to occur in the cultivar Waltham Cross (Dattier) in South Africa (Boyes and Villers, 1933) and in the cultivar Muscat of Hambourg in Israel (Lavee, 1959). Lavee (1959), working on dry drop of Muscat of Hambourg and Dabouki grapes concluded that it is physiologically controlled.

As shown by many investigators, the flesh of avocado, olive, lemon (Milborrow, 1967), pear (Rudnicki *et al.*, 1969), sour cherry (Kaska, 1970b, 1971), grape (Niimi *et al.*, 1978; Bhullar and Dhillon, 1979; During *et al.*, 1979; Paleswals *et al.*, 1980) and seeds of apple (Rudnicki, 1969; Kaska, 1970a, sour cherry (Kaska, 1970b), grape (Lott, 1968; Eris and During, 1978; Broquedis and Bouard, 1986, Broquedis, 1987) all contain ABA or ABA-like substances. Their concentrations could be important in the regulation of cell separation in berry drop.

This chapter reports the effects of naphthaline acetic acid (NAA) and and its acetamide (NAAm) on post harvest shatter and on the abscisic acid content of the variety Tarsus Beyazi, which is one of the early maturing grapes grown in large areas of southern Turkey. This white table grape has medium sized and moderately loose clusters of weakly attached berries that mature in about the third week of June depending on the growing site. The attachment of the berries is so weak that one third of them may fall-off when the cluster is shaken. Since this variety is important in the export trade of fresh grapes from Turkey, an attempt has been made to correct this undesirable post-harvest berry drop.

MATERIALS AND METHODS

Fruit Clusters of the variety Tarsus Beyazi (*V. vinifera* L.) were dipped into 50, 100 and 200 ppm solutions of NAA and 10, 50 ppm solutions of NAAm, for one minute, 30, 15, 7 or 3 days before harvest. The samples were separated into flesh and seed before fragmenting or crushing and extracting three times with 80% (v/v) methanol. The extraction, purification and assay procedures were those of Kaska (1970b), Milborrow and Mallaby (1975), Jones and Lacey (1968) and Beyers *et al.* (1972). Penjame-62 wheat was used for the bioassays. Methylated ABA was measured using a gas chromatograph equipped with ^{63}Ni electron capture detector (Schlenk and Gellerman, 1960; Stowe and Schilke, 1964; Powell, 1964.

At harvest, the percentage total soluble solids and total acidity were determined with a hand refractometer and pH meter. In addition, the force required to remove berries from treated and untreated clusters were measured by a Chattillon Dynamometer, sensitive to 5 grams.

RESULTS AND DISCUSSION

The effectiveness of plant growth regulators in reducing berry drop to an acceptable level depends upon the chemical used, its concentration and physiological status of fruit at the time of application. In this work, changes in ABA-like substances of the berry (including seed) extracts were linked to berry adherence as measured with the Dynamometer.

The results of analysis of grape flesh and seeds of the variety Tarsus Beyazi show that activity of ABA or ABA-like inhibitors was contained in zones on thin layer plates with Rf values of 0.65-0.80 and 0.7-0.9 and the amount of inhibitor in the ABA zone of flesh extracts was proportionally higher than from seeds. Treatments with plant growth regulators changed the amounts of inhibitor in both tissues (data not shown) and the pull force required to remove berries (Table 1). Naphthalene acetic acid was more effective than NAAm (Tables 1 and 2).

TABLE 1. Pull force of berries from NAA and NAAm applied to clusters at harvest time

	Pull force (g)	
Treatment	1st year	2nd year
Control	237.00 b*	155.38 b
Auxins (NAA/NAAm)	265.50 a	184.47 a**

Different letters (a, b) indicate significant differences within each column.
(*) p = 0.05, (**) p = 0.01

Although there was no significant difference between the effectiveness of NAA at 100 ppm and 200 ppm in the first year, both concentrations were more effective than 50 ppm NAA. Yet, in the second year NAA applications were less effective and 200 ppm was the most effective concentration.

TABLE 2. Effects of NAA and NAAm applications on the pull force of berries at harvest time

Treatment	Pull force (g)	
	1st year	2nd year
NAA, all concentrations	280.84 b**	192.96 b*
NAAm, all concentrations	242.50 a	171.74 a

Different letters (a, b) indicate significant differences within each column.
(*) p = 0.05, (**) p = 0.01

NAAm applications, which were less effective on increasing the pull force of berries than was NAA applications, showed no significant difference between 10 ppm and 50 ppm at harvest.

TABLE 3. Effects of NAA and NAAm application times on the pull force of berries at harvest

Treatment	Pull force (g)	
	1st year	2nd year
30 days before harvest	265.23 ab	188.27 a
15 days before harvest	274.71 a	193.50 a
7 days before harvest	248.98 c	170.37 b
3 days before harvest	254.08 bc	166.36 b
L.S.D. (0.01)	16.03	10.90

Different letters (a, b) indicate significant differences within each column.

Application time of the growth regulators is important if the maximum benefit is to be achieved. Applications 15 days before harvest were found to be the most effective (Table 3). However the reduction of berry drop by treatments a month before harvest was also found acceptable which is consistent with an earlier report (Rao and Kololgi, 1971).

The first report of applications of plant growth regulators to grapes was made by Pentzer (1941) who attempted to improve berry adherence. He found that 50 ppm NAA applied to several varieties of *V. vinifera*, *V. labrusca* and *V. rotundifolia* either by spraying or dipping 6 days before harvest had little effect. Similar results were also obtained by Lavee (1959), who sprayed bunches and surrounding canopy of Dabouki grapes with 10-20 ppm NAA 10 days after set or 4 days before picking. Although these results support some of our findings, there are other reports that differ from ours. For instance, it was reported that 100 ppm NAA applications 7 and 3 days before harvest reduced berry shattering significantly in Bangalore Blue grapes (Narasimhan *et al.*, 1967) and Aneb-e-Shahi grapes (Rao *et al.*, 1968). These results indicate that different varieties, different species, different locations or years can influence the effect of plant growth regulators. In addition, environmental conditions will also be an important determinant of effectiveness (Ulrich, 1958).

The variation between years seen in this work may be explained by water deficiency in the soil just before harvest (Beyers, 1935; 1937), hot and dry air at the time of maturity (Beyers, 1937), variation in seed number per berry (Beyers, 1936; 1938), or the number of flowers per cluster (Schneider and Staudt, 1978).

Treatment with NAA and to a lesser extent NAAm reduced the content of soluble solids and total acidity of berries. This effect may be associated with a delay in ripening and negatively correlated with the ABA content of fruit.

Our work indicates that to understand more thoroughly the involvement of ABA in both "wet" and "dry" drop of grape fruit, a histological study of berries from the Tarsus Beyazi variety in which adherence was improved by the application of NAA treatments, should be undertaken before this variety disappears from commerce.

REFERENCES

Ağaoğlu YS, Çelik S (1978) An investigation on the method of measuring berry drop and separation force and their use (in Turkish). A U Zir Fak Yilliği 28: 60-71

Beyers E (1935) Resumé of drop berry investigations. Low Temp Res Lab Capetown Ann Rep 1934 5: 114-121

Beyers E (1936) "Drop berry" in Waltham Cross grapes. Low Temp Res Lab Capetown Abb Rep 1935 6: 187-199

Beyers E (1937) Crop berry and desiccation of stalks in Waltham Cross grapes. Low Temp Res Lab Capetown Ann Rep 1936, 7: 91-101
Beyers E (1938) Relationship between "Drop" and seedlessness in Waltham Cross grapes. Low Temp Res Lab Capetown, Ann Rep 1937, 8: 87-90
Beyers RE, Emerson PH, Dostal HC (1972) Effect of SADH on the endogenous level in Redskin peach mesocarp of gibberellin-like material. HortSci 7: 386-387
Bhullar JS, Dhillon BS (1979) Auxin activity in developing berries of Perlette and Anab-e-Shahi grapes. Hort Abs 49/11
Boyes WW, Villers JR (1933) Effect of delayed storage on quality of table grapes. Low Temp Res Lab Capetown Ann Rep 1933, 96-99
Broquedis M, Bouard J (1986) Effect of ringing on the quality of grape seeds, consequences for their abscisic acid contents and their germination potential. Hort Acts Vol 56/2
Broquedis M (1987) Interconversion of β-D-glycopyranosyl abscicate into free abscisic acid during the stratification of grape seeds. Hort Abs Vol 57/10
Cooper WC, Rasmussen GK, Rogers BJ, Recee PC, Henry WH (1968) Control of abscission in agricultural crops and its physiological basis. Plant Physiol 43: 1560-1576
Craker LE, Abeles FB (1969) Abscission: Quantitative measurement with a recording abscissor. Plant Physiol 44: 1139-1143
During H, Alleweldt G, Kock R (1979) Studies on hormonal control of ripening in berries of grapevines. Hort Abs Vol 49/11
Eris A, During H (1978) Determining the effect of stratification on ABA level of Muscat of Hambourg seeds by HPLC (in Turkish). A U Zur Fak Yilliği 1977. Cilt 27 (3/4) S 490-498
Fidan Y, Ağaoğlu YS, Çelik S, Alleweldt G, During H (1984) Untersuchungen über die ursachen und die vorbeugung des abbeerens bet (*Vitis vinifera* L.) sorten. Forschung zur Entwicklung der Turkischen Landwirthschaft-Deutsch-Türkisches Symposium v. 19.-23.11.1980 in Ankara Herausgeber: Die Partnerschaftausschüsse der Landwirtschaftlichen Fakultäten Göttingen und Ankara, Göttingen 1984
Jones OP, Lacey HJ (1968) Gibberellin-like substances in the transpiration stream of apple and pear trees. J Exp Bot 19: 526-531
Kaska N (1970a) Changes in the levels of growth promoters and inhibitors in apple leaves, buds, fruits and seeds during the growing period. Univ. Ank Yearbook of Fac of Agric. 1969, pp 67-83
Kaska N (1970b) Investigations on the changes of ABA content of the seeds from Apricot and Kütahya sour cherry during stratification (in Turkish). A U Zir Fak Yayinlari 431 s 104
Kaska N (1971) Investigations on the endogenous growth regulators in the sour cherry (in Turkish). A U Zir Fak Yilliği 1970, s 579-596
Lavee S (1959) Physiological aspects of post harvest berry drop in certain grape varieties. Vitis 2: 34-49
Lott H (1968) Uber den machweis von ascisinsaurs in samen von reben. Vitis 7: 221-222
Milborrow BW (1967) The identification of (+)-abscisin II [(+)-dormin] in plants and measurement of its concentrations. Planta 76: 93-113
Milborrow BW, Mallaby R (1975) Occurrence of methyl (+)-abscisate as an artifact of extraction. J Exp Bot 26: 741-748
Narasimham P, Rao M, Nagaraja N, Anandaswamy B (1967) Effect of pre-harvest application of growth regulators on the control of berry drop in Bangalore Blue grapes (*V. labrusca* L.). J Food Sci Tech 4: 162-164
Nickell LG (1984) Plant growth regulating chemicals. Vol 1 CRC Press, Boca Raton, p 183
Niimi Y, Ohkawa M, Torikata H (1978) Changes in auxin and abscisic acid-like activities in grape berries. Hort Abs Vol 48/5

Ozbek AS (1977) Fruit drops. In: General fruit growing (in Turkish) Ç U Zir Fak Yayinlari 111 Ders kitabi 6, Ankara, s 386

Paleswala VA, Parikh HR, Modi VV (1986) The role of abscisic acid in the ripening of grapes. Hort Abs Vol 65/4

Pentzer WT (1941) Studies on the shatter of grapes with special reference to the use of solutions of naphthalene acetic acid to prevent it. Proc. Amer Soc Hort Sci 38: 397-400

Powell LE (1964) Preparation of indole extracts from plants for gas chromatography and spectrophotofluorimetry. Plant Physiol 39: 836-842

Rao MM, Narasimham N, Nagraja N, Anandaswamy B (1968) Effect of pre-harvest spray of α-naphthalene acetic acid and para-chlorophenoxyacetic acid on control of berry drop in Aneb-e-Shahi grapes (*V. vinifera* L.). J. Food Sci Tech 5: 127-128

Rao MM, Kololgi SD (1971) Further studies on the control of post-harvest berry drop in Aneb-e-Shahi grapes (*V. vinifera* L.). Indian Food Packer 25: 67-69

Rudnicki RJ, Pieniazek J, Pieniazek N (1968) Abscisic II in strawberry plants at two different stages of growth. Bull Acad Polon Sci 16: 127-130

Rudnicki R (1969) Studies on abscisic acid in apple seeds. Planta 86: 63-68

Schlenk H, Gellerman JL (1960) Esterification of fatty acids with diazomethane on a small scale. Anal Chem 32: 1412-1414

Schneider W, Staudt G (1978) The relationship between berry drop in *Vitis vinifera* and environment and genome. Vitis 17: 45-53

Stowe BB, Schilke JF (1964) Submicrogram identification and analysis of indole auxins by gas chromatography and spectrophotofluorimetry. In: Regulateurs naturels de la crossance végétale. Intern Centre Natl Rech Sci, Paris

Ulrich R (1958) Post harvest physiology of fruits. Ann Rev Plant Physiol 9: 385-416

Winkler AJ, Cook JA, Kliewer WM, Lider AA (1974) General viticulture. Univ. Calif Press, Berkeley Los Angeles London, p 710

OLIVE INFLORESCENCE, FLOWER, FRUIT AND LEAF ABSCISSION WITH CHEMICALS USED FOR MECHANICAL HARVEST

George C. Martin
Department of Pomology
University of California
Davis, California 95616
U.S.A.

INTRODUCTION

All crops are harvested when judged horticulturally mature, a stage of development that meets an agreed-upon standard established in the market place. For the California olive fruit, horticultural maturity occurs about four months prior to physiological maturity. By harvesting early, olive fruit can be processed by the California black-ripe or green-ripe procedures to achieve the defined industry standard. Recent olive harvest timing research shows that there is a 14-day period in October for harvest when maximum crop value can be obtained (Sibbett *et al.*, 1986).

Growers eager to take advantage of this information have been frustrated by the slow rate of the olive harvest using hand labour. The best laborers can harvest 230 kg of olive fruit per day. Good orchards will average about 6.6 metric tons of olive fruit per hectare; thus, in this example 29 man-days are required to harvest a hectare of fruit. In the case of California, the defined economic unit is a size of orchard that can support and be managed by a family of four. This is about 40 hectares. It would require 29 workers working 40 days to complete harvest. These estimates mean that harvest would exceed the period for maximum return by at least 26 days. Many growers are able to complete harvest in about 21 days, but, as shown by Sibbett *et al.* (1986), they still loose much crop value.

Recent developments in labour management and availability in the USA make it less likely that olive growers can achieve optimum harvest by hand. Research in California has shown that mechanical devices that position people in the orchard - such as picking platforms - are not

NATO ASI Series, Vol. H 35
Cell Separation in Plants
Edited by D. J. Osborne and M. B. Jackson

satisfactory choices for improving harvest rates. Fridley (1969) shows that these devices improve the picking speed of the slowest workers only, and that the best workers, however, may be slowed down. The latter situation occurs because the picking aid (platform, ramps, etc.) moves at the rate of the slowest picker. Well motivated workers can do better without the harvest aid. Crew matching, therefore, becomes critical. Although there is little or no interest among olive growers in harvest aids, there is great interest in mechanical harvest. The reasons for grower interest in mechanical olive harvest are predictable. McKibben (1953) analyzed the history of mechanical harvest in the USA and reveals many factors contributing to such interest, among them are the high cost of and the shortage of labour. Currently, the olive industry has both of these problems, and this has already motivated a few growers to change to mechanical olive harvesting.

The specific problem of maximum chemical-induced fruit loosening with minimal leaf loss has necessitated investigations into the abscission process for many organs, from inflorescences and flowers, to fruit, and to leaves.

THE ABSCISSION PROCESS

The earliest published observations on olive leaf abscission are those of Hewitt (1938). His work identified entry points for olive knot and did not concern anatomical details. However, Polito and Lavee (1980) studied the anatomy of leaf abscission following treatment with the ERC 2-chloroethylphosphonic acid (ethephon). In this study, abscission zones from water-treated controls were directly compared to those treated with ethephon.

External examination reveals a clearly defined abscission zone at the base of mature olive leaves. Polito and Lavee (1980) describe leaf abscission events as similar to those for other crops. For olive, ethephon application led to onset of separation within 36 to 60 hours following treatment. In another 8 to 12 hours, separation was evident in the abaxial cortical cells adjacent to the vascular system, with separation proceeding from these interior cells toward the epidermis. Next, separation in the adaxial side occurred in the same manner, that is, from inside-out. The last point of separation occurred in the vascular region and the epidermal cells of the region of indentation that can be seen at the leaf petiole base. During the abscission process there was no cell division in the abscission zone, which contained from 10 to 12 rows of isodiametric cortical parenchyma cells. The line of separation during abscission was marked by cell wall swelling and dissolution of the middle lamella.

Histochemical examination of the olive leaf abscission zone by Polito and Lavee (1980) reveals the following: 1) starch localization in abundance near the vascular region, with no evident changes in starch content during abscission; 2) a noteworthy increase in cell protein content prior to and following abscission; 3) a loss of intercellular pectic substances; 4) little loss of cellulose; 5) no tylose formation, and 6) no lignification of abscission zone cells.

Olive fruit abscission represents an even more complicated process (Reed and Hartmann, 1976). These authors describe two abscission zones, namely, at the peduncle:pedicel junction and at the pedicel:fruit junction. The peduncle:pedicel abscission zone is much like that of the leaf: it has an easily identifiable separation zone with isodiametric cells which are smaller than the adjacent elongated cells. In contrast, prior to abscission the pedicel:fruit junction does not contain an identifiable zone of abscission layer cells.

To induce fruit abscission, Reed and Hartman (1976) used the ERC 2-chloroethyl-tris-(2-methoxyethoxy)-silane (Alsol). With Alsol treatment, fruit separation occurred most frequently in the pedicel:fruit zone, but there was abscission at both junctions, i.e. peduncle:pedicel and pedicel:fruit. The separation process they describe for fruit is, in general, similar to that for olive leaves. Cell plasmolysis occurs first in the pedicel pith and cortical regions and then moves toward the phloem. However, there is no xylem disruption evident. Closer to the detachment event, there was both loss of pectin and cell wall polysaccharides and presence of starch grains. Cell wall and middle lamella disintegration occurred as plasmolysis proceeded. No cell division was evident in the abscission zone during separation. These authors mentioned that two other ERC's, cycloheximide and ethephon, effected an identical anatomical separation process.

It is, however, the olive inflorescence that contains the most complex assembly of abscission zones formed in response to ethephon treatment or the natural separation process. Weis *et al.* (1988) describe eight abscission zones in the inflorescence (Fig. 1). These were identified by subjecting sampled inflorescence to ethephon or ethylene gas treatment weekly beginning eight weeks before and until the time of anthesis. Samples of abscission zones were taken for anatomical investigations. As was the case in the leaf, the abscission zones of the inflorescence axis are preformed early in their development. The general cellular characteristics of inflorescences are similar to those of

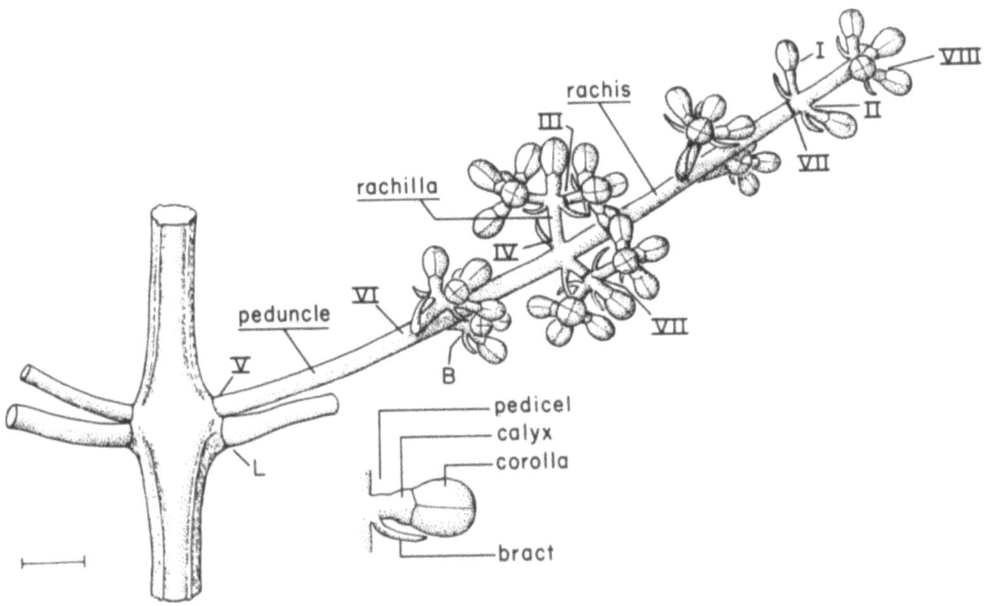

FIGURE 1. Morphology of the olive inflorescence (idealized) showing all possible patterns of branching. Examples of all possible abscission zones types are indicated. Inset shows olive floral morphology (inflorescence abscission zones I-VIII, bract B, leaf L). Scale bar = 0.5 cm. Taken from Weis et al., 1988.

leaves: abscission zone cells had small cell size and dense cytoplasm and there were differences in cell wall staining when compared to cells outside this zone (Weis, personal communication). Whereas the leaf abscission zone is only from two to five cells wide and distinct, the inflorescence abscission zone is many cells wide and diffuse.

The separation characteristics of the inflorescence, whether induced naturally or by ethephon, are similar to those of leaves. As nearly as can be determined by the histochemical evidence, leaf, inflorescence, flower or fruit abscission as induced by ERC, is similar to that induced naturally. Treatment with ERC or ethylene gas will of course accelerate the onset of separation. Both Lavee and Martin (1981b) and Weis et al. (1988) show that inflorescence abscission occurs after that of leaves when experimental units are treated with ethephon. The reason for this difference in response rate is not clear. Lang and Martin (1987) find that fruit abscise at approximately half the concentration of ethylene gas required for leaf abscission. When treated with ethylene gas, rather than with ethephon fed through the cut stem, Weis et al. (1988) found inflorescence abscission occurred before that in leaves. These apparent contradictory findings are difficult to resolve without determining the amount of ethylene reaching the abscission site. From a practical standpoint, there is a more rapid abscission response from fruit than there is from leaves when ethephon is applied in the field. Field treatments can be compromised by many factors, including: 1) the environmental conditions at treatment time, 2) the length and the dosage of ethylene exposure, and 3) the type of ERC and its ethylene releasing characteristics. These will be covered next.

CHARACTERISTICS OF ETHYLENE ACTION

Environment. When used under field conditions, ERC's result in variable plant responses. The variability is probably the result of temperature and moisture conditions and possibly the result of O_2 and CO_2 concentrations and light conditions. As for this last factor, fruit are either in the light or shade. While light may be a subtle affecting factor, as was shown with cherry explants by Wittenbach and Bukovac (1973), it probably does not exert a major influence on abscission in the field. Similarly, while low-O_2 (0%) and high-CO_2 (5%) concentrations can reduce abscission *in vitro*, these conditions are irrelevant under field conditions (Carnes *et al.*, 1951, Wittenbach and Bukovac, 1973). In contrast to the situations with light and gas concentration, temperature and moisture conditions play major roles in the action of ERC's. Klein *et al.* (1978) show that at 95% RH, ethephon decomposition rate is reduced, compared to lower RH even when temperature was increased from 20° to 30°C. However, when kept at 37% or 70% RH -- conditions common in the field -- the ethephon decomposition rate increased as temperature was increased from 20° to 50°C.

Tree moisture status at ERC treatment time is also important (Hartmann et al., 1972; Lavee and Haskal, 1976; Martin *et al.*, 1981). These authors show that ERC's induce more fruit removal with less leaf loss when the treated orchard is at the end of the normal irrigation cycle. This does not necessarily imply that the trees are stressed but, rather, that irrigation is due. Careful laboratory studies and field treatment of non-irrigated trees by Klein *et al.* (1978) corroborates the previously mentioned field research. These authors report more fruit abscission with modest leaf loss in non-irrigated than in irrigated trees when treated with ethephon. In the laboratory, they show that partially desiccated olive shoots take up more ^{14}C-ethephon and translocate less of it than do similar shoots with sufficient moisture. In this laboratory system, the fruit contained a similar amount of ^{14}C-ethephon whether partially desiccated or moist; however, the rest of the shoot, when desiccated, contained about half the ^{14}C-ethephon when compared to a similar, but moist, shoot. These results show that moist shoots distribute ^{14}C-ethephon that may reach target leaf cells better and more easily than is the case in desiccated shoots.

My own experience in the field has shown that ethephon application provides the greatest fruit separation with the least leaf loss when the temperature is between 16° and 30°C and when trees require irrigation (Martin *et al.*, 1986). All of these data corroborate the importance of the amount and duration of ethylene exposure at the abscission zone.

Ethylene Exposure Time and Concentration. For experimental purposes, excised olive shoots fed an ERC via the stem are convenient for study under controlled laboratory studies. No experimental laboratory system can completely approximate conditions in the field; however, without use of the excised shoot system, experimental study of olive abscission is difficult -- if not impossible -- and is compromised by the variable and uncontrolled environmental factors previously discussed. In spite of the potential and real shortcomings involved in using the excised olive shoot system, in studies where direct comparisons were made with field treatments, results were remarkably similar (Lang and Martin, 1985).

Using the excised olive shoot system, Lavee and Martin (1981a) made several comparisons of leaf response to ethephon treatment. These authors showed a time and concentration-dependent effect in use of ethephon, regardless of leaf location. The leaves that were first to begin significant release of ethylene gas following ethephon treatment were the first to abscise. Whereas 90% leaf abscission was induced with 2 hours of an ethephon pulse at 250 mg l^{-1} only 1 hour was required at 500 mg l^{-1}. In these experiments there was a minimum induction period following treatment of 50 hours that led to leaf abscission. They reported that untreated leaves in the laboratory system were stable and showed little or no ethylene evolution for about seven days and that stem feeding ethephon offset the variability found in use of foliar sprays. In an interpretation of results, these authors suggest that ethephon arrived at the leaves long before decomposition, with subsequent release of ethylene. Verification of their logic begs further experimentation, some of which is underway in my laboratory.

In subsequent experiments, Lavee and Martin (1981b) compared leaf, inflorescence, and fruit abscission following ethephon treatment. These authors were the first to show that leaves are actually more responsive to ethephon-induced abscission than are fruit when fed through the cut stems. In field treatments of ethephon foliar sprays, fruit abscission is always greater than leaf abscision. The former may result from more ethephon collecting at the fruit than on the leaf. Still, in either foliar- or stem-fed treatments, we do not know the exact concentration of the chemical at the abscission zone. These authors also showed that the amount of ethylene evolution from organs following ethephon stem-feeding treatment does not indicate abscission rate. Leaves and inflorescences evolved similar amounts of ethylene, but there was far greater leaf abscission than there was inflorescence abscission. Fruit evolved low amounts of ethylene and showed variable abscission response.

In an effort to improve the excised olive shoot system for experimental use, Lang and Martin (1985) compared droplet application in the pedicel:fruit cavity on excised shoots in the laboratory with the same treatment to unaltered shoots in the field. The objective was to eliminate the uptake and transport variable inherent in delivery of treatments via cut stems in the excised shoot system. A 2.5 μl droplet of 69 mM ethephon applied to the pedicel:fruit cavity led to similar abscission response curves for field or laboratory treatments. While abscission occurred earlier in the laboratory it was thought that this was due to the constant high temperature about 25°C. In the field, this temperature was achieved for only about 8 hours per day with much cooler nights. These results agree with those of Klein *et al.* (1978), who illustrated an increased rate of ethylene release from ethephon with increase in temperature.

Conjecture has led many to believe that release characteristics from ERC's could be important with respect to leaf vs. fruit abscission comparisons. Using the droplet method, Lang and Martin (1985) measured ethylene release from both CGA-15281 and ethephon. CGA-15281 gave maximum ethylene peaks when the first measurement was made 2 hours after treatment, whereas ethephon achieved its maximum ethylene release 13 to 18 hours after treatment. Fruit abscision from CGA-15281 began 7 to 12 hours after treatment, whereas for ethephon it began 19 to 25 hours after treatment. These data substantiate ERC field trials which suggest that large quantities of ethylene applied for a short time might enhance fruit abscission but minimize leaf abscission (Ben-Tal and Lavee, 1976; Lang, 1983; Martin *et al.*, 1981). A time-concentration constraint does exist, as laboratory work by Lavee and Martin (1981c) using ACC shows that very short ethylene bursts of saturating quantities do not induce leaf abscission.

The above data highlight the importance of the duration and concentration of ethylene to induce differential abscission of fruit and leaves. Lang and Martin (1989) designed a test system to quantify the dosage and time of ethylene gas exposure which would define the range of fruit and leaf abscission. In their system, olive shoots with fruit were positioned with cut bases in vials of water and then placed in 10-l jars through which ethylene gas or pure air could flow. Over 100 combinations of ethylene gas concentrations and durations were tested for leaf and fruit abscission. These data were assembled to construct three-dimensional response surface plots. Ethylene concentrations greater than 10 $\mu l\ l^{-1}$ for more than 20 hours led to both excessive leaf and fruit abscission. A working range of 4 to 7 $\mu l\ l^{-1}$ ethylene gave the highest fruit abscission with the lowest leaf abscission. For optimizing fruit abscission above 90%, greater than 3 $\mu l\ l^{-1}$ ethylene are required for at least 28 hours. To restrict leaf abscission to below 25%, the ethylene content should be below 7.5 $\mu l\ l^{-1}$ for less than 38 hours. From these data, an ideal three-dimensional response surface plot was calculated showing fruit/leaf (F/L) abscission ratios greater than 3.6 (i.e., 90% or greater fruit abscission divided by 25% or less leaf abscission). The most favourable combinations resulted in F/L ratios up to 13.3 and occurred with ethylene doses of from 3 to 5 $\mu l\ l^{-1}$ for 28 to 34 hours (Fig. 2).

Ethylene Pulse Treatments. In the field, ethylene release from ERC is affected by temperature (Klein *et al.*, 1978; Lang and Martin, 1985). The results from continuous ethylene treatment mentioned above, while useful, do not represent the conditions of ethylene release in the field, where because of lower temperature much less ethylene is released at night. Thus, the system used by Lang and Martin (1989) and described above was used, but, instead of continuous ethylene exposure, pulses of ethylene were delivered. The protocol involved 8 to 16 hours of ethylene, followed by air for 12 to 20 hours, ethylene for 12 to 32 hours followed by leaf and fruit abscission determination. The pulse treatment approach allowed for longer periods of ethylene exposure with less leaf abscission and thus far superior F/L ratios than was the case with continuous ethylene treatment (Fig. 3). These preliminary experiments make evident the possibilities of "tailoring" an ERC with certain release characteristics. Resulting in about 98% fruit abscission with less 6% leaf abscission, the ideal protocol was 8 hours ethylene, 16 hours air, followed by an additional 24 to 32 hours of ethylene (Fig. 3). The most successful combinations of ethylene concentration and time of exposure had similar results, whether using continuous ethylene or pulse ethylene treatments;

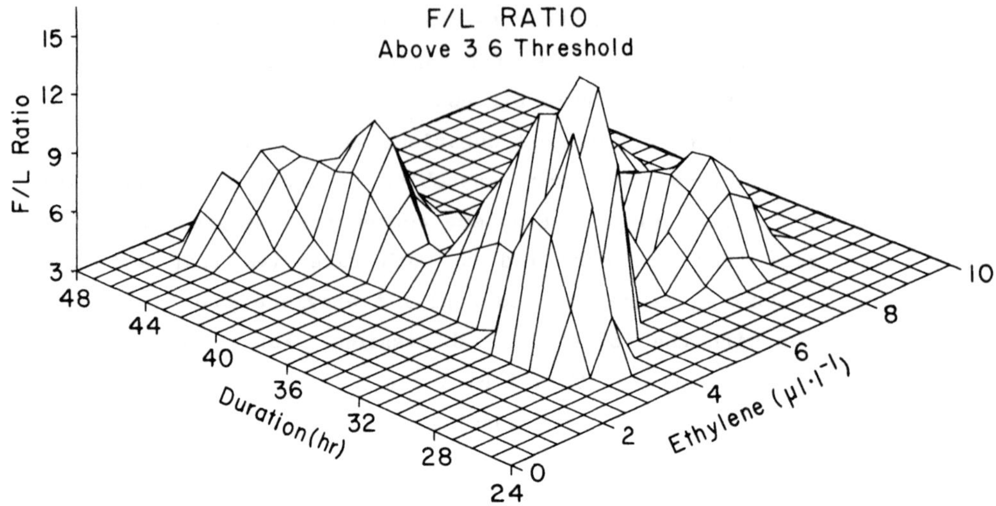

FIGURE 2. Olive fruit-to-leaf (F/L) abscission ratio response surface to ethylene concentration and duration: portion of surface depicting doses which induce F/L ratios higher than 3.6. Taken from Lang and Martin, 1989.

however, the optimum with maximum fruit abscission and minimum leaf abscission was achieved with pulse treatments. Compare Fig. 2 with Fig. 3.

Enhanced Ethylene Production and Leaf Abscission. Several scientists have shown ethephon to generate two peaks of ethylene (Ben-Tal and Lavee, 1976; Hartmann *et al.*, 1970; Lang and Martin, 1987). It has been suggested that the second ethylene peak is autocatalytic and arises from the initial release of ethylene from ethephon (Ben-Tal and Lavee, 1976; Daniell and Wilkinson, 1972). As well, Lang and Martin (1987) showed that pulse treatments of ethylene have physiological significance (discussed above). It was of interest to determine the influence of

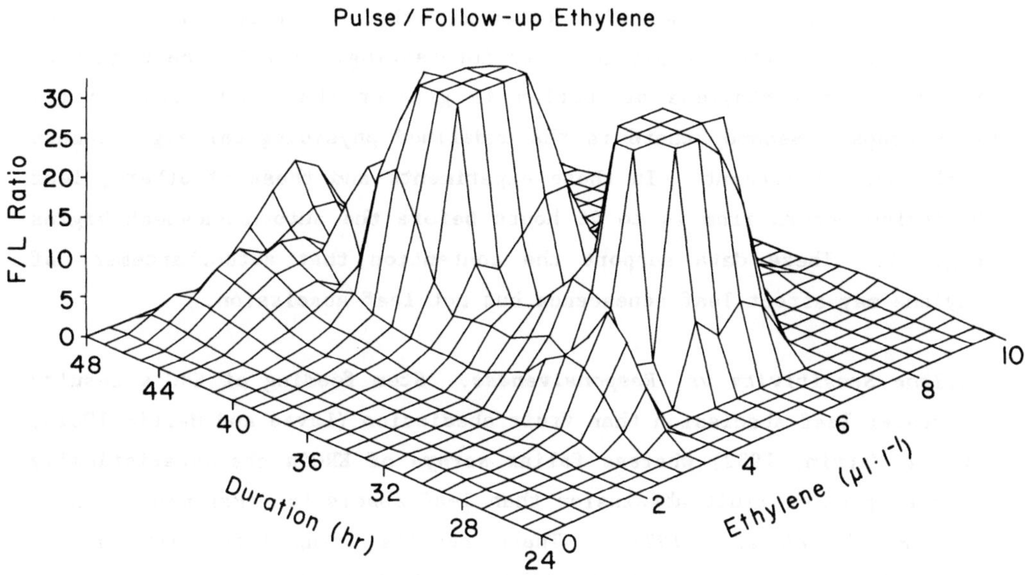

FIGURE 3. Response surface model of olive fruit-to-leaf abscission ratios. The effect of pulse/follow-up ethylene treatments. Taken from Lang and Martin, 1987.

external ethylene on subsequent endogenous ethylene production and whether regulation of leaf abscission is involved. To that end Goren *et al.* (1988) made a test using the excised olive shoot system in containers with adjusted flow of air or ethylene gas. In the case of fruit they found, as have others, low natural ethylene production. When fruit were treated with ethylene the enhanced ethylene production resulted from external ethylene trapped in the fruit; thus, there was no autoenhancement of ethylene evolution in fruit. By contrast, leaves, when supplied exogenous ethylene, showed autoinhibition of ethylene production during the first 24 to 48 hours but autoenhancement of ethylene by 72 hours. The autoenhancement of ethylene production by leaves was 100- to 400-fold, irrespective of leaf age or time of year when supplied. The autoenhancement of ethylene production in intact or detached olive leaves was positively correlated with external ethylene concentration. Two aspects of the work are particularly interesting. First, the magnitude of autoenhanced ethylene production is greater than that reported for other crops. Second, there is the undefined physiological significance of the autoenhancement. In these experiments and those of others, leaf abscission occurs from 24 to 48 hours before the autoenhancement begins (Fig. 4). These data support the contention that autoenhancement of ethylene may affect leaf senescence but not leaf abscission.

Ethylene Sensitivity vs. Responsiveness. Stem feeding of ERC's results in greater leaf abscission than fruit abscission (Lavee and Martin 1981b; Sun and Martin, 1982) whereas foliar sprays of ERC's characteristically leads to greater fruit abscission than leaf abscission (Hartmann *et al.*, 1970; Martin *et al.*, 1981). These results bring into question the sensitivity and/or responsiveness of treated olive organs.

Foliar Spray vs. Stem-Fed Treatments. Sun and Martin (1982) showed that 600 ppm CGA-15281 in foliar spray resulted in 100% fruit abscision and 7% leaf abscission. When fed through the stem the same concentration of CGA-15281 resulted in 38% fruit abscission and 26% leaf abscission. Lavee and Martin (1981b) found stem feeding of ethephon to induce 100% leaf abscission but as low as 5% fruit abscission. Perhaps, with either ethephon or CGA-15281, the spray treatment results in collection of treatment chemical in the pedicel:fruit cavity, whereas no such collection basin exists in olive leaves. Confirmation of this possibility came from treatments with 2.5-μl droplets of 69-mM ethephon in the pedicel:fruit cavity of olive, which induced 100% abscission (Lang and Martin, 1985).

The issue of sensitivity to ethylene is not resolved with these data. Whether by foliar spray or stem feeding, the amount of ethylene exposure at the abscission zone cannot yet be determined. The ethylene gas treatments show separation starts first in fruit, well ahead of that in leaves; still, it may be that the gas penetrates to the fruit abscission zone faster than it does to that of leaves. Certainly the fruit are more responsive to ethylene gas or ERC spray than are leaves.

Currently we are working on a method to determine where ethephon is transported following stem feeding. Perhaps with this procedure we can gather data to determine the comparative ethylene sensitivity between olive leaves and fruit.

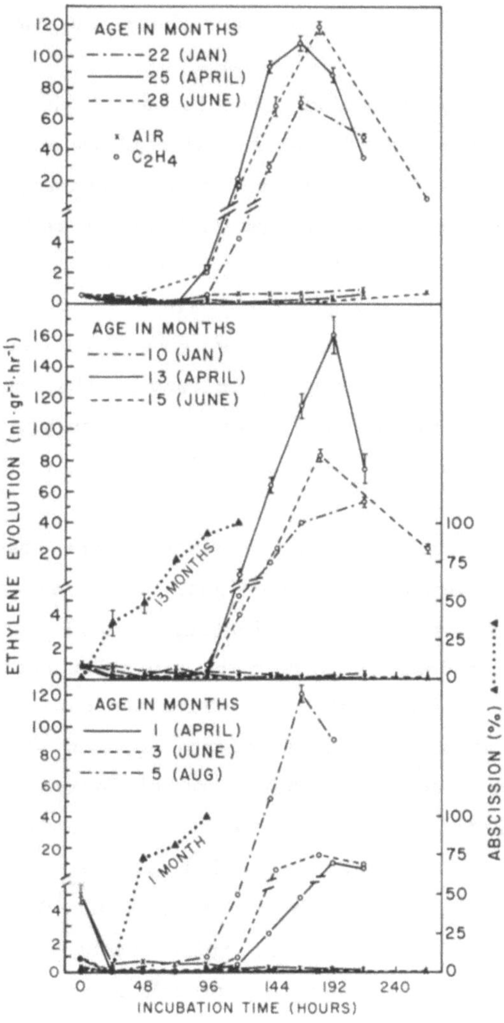

FIGURE 4. Endogenous ethylene evolution from olive leaves and abscission of leaves at different ages following exogenous ethylene treatment at 10 $\mu l\ l^{-1}$. Error bars ± SE of the mean (n = 12). Taken from Goren et al., 1988.

Future Directions for Olive Abscission Research. We are at a stage in olive abscission research where the search for a "good idea" remains most important. I do not discourage empirical studies where candidate abscission chemicals are tested; however, 40 years worth of these studies has given us precious little useful information. Firstly, we need to understand the plant better. However, some factors involved in mechanical harvest of olive are already clear, and these are summarized below.

Olive Abscission Recapitulation. The forces which dictate mechanization are the combined problems of labour availability and economics. We have no control over these factors, which override reason, desire or politics. The principles of mechanization are well documented and require no more than touch-up research -- a tailoring of solution to application. These mechanization mechanics demand a certain tree form, which can be described in detail by engineers. The tree form would transmit shaking energy in absolute terms at high efficiency to the fruit -- presto, fruit separation. Proper management, however, can only slightly modify tree form, and geneticists have even less influence on tree form at this time.

The problem of fruit separation rests in the abscission zone. Here we find a mystery wrapped in an enigma. Abscission research the world over has unveiled one of the fascinating mysteries of life. But romanticizing abscission does not answer the central question, namely, what is the mechanism of abscission? For olive we can describe some of the anatomy of separation and can elucidate aspects of its histochemistry. For instance, we know that there are preformed abscission zones of similar configuration on the leaf and at the peduncle:pedicel junction of the inflorescence and that separation events for both are similar, whether occurring naturally or by ERC treatment. Even so, study of these abscission zones and the similar, inflorescence and flower abscission zones have not revealed a research approach to exploit.

Ethylene is surely a major feature of the abscission process. Application of ethylene via ERC's provides a possible solution one to which we cling. Still, the problem remains that ERC treatment, although resulting in excellent fruit removal also leads to excessive leaf loss. The leaf loss results in outbreaks of olive knot and reduced cropping.

We have learned a great deal about ERC's and how ethylene release is affected by moisture and temperature. We have learned that models for ethylene exposure for specific periods and at prescribed concentrations can provide us with a solution for complete fruit separation with negligible leaf loss. However, perhaps such solutions are no more practical than those of the engineers, who model optimum olive tree form or maximum shaker energy conduction. So where do we go next?

Abscission Research Directions. The choices should be apparent. We must design the proper tree form to conduct shaker energy, make the tree behave by releasing its fruit on schedule, and design an ERC with specific ethylene release characteristics. Are there other choices? Probably many others, but what are they? Of the choices we have, which should be investigated and supported with priority over the others?

Restructuring the olive tree seems to be a formidable task and a project for genetic engineering in the 21st century. We can start on that journey by supporting basic studies in genetic engineering. In this area, gene trait identification, gene manipulation, and controlled gene expression must be perfected. Then regeneration of whole plants from genetically altered cells must be under control. These problems, while daunting, will yield to time -- a great deal of time -- and to research effort - a great deal of effort.

Making the plant release its fruit at a prescribed time may actually be a part of the previous choice. Here, instead of restructuring the entire tree, we are interested in controlling the tree's physiology, the physiology of separation. To control abscission artificially we must understand how it works. The choice here is clear. In spite of all our blundering, we <u>must</u> continue to research the abscission process. But how? Perhaps a fresh outlook is needed. The history of molecular biology shows us that the teaming of natural products chemists, physicists, and physical chemists of the 1930's and 1940's worked toward cracking the genetic code. They entered the research with a fresh approach as it was a different area of research for them. Perhaps the time is right for a fresh approach in abscission physiology. Lacking this, we must continue our quest with what we have.

The final choice in controlling abscission intrigues me. The dosage-time model previously described (Figs. 2,3) outline the conditions for ethylene release that effect maximum fruit removal with minimum leaf loss. I believe an unrelenting effort should be mounted to tailor an ERC which can behave consistently over several prescribed parameters of temperature, light, RH and pH. The technology to meet these requirements is available. The chemical tailoring solution may cost more than chemical companies are willing to invest.

Controlling abscission seems to be an overwhelming task as we now begin to understand the problem. Yet, we have now entered a stage of research capability when we have greater means for finding solutions. The time is ripe and the future bright in abscission control research. As we gradually peel back the layers of our ignorance concerning plant growth, the solution will appear. Time coupled with continued research effort will prevail.

REFERENCES

Ben-Tal Y, Lavee S (1976) Increasing the effectiveness of ethephon for olive harvesting. HortSci 11:489-490

Carnes HR, Addicott FT, Lynch RS (1951) Some effects of water and oxygen on abscission in vitro. Plant Physiol 26:629-630

Daniell JW, Wilkinson RE (1972) Effect of ethephon-induced ethylene on abscission of leaves and fruits of peaches. J Am Soc Hort Sci 97:682-685

Fridley RB (1969) Tree fruit and grape harvest mechanization progress and problems. HortSci 4:235-237

Goren R, Nishijima C, Martin GC (1988) Effects of external ethylene on the production of endogenous ethylene in olive leaf tissue. J Am Soc Hort Sci 113:778-783

Hartmann HT, Tombesi A, Whisler J (1970) Promotion of ethylene evolution and fruit abscission in the olive by 2-chlorethanephosphonic acid and cycloheximide. J Am Soc Hort Sci 95:635-640

Hartmann HT, El-Hamady M, Whisler J (1972) Abscission induction in the olive by cycloheximide. J Am Soc Hort Sci 97:781-785

Hewitt WB (1938) Leaf-scar infection in relation to the olive-knot disease. Hilgardia 12:41-71

Klein I, Epstein E, Lavee S, Ben-Tal Y (1978) Environmental factors affecting ethephon in olive. Scientia Hortic 9:21-30

Lang GA (1983) Ethylene-induced olive (Olea europaea L) organ abscission: a system to model fruit and leaf response to ethephon and CGA-15281 preharvest sprays. MS Thesis, Univ of California, Davis

Lang GA, Martin GC (1985) Ethylene-releasing compounds and the laboratory modelling of olive fruit abscission versus ethylene release. J Am Soc Hort Sci 110:207-211

Lang GA, Martin GC (1987) Ethylene-induced olive organ abscission: ethylene pulse treatments improve fruit-to-leaf abscission ratios. Acta Hort 210:43-52

Lang GA, Martin GC (1989) Olive organ abscission: fruit and leaf response to applied ethylene. J Am Soc Hort Sci 114:134-138

Lavee S, Haskel A (1976) Further field studies of the mode of application and efficiency of various ethylene-releasing compounds to facilitate olive fruit harvest. Riv Ortoflorofruttic Ital 60:166-175

Lavee S, Martin GC (1981a) In vitro studies on ethephon-induced abscission in olive. I The effect of application period and concentration on uptake, ethylene evolution, and leaf abscission. J Am Soc Hort Sci 106:14-18

Lavee S, Martin GC (1981b) In vitro studies of ethephon-induced abscission in olive. II The relation between ethylene evolution and abscission of various organs. J Am Soc Hort Sci 106:19-26

Lavee S, Martin GC (1981c) Ethylene evolution following treatment with 1-aminocyclopropane-1-carboxylic acid and ethephon in an in vivo olive shoot system in relation to leaf abscission. Plant Physiol 67:1204-1207

Martin GC, Lavee S, Sibbett GS (1981) Chemical loosening agents to assist mechanical harvest of olive. J Am Soc Hort Sci 106:325-330

Martin GC (1986) Olive harvest in California, United States of America. Olivae 3:11-20

McKibben EG (1953) The evolution of farm implements and machines. Agri Engineer 34:91-93

Polito VS, Lavee S (1980) Anatomical and histochemical aspects of ethephon-induced leaf abscission in olive (Olea europaea L). Bot Gaz 141:413-417

Polito VS, Stallman V (1981) Localized cell growth in ethephon-treated olive leaf abscission zone. Scientia Hortic 15:341-347

Reed NR, Hartmann HT (1976) Histochemical and ultrastructural studies of fruit abscission in olive after treatment with 2-chloroethyl-tris-(2-methoxyethoxy)-silane. J Am Soc Hort Sci 101:633-637

Sibbett GS, Freeman MW, Ferguson L, Anderson D, Welch G (1986) Timing Manzanillo olive harvest for maximum profit. Calif Agric 40:19-22

Sun FZ, Martin GC (1982) Evolution of (2-chloroethyl)methylbis(phenylmethoxy)silane (CGA-15281) as a chemical fruit abscising agent for olive using detached shoots. HortSci 17:957-958

Weis KG, Goren R, Martin GC, Webster BD (1988) Leaf and inflorescence abscission in olive. I Regionation by ethylene and ethephon. Bot Gaz (in press)

Wittenbach VA, Bukovac MJ (1973) Cherry fruit abscission: effect of growth substances, metabolic inhibitors and environmental factors. J Am Soc Hort Sci 98:348-351

SECTION VII

Additional Contributions

PUTATIVE ETHYLENE BINDING PROTEIN(S) FROM ABSCISSION ZONES OF PHASEOLUS VULGARIS

Cathal P. Connern, A.R. Smith R. Turner and M.A. Hall
Department of Biological Sciences
University College of Wales
Aberystwyth, Dyfed, SY 23 3DA
Wales, U.K.

INTRODUCTION

Ethylene binding sites exist in abscission zones of *Phaseolus vulgaris* (Hall *et al.*, 1987). However, attempts to characterise these binding sites have been unsuccessful due to both the low concentrations of binding sites present in a zone two cells thick and the low maximum specificity of ^{14}C-ethylene. An alternative approach has now been taken using an ethylene binding protein (EBP) purified from developing cotyledons of *Phaseolus vulgaris* as previously described by Thomas *et al.* (1985) and Williams (1988). Polyclonal antibodies (Anti-EBP) were raised against the EBP and characterised in terms of immune response, sensitivity and specificity (Western blotting) using a colloidal gold labelled goat-anti-rabbit secondary antibody. Antibodies, specific to the EBP were used to probe abscission zones from the same tissue. In this paper, evidence will be presented for the existence of putative EBP (cross reacting antigens) in abscission zones from *Phaseolus vulgaris.* However, the nature and biological function of these proteins is unknown at present.

MATERIALS AND METHODS

Phaseolus vulgaris L. cv Canadian Wonder was grown in 35x20 cm seed trays containing Levington Universal compost with 25 plants per tray. Plants were maintained in a heated greenhouse at a daytime temperature of 20°C and a night temperature of 15°C with a 16 hour photoperiod supplemented by 400 HP sodium lamps. Treatments were carried out on 14-day-old plants.

Chemical Treatments. Ethephon, supplied as Ethrel-C (A.H. Marks, Bradford, U.K.) was diluted with water to a final concentration of 250 mg l^{-1} and applied as a foliar spray to run off.

Deblading Treatments. A cut was made with a scalpel to remove the leaf blade from the petiole of the first two leaves.

NATO ASI Series, Vol. H 35
Cell Separation in Plants
Edited by D. J. Osborne and M. B. Jackson

Tissue Preparation. Distal abscission zones and 1 cm segments of petiole tissue adjacent to these zones were excised 24 hours after treatment. Homogenates from both tissues were fractionated by differential centrifugation as described previously (Williams, 1988) and integral membrane proteins solubilised with 30 mM octyl-glucoside.

Dot Blotting. Dot blotting of plant material was carried out using colloidal gold labelled goat-anti-rabbit secondary antibody. Antigen-antibody complexes were visualised with silver enhancement (Janssen Pharmaceutical).

RESULTS AND DISCUSSION

Homogenates were prepared from distal abscission zones and 1 cm petiole segments of *Phaseolus vulgaris* plants which had been debladed to induce abscission. Both samples were screened for cross reacting antigens using anti-EBP by immuno dot blotting (Fig. 1). Cross reacting antigens were detected in both abscission zone and petiole tissue homogenates. No comparison in the level of cross reactivity between abscission zones and petiole tissue could be made at this stage due to the high background staining detected with non-immune serum.

Both homogenates were fractionated by differential centrifugation into their respective cytosol (S_2) and membrane proteins (P_2). Fractions S_2 and P2 from both tissue samples were screened for immuno reactive antigens as before (Fig. 2). No immuno reactivity was detected in S_2 fractions from either sample. Cross reacting antigens were detected in both membrane preparations P_2. However, the high background staining rendered it impossible to make an absolute comparison in the degree of cross reactivity between abscission zone membrane P_2 and petiole membrane P_2.

Solubilised integral membrane proteins (S_4) from petiole and abscission zones were prepared from 14-day-old seedlings of *Phaseolus vulgaris* in which abscission had been incuded by deblading/ethephon treatment. Immuno dot blotting was carried out as before (Fig. 3). No background staining was detected with non-immune serum (data not shown). This allowed a direct comparison in the degree of cross reactivity between abscission zones and petiole tissue. From the density of staining, there appeared to be a greater level of cross reactivity in abscission zones that in petiole

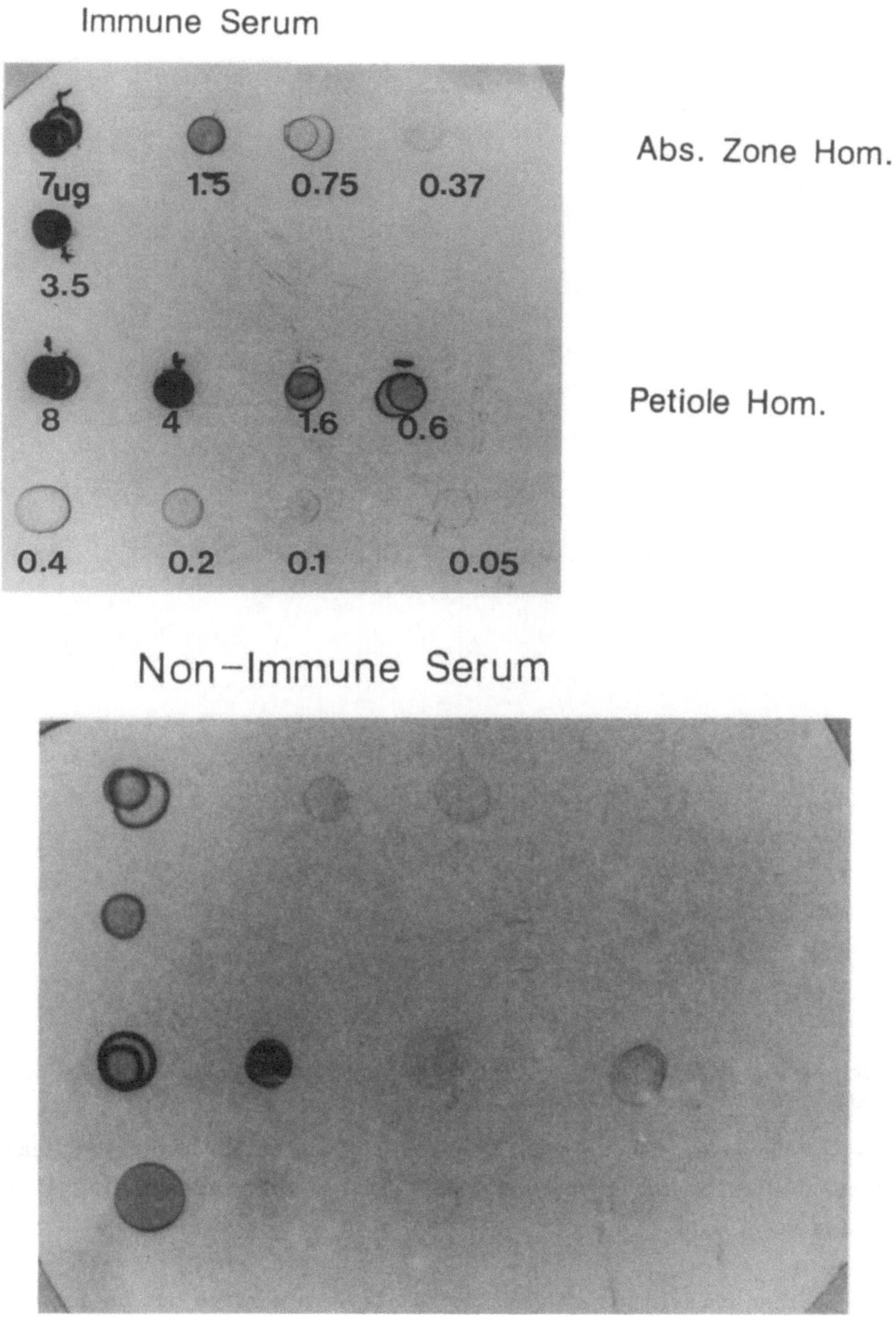

FIGURE 1. Abscission zone homogenate and petiole homogenate were applied to nitrocellulose membrane strips over the protein range 7 μg- 0.05 μg and 8 μg-0.05 μg respectively. Antigen-antibody complexes were visualised by using colloidal gold labelled goat anti-rabbit 2° antibody followed by silver enhancement.

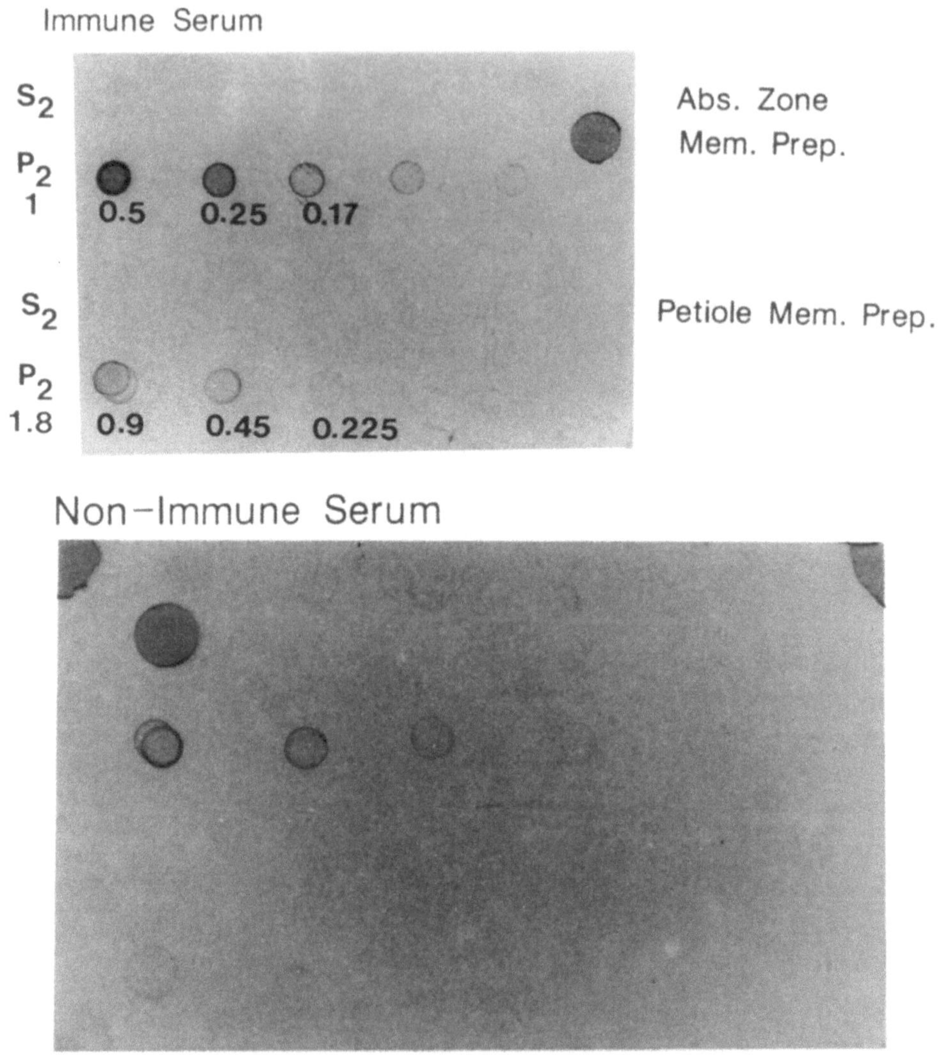

FIGURE 2. Cytosol S_2 and membrane P_2 fractions from both abscission zones and petiole tissue, were screened by dot blotting - previously described.

tissue. Thus cross reacting antigen could be detected down to at least 63 ng from abscission zones whereas from petioles the corresponding limit of detection was 201 ng.

In further experiments, immuno dot blots were carried out on solubilised abscission zones integral membrane proteins in which abscission had been induced by deblading only and those from abscission zones and petiole tissue in which abscission had been induced by a combination of deblading and ethephon (Fig. 4).

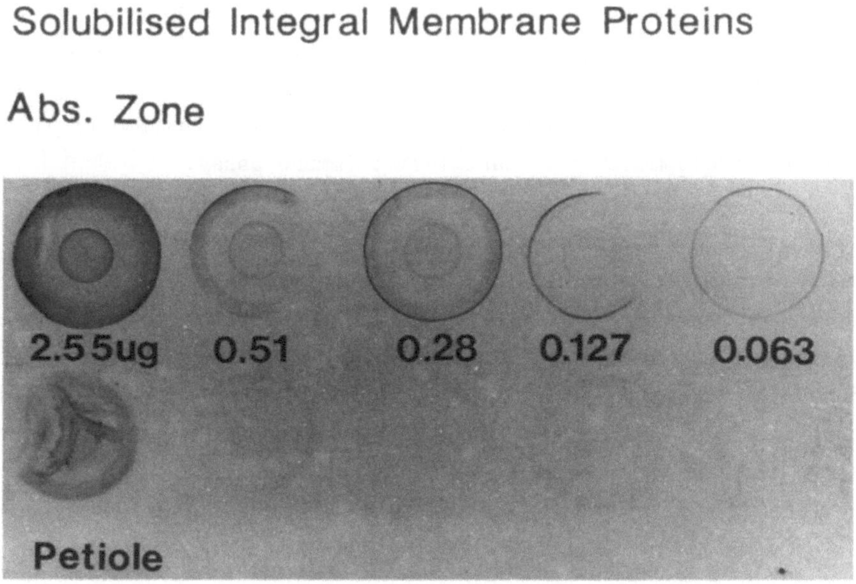

FIGURE 3. Solubilised integral membrane proteins debladed-ethephon induced abscission zones and petioles screen by dot blotting

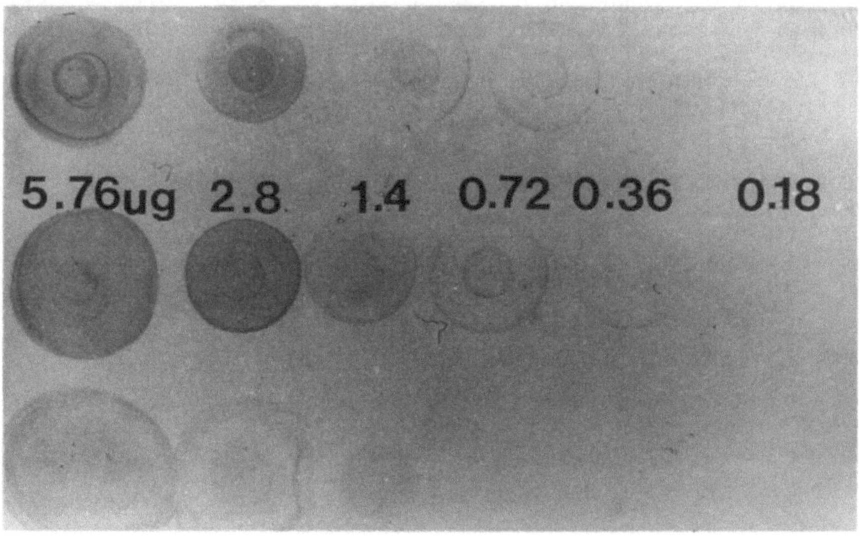

FIGURE 4. Comparison between the levels of cross reactivity detected by immuno dot blotting of solubilised integral membrane proteins prepared from abscission zones in which abscission had been induced by deblading only (top row) and those from abscission zones (middle row) and petiole segments (bottom row) in which abscission had been induced by a combination of deblading and ethephon treatment

Again, results indicated that higher levels of immuno reactive antigen were present in abscission zones that in petiole tissue. There is some indication that more cross reacting antigen is present in the ethephon treated samples than in the debladed samples but confirmation of this will have to await development of a quantitative immuno assay.

Osborne *et al.* (1985) have shown that unique protein determinants exist in induced abscission zones relative to petiole segments. Work presented in this paper has shown by qualitative immuno dot blotting that a polyclonal antibody raised against an ethylene binding protein from developing cotyledons of *Phaseolus vulgaris* cross reacts with integral membrane protein(s) from petiole tissue. Much more work needs to be carried out in this area. Caution is necessary in the interpretation of these results and Western blotting will need to be carried out on induced abscission zones and on petiole segments to determine the specificity of the antisera for these proteins.

REFERENCES

Hall MA, Howarth CJ, Robertson D, Sanders IO, Smith AR, Smith PG, Starling RJ, Tang Z-D, Thomas CJR, Williams RAN (1987) Ethylene binding proteins. In: Jackson MB, Fox JE (eds) Molecular biology of plant growth control, Alan R Liss, New York, pp 335-344

Osborne DJ, McManus MT (1985) Target cells for ethylene action. In: Roberts JA, Tucker GA (eds) Ethylene and plant development. Butterworths, London, p 197

Thomas CJR, Smith AR, Hall MA (1985) Partial purification of an ethylene binding site from *Phaseolus vulgaris* L. cotyledons. Planta 164: 272-277

Williams R (1988) Purification of an ethylene binding protein from *Phaseolus vulgaris* L. cotyledons. PhD Thesis, University College of Wales, Aberystwyth

DIFFERENTIAL ABSCISSION AND RIPENING RESPONSES TO ETHYLENE BY TABASCO PEPPER LEAVES AND FRUIT: PROTEIN "MARKER EVENTS" AS PROBES

Gregory A. Lang
Department of Horticulture
Louisiana State University
Baton Rouge, Louisiana 70803
USA

INTRODUCTION

The cultivated pepper, Capsicum spp., is a major horticultural crop worldwide. The most generally grown species is C. annuum, which includes bell, chili, jalapeno, pepperoncini, wax, pimento, cayenne, and cherry peppers, among others (Smith *et al.*, 1987). However, the second-most valuable pepper in the U.S. (on a dollar per ton basis [Andrews, 1984]) is the tabasco pepper, C. frutescens, a short-lived perennial plant of the tropics which is grown in Louisiana and used for making hot pepper sauce. One striking difference between most cultivated C. annuum cultivars and C. frutescens 'Tabasco' is the site of the mature fruit abscission zone (AZ): abscission occurs at the pedicel/stem AZ in C. annuum and at the pedicel/receptacle AZ in C. frutescens. Abscission at the pedicel/receptacle is desirable for the pepper sauce process, since there is no attached pedicel to be ground up with the harvested fruit.

Studies of ethylene-induced ripening (development of red coloration, primarily capsanthin and other carotenoids) in C. annuum spp. date back to the early 1970's when 2-chloroethylphosphonic acid (ethephon) first became available (Love *et al.*, 1971; Sims *et al.*, 1970). Ethephon has proved to be an efficient inducer of red colouration, but the range of dose concentrations which induce colouration without leaf or fruit abscission is narrow. When used for colour induction of tabasco peppers, ethephon also has caused fruit abscission at the pedicel/stem AZ unless modified with calcium hydroxide

NATO ASI Series, Vol. H 35
Cell Separation in Plants
Edited by D. J. Osborne and M. B. Jackson

(Conrad and Sundstrom, 1987). Recent commercial efforts to use ethephon+calcium have resulted in complete defoliation of tabasco peppers (FJ Sundstrom, unpublished). The leaf and the pedicel/stem AZs appear to be similar anatomically and responsively, except under water stress (Lang, unpublished), suggesting that research on preventing the induction of one may help prevent induction of the other as well.

Recent work in other fruit crops on ethylene-induction of fruit abscission without leaf abscission (Lang and Martin, 1985, 1989; Perry and Gianfagna, 1987), and fruit colouration without softening (Murphey and Dilley, 1988), suggests that differential induction of physiological responses may be possible by modification of the ethylene dose timecourse. There is also evidence from olive (Lang and Martin, 1987) and tabasco pepper (Lang, unpublished) that a cycle of ethylene treatments of modest concentration and short duration with ethylene-free intervals may effectively induce fruit abscission with very little leaf abscission. Consequently, the research programme outlined here focuses on ethylene as an inducer/enhancer of multiple responses in a single plant, and how to induce differentially and specifically the responses of choice.

RESEARCH QUESTIONS AND DIRECTIONS

Intriguing (and, with current technology, generally unanswerable) questions surround the possibility of differential induction of responses among tissues exposed to the same ethylene dose. How are observed differential responses manifested? Perhaps differential responses are due to different rates of ethylene diffusion to target cells. Other possibilities include differences among tissues in ethylene sensitivity resulting from different receptor-site concentrations and/or receptors with different binding characteristics. Does one "generic" binding site exist for a class of reactions or are there several unique sites, with different binding affinities, for different responses? Is the theoretical binding dissociation constant similar for each ethylene-induced response? Is turnover of the ethylene/receptor complex similar in all tissues? How is the ethylene signal transduced from the receptor complex to gene expression or enzyme activity modification? As the duration of ethylene exposure can be critical for the partial or full induction of responses, questions exist regarding how this durational factor is accounted for - e.g., for a particular response, does ethylene directly act on only one biochemical step in a "cascade of events", or does it have various roles at several points along the binding-turnover-transduction-transcription-translation-secretion-activation response pathway? What steps along this pathway are reversible/irreversible? At what levels do modifying factors (e.g., auxin, water stress, ethylene pre-treatment) affect ethylene-induced responses?

To understand better the potential for differential induction of ethylene-mediated events, the research approach we are developing is to document "threshold doses" for four physiological events of agricultural interest - fruit colouration, fruit abscission at the pedicel/receptacle AZ and at the pedicel/stem AZ, and leaf abscission. We are currently investigating the use of both whole plants and shoot explants for this documentation. Using the threshold dose to predictably induce each event, the mRNA profiles will be characterized before, during, and after ethylene induction to probe "marker" genetic events associated with the ethylene-mediated response(s). Important experimental questions to be addressed include: which ethylene-inducible proteins cease or continue to be synthesized if ethylene treatment is stopped after induction? Which proteins appear to be indirectly induced by ethylene - that is, do not appear until well after the initial exposure to ethylene has occurred?

The experimental procedure will be to extract total cellular RNA from the plastids of the fruit pericarp (Hadjeb *et al.*, 1988), and from the leaf and fruit AZs (GA Tucker, pers. comm.), over a timecourse of ethylene treatment. Extracts will be fractionated into poly(A)+ RNA and translated in vitro. Polypeptides synthesized in vitro will be separated by one- and two-dimensional gel electrophoresis and visualized by fluorography. The mRNA populations will be compared for dramatic increases or decreases in similar and novel mRNAs over the timecourse of ethylene exposure and development of physiological responses to determine which mRNAs appear to be absolute requirements for response and which merely accompany ethylene treatment. Nuclear run-off techniques also will be used to determine which changes in polypeptide synthesis are regulated at the level of transcription.

CONCLUSIONS

The long-range goal is to contribute to the eventual identification of response-specific, ethylene-induced genes for insertion into, or deletion from, the genome of Capsicum spp., or for nullifying specific ethylene-induced responses by inserting genes that code for antisense RNA (Smith *et al.*, 1988). This may allow easy differential manipulation of ethylene-induced responses in specific organs of agriculturally-important crops. In the case of tabasco pepper, the goal will be to use an ethylene-releasing compound, such as ethephon, to induce the expression of genes for fruit coloration and abscission at the pedicel/receptacle AZ, while reducing or eliminating ethylene-induced expression of the genes for abscission at the leaf and pedicel/stem AZs.

REFERENCES

Andrews J (1984) Peppers, the domesticated capsicums. Univ of Texas Press, Austin

Conrad RS, Sundstrom FJ (1987) Calcium and ethephon effects on Tabasco pepper leaf and fruit retention and fruit color development. J Am Soc Hort Sci 112:424-426

Hadjeb N, Gounaris I, Price CA (1988) Chromoplast-specific proteins in Capsicum annuum. Plant Physiol 88:42-45

Lang GA, Martin GC (1985) Ethylene-releasing compounds and the laboratory modeling of olive fruit abscission vs. ethylene release. J Am Soc Hort Sci 110:207-211
Lang GA, Martin GC (1987) Ethylene-induced olive organ abscission: ethylene pulse treatments improve fruit-to-leaf abscission ratios. Acta Hort 201:43-52
Lang GA, Martin GC (1989) Olive organ abscission: fruit and leaf response to applied ethylene. J Am Soc Hort Sci 114:134-138
Love JE, Fontenot JF, White JW (1971) Ripening hot peppers with ethrel. Louisiana Agric 14:14-15
Murphey AS, Dilley DR (1988) Anthocyanin biosynthesis and maturity of 'McIntosh' apples as influenced by ethylene-releasing compounds. J Am Soc Hort Sci 113:715-723
Perry SC, Gianfagna TJ (1987) Effect of Silaid and Ethrel on peach leaf and fruit abscission in relation to the kinetics of ethylene release. Acta Hort 201:157-163
Sims WL, Collins HB, Gledhill BL (1970) Ethrel effects on fruit ripening of peppers. Calif Agric 24:4-5
Smith CJS, Watson, CF, Ray J, Bird CR, Morris PC, Schuch W, Grierson D (1988) Antisense RNA inhibition of polygalacturonase gene expression in transgenic tomatoes. Nature 334:724-726
Smith PG, Villalon B, Villa PL (1987) Horticultural classification of peppers grown in the United States. HortSci 22:11-13

CHIMERIC ANALYSIS OF CELL LAYER INTERACTIONS DURING DEVELOPMENT OF THE FLOWER PEDICEL ABSCISSION ZONE

Eugene J. Szymkowiak and Ian M. Sussex
Department of Biology
Yale University
New Haven, Connecticut 06511
USA

INTRODUCTION

The abscission zone that forms on the pedicel of tomato is composed of 5-10 tiers of unexpanded cells that traverse the pedicel at a point where a groove or indentation on the surface is visible. Cell separation occurs between adjacent cells within the abscission zone in response to ethylene (Roberts *et al.*, 1984). The cells of the abscission zone are morphologically distinct prior to separation and are derived from three distinct cell lineages in the shoot apical meristem. In the case of foliar abscission in *Impatiens sultani*, cells derived from all three lineages including the epidermal cells of the abscission zone produce their own wall degrading enzymes during the abscission process (Sexton, 1976).

The shoot apical meristem of *L. esculentum* and other members of the Solanaceae is composed of three cell layers maintained in the meristem by predominantly anticlinal cell division patterns in the first and second cell layers, the L1 and L2. The term periclinal chimera is applied to plants having an apical meristem with one cell layer genetically different from the other layers. Colchicine-induced cytochimeras of *Datura stramonium* that are polyploid in one cell layer (Blakeslee *et al.*, 1940) and plastid chimeras of tobacco that are albino in one cell layer (Burk *et al.*, 1964) have been used to identify the contribution of the derivatives of each of the meristem cell layers to the organs of the plant. The L1 contributes a single cell layered thick epidermis covering the entire shoot and L2 and L3 each contribute a variable amount of internal tissue depending on the organ type. The pedicels of tomato and related species have an L1 derived epidermis, a one or two cell thick sub-epidermal layer derived from L2, and the remaining tissue derived from L3.

Using a grafting technique similar to that of Winkler (1907), we have generated genetic periclinal chimeras to see whether information is communicated between cells of the three lineages which form the abscission

NATO ASI Series, Vol. H 35
Cell Separation in Plants
Edited by D. J. Osborne and M. B. Jackson

zone or whether cells of different lineages behave in a cell autonomous manner. In these chimeras one of the cell layers of the shoot apical meristem differs genetically from the other cell layers, either in the ability to form an abscission zone or in the position of the abscission zone on the pedicel. Interspecific chimeras were generated between *Lycopersicon esculentum*, in which the abscission zone develops at a point on the pedicel approximately midway between the flower and the peduncle, and *L. pennellii*, in which the abscission zone develops at the base of the pedicel. Other chimeras were generated in which tomato mutant for *jointless*, which does not form an abscission zone (Butler, 1936), contributed the L1 of an intraspecific chimera with tomato wildtype for *jointless*, or the L2 and L3 of interspecific chimeras with *L. peruvianum*, a species which forms abscission zones similar to those of tomato.

Analysis of the chimeras indicates that the cells derived from L3 play an important role in determining the development and the position of the pedicel abscission zone and that L1 and L2 derived cells can respond to signals from the L2 and differentiate accordingly.

METHODS

All seed stocks were obtained from the Tomato Genetic Stock Center, Dept. Veg. Crops, University of California-Davis, Davis, CA 95616.

To facilitate the identification of chimeras, the tomato mutants *Xanthophyllic 2, Xa-2*, which has reduced chlorophyll content and a yellow appearance, and *anthocyanin gainer, ag.* which has reduced anthocyanin pigmentation, were used to mark genetically the L2 and L3 cell layers. The mutation *hairless, h*, which has a greatly reduced number of long trichomes on the epidermis, was used to genetically mark the tomato L1. All of these mutations have been shown to be cell layer autonomous in their expression (unpublished results). The epidermis of *L. pennellii* and *L. peruvianum* are sufficiently different from tomato epidermis to easily identify the plant that contributes the L1 layer to the chimera.

Six week old plants of *L. esculentum* that carried the cell layer markers *Xa-2, h,* and *ag*, and that were either mutant or wild type for jointless, were reciprocally slant grafted to *L. esculentum, L. pennellii* or

L. peruvianum. After 30 days, a cut was made through the graft junction leaving exposed tissue from both stock and scion. Callus formed at the cut graft junction and shoot meristems were formed from the callus cells. Occasionally cells from both stock and scion were incorporated into a newly forming shoot meristem. These shoots were identified by having novel combinations of the genetic cell layer markers, i.e. a yellow shoot with *L. pennellii* type epidermal hairs arising from a graft of a *Xa-2* tomato and *L. pennellii*. Such shoots were generally sectorial in nature and were manipulated by continual pruning to force periclinal buds within the chimeric sector to develop. Pedicels of flowers just prior to anthesis were fixed in FAA for both light microcopy and scanning electron microscopy.

RESULTS

The chimeras generated and their phenotypes are summarized in Table 1. From grafts of *L. esculentum* and *L. pennellii* two different chimeras were obtained. Chimera 1A1 had the L1 of *L. pennellii* and the L2 and L3 of *L. esculentum* and chimera 1B1 had the L1 and L2 of *L. pennellii* and the L3 of *L. esculentum*. The abscission zone formed on the pedicel of both chimeras at the same position as in *L. esculentum*. In the chimera 1B1 only the cells derived from the L3 layer are genetically "programmed" to form the abscission zone at the tomato position, yet the *L. pennellii* cells derived from L1 and L2 layers in the chimera appear as abscission zone type cells in the tomato position and not at the base of the pedicel where *L. pennellii* normally forms an abscission zone. Although no experiments have yet been performed to determine if the *L. pennellii* cells in these chimeras proceed through all stages of abscission, the chimeric flowers do abscise if pollination is prevented or if the flower is wounded prior to anthesis.

TABLE 1. Graft partners (plant 1, plant 2) and meristem layer constitution of chimeras.

Plant 1	Abs. zone	Plant 2	Abs. zone	Chimera	L1L3L3	Abs. zone
L. esculentum	mid pedicel	L. pennellii	base pedicel	1A1	2 1 1	mid pedicel
L. esculentum	mid pedicel	L. pennellii	base pedicel	1B1	2 2 1	mid pedicel
L. esculentum	jointless	L. esculentum	mid pedicel	7A1	1 2 2	mid pedicel
L. esculentum	jointless	L. peruvianum	mid pedicel	3A2	2 1 1	jointless*
L. esculentum	jointless	L. peruvianum	mid pedicel	3B2	2 1 2	mid pedicel*

*incomplete floral development

Tomato plants mutant for *jointless*, in which the pedicel develops without an abscission zone and unpollinated flowers are not abscised, were grafted with tomato plants wildtype for *jointless*. From these grafts the chimera 7A1 was obtained which had a jointless L1 and a wild type L2 and L3. The chimera produced normal abscission zones. SEM examination of the epidermal cells of the abscission zone revealed cell morphologies similar to those of wildtype abscission zones. This chimera suggests that the underlying cells of the L2 and L3 are capable of inducing mutant L1 cells to develop phenotypically as wild type epidermal cells in the region of the abscission zone.

From grafts of *L. peruvianum* with jointless *L. esculentum*, two chimeras were obtained. Chimera 3A2 had the L1 of *L. peruvianum* and the L2 and L3 of jointless *L. esculentum*. Chimera 3B2 had the L1 and L3 of *L. peruvianum* and the L2 of jointless *L. esculentum*. Although in both of these chimeras vegetative growth proceeds normally, floral development does not proceed to maturity. This result will be addressed in a future publication. However the pedicel did begin to develop and in chimera 3B2, where only the L2 is *jointless*, an abscission zone began to form with the characteristic surface indentation being very apparent. No such indication of abscission zone development was observed in chimera 3A2 where both the L2 and L3 were *jointless*. These results indicate, as seen in 3B2 where only the L2 is mutant, that the mere presence of jointless tissue is not sufficient to suppress abscission zone development and in this case the mutant L2 cells develop morphologically as wildtype cells. In addition, as seen in 7A1 and 3A2, the morphology of epidermal cells in the region of the abscission zone depends not on their own genotype but on the genotype of cells internal to them.

DISCUSSION

The periclinal chimeras analyzed in this study indicate that communication takes place between the three cell lineages which form the abscission zone. This information can be transferred from wildtype to mutant cells (chimera 7A1, 3A2 and 3B2), and also between tomato cells and the cells of two related species (chimeras 1A1, 1B1, 3A2 and 3B2). The L1 and L2 derived cells do not determine the location of the abscission zone on the pedicel. Abscission zone location was determined by the L3 in the chimeras, and the L1 and L2 responded to the L3 by forming normal abscission zone type cells. To discover whether tomato tissue need only be present in the pedicel to determine the position of the abscission zone, regardless of which cell layer it is in, the remaining four chimeric cell layer combinations will need to be generated. These include, L1=pennellii, L2-esculentum, L3=pennellii or PEP, EPP, EEP and EPE.

Although the molecular basis of the jointless mutation is unknown, the chimeras containing this mutation in one or more cell layers indicate that the cells of the L1 and L2 can develop morphologically according to the genotype of the cell layers beneath them and not according to their own genotype with respect to abscission zone formation. In order to avoid the complications of abnormal floral development encountered in chimeras with *L. peruvianum* and to determine the extent and direction of these interactions a full set of intraspecific chimeras using the *jointless* mutation will need to be generated so that the abscission zone development can be followed throughout the development of the pedicel and subsequent abscission processes.

The kinds of cell layer interactions discussed here are similar to those observed in some other genetic chimeras. In maize, the *Knotted* mutation causes abnormal epidermal divisions in the leaf. X-ray-induced sectors chimeric for *Knotted* were used to show that it is the genotype of the internal tissue and not the epidermal L1 tissue that causes the L1 cells to divide (Hake and Freeling, 1986). The chimera *Camellia + 'Daisy Eagleson'*, which has the L1 of *C. sasanqua*, a self fertile species, and the L2 and L3 of *C. japonica*, a sterile species, produces self fertile flowers (Steward *et al.*, 1972). In this case it is the epidermis which induces internal L2 and L3 derived cells to form organs they normally do not form.

SUMMARY

The morphologically distinct abscission zone develops at a point midway on the pedicles of *Lycopersicon esculentum* and *L. peruvianum*. In *L. pennellii* the abscission zone forms at the base of the pedicel near the peduncle. In the tomato mutant *jointless* no abscission zone develops on the pedicel. Genetic chimeras were generated to determine if abscission zone development is a cell autonomous process. Periclinal chimeras, having one of their three stable meristematic cell layers, L1, L2 and L3, genetically different from the remaining cell layers, were generated by grafting and bud regeneration of *L. esculentum* with *L. pennellii*, *L. esculentum* with jointless *L. esculentum*, and *L. peruvianum* with jointless *L. esculentum*. These chimeras indicate that the L1 and L2 derived cells do not develop autonomously during abscission zone formation but respond to signals from internal L3 derived cells.

REFERENCES

Blakeslee AF, Satina S, Avery AG (1940) Utilization of induced periclinal chimeras in determining the constitution of organs and their origin from three germ layers in *Datura*. Science 91:423.

Burk, LG, Stewart RN, Dermen H (1964) Histogenesis and genetics of a plastid-controlled chlorophyll variegation in tobacco. Am J Bot 51:713-724

Butler L (1936) Inherited characters in the tomato. J Herd 27:25-26

Hake S, Freeling M (1986) Analysis of genetic mosaics shows that the extra epidermal cell divisions in *Knotted* mutant maize plants are induced by adjacent mesophyll cells. Nature 320:621-623

Roberts JA, Schindler CB, Tucker GA (1984) Ethylene-promoted tomato flower abscission and the possible involvement of an inhibitor. Planta 160:159-163

Sexton R (1976) Some ultrastructural observations on the nature of foliar abscission in *Impatiens sultani*. Planta 128:49-58

Stewart RN, Meyer FG, Dermen H (1972) Camellia + 'Daisy Eagleson,' A graft chimera of *Camellia sasanqua* and *C. japonica*. Am J Bot 59:515-524

Winkler H (1907) Ueber pfropfbastarde und pflanzlich chimaeren. Ber Deut Bot Ges 25:568-576

THE ROLE OF ETHYLENE IN PETAL ABSCISSION OF RED RASPBERRY

J. N. Burdon, R. Sexton
School of Molecular and Biological Sciences
University of Stirling
Stirling FK9 4LA
Scotland, U.K.

INTRODUCTION

A great deal of evidence has accumulated which implicates ethylene in the abscission of leaves, fruits and flower buds (Sexton *et al.*, 1985). Until recently little attention had been paid to petal fall although one of the original observations that ethylene accelerated abscission was made by Fitting in 1911 with geranium petals. These observations have since been repeated (Sexton *et al.*, 1983) and extended to other species (Halevy and Mayak, 1981).

The hypothesis that ethylene is the natural regulator of petal abscission is supported by the observations of Wallner *et al.* (1979) and Stead and Moore (1983). Both groups showed correlations between elevated rates of ethylene formation and petal/corolla fall. In snapdragon (Wang *et al.*, 1977) and foxglove (Stead, 1985) the inhibition of ethylene synthesis with aminoethoxyvinylglycine (AVG) retarded corolla abscission. Halevy *et al.* (1984) have shown that increased ethylene sensitivity is a critical factor in the shedding of cyclamen corollas. There are also cases where there is no clear evidence of a relationship between ethylene production and petal fall (Armitage *et al.*, 1980).

PETAL FALL IN RASPBERRIES

During work on the role of ethylene in the shedding of raspberry fruit (*Rubus idaeus* cv. Glen Clova) it was noted that very immature fruits had higher ethylene production rates than the enlarging green berries (Burdon, 1987). An examination of flower development revealed a burst of ethylene production associated with the later stages of flower opening and petal fall. The vagaries of the climate and visits by pollinators made a field investigation difficult. Hence a laboratory based method was devised to examine the role played by ethylene in petal fall.

NATO ASI Series, Vol. H 35
Cell Separation in Plants
Edited by D. J. Osborne and M. B. Jackson

MEASURING ETHYLENE PRODUCTION RATES DURING FLOWER OPENING

It was found that flower buds brought into a laboratory would open and shed their petals over the subsequent 36 to 48 h. This time course seemed similar to that of flowers in the field. To ensure consistency, buds were gathered when the sepals had just started to part to reveal the petals. The pedicel of each bud was threaded through a 1 cm diameter disc of 1 mm thick expanded polystyrene which acted as a floatation collar. The bud was then floated on water in a vial (2 cm x 5 cm) which could be sealed for a short time to allow ethylene to accumulate. In this way the ethylene production rates could be measured at different stages of development.

The sequential stages of flower opening were thus categorized. From a fully expanded bud (stage 1) the sepals start to separate along their margins (2) until they are no longer attached at their tips (4). The sepals then move through the vertical (5) to the horizontal (9) and finally curl back towards the pedicel (10). At this time the petals are closed around the stigmas (11) and gradually open until almost horizontal (15). Once the first petal has been shed (16) the rest follow over the next 3-4 h until none remain (20). At the time of shedding the petals show no visible signs of senescence.

ETHYLENE AND PETAL ABSCISSION

The overall pattern of ethylene production (Fig. 1) seemed similar to that of field harvested flowers. During the stages where the sepals reflexed (stages 1 - 10) the rates were low, increasing during petal opening (stages 11 - 15) and decreasing during petal abscission (stages 16 - 20). The peak in ethylene production preceded petal abscission by approximately 3 h. It is known that pollination induces increased ethylene production (Burg and Dijkman, 1967; Stead, 1985; Hill et al., 1987). The addition of pollen to raspberry flowers opened in the laboratory had no effect on ethylene production. However, SEM observations showed that the outer whorls of stigmas were self-pollinated.

To investigate if the ethylene burst during flowering was involved in subsequent petal fall, buds were allowed to open while transpiring aminoethoxyvinylglycine (AVG). This treatment prevented the normal increase in ethylene production (Fig. 1) but did not stop the flowers

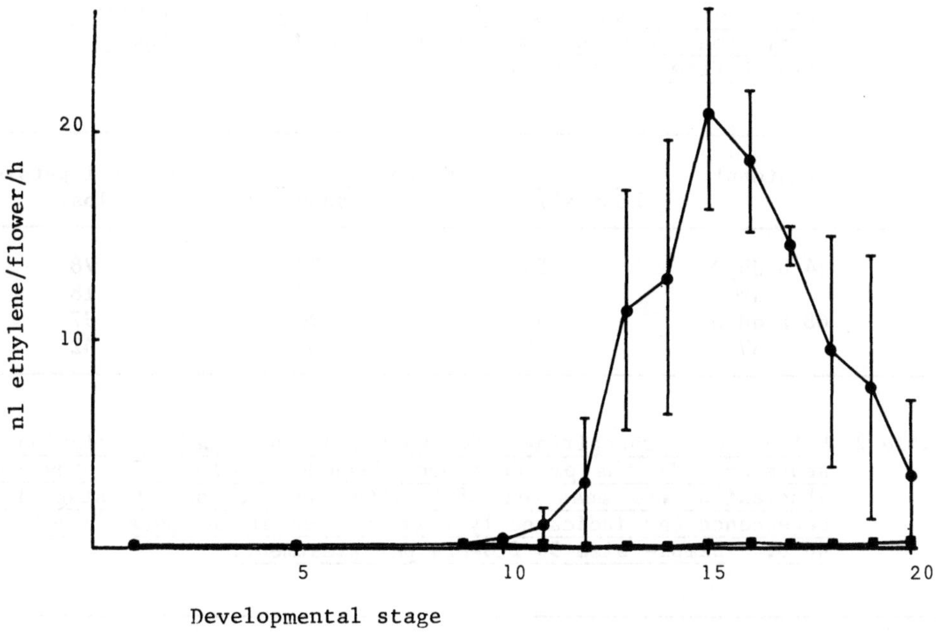

Fig. 1 Ethylene production rates of individual flowers at different stages of development. One group was allowed to transpire 0.4 mM AVG (■), the other distilled water (●). Sample size 20 flowers per stage, ± S.E.

opening and shedding their petals. The period over which petals were lost however was greatly extended. To examine the role of ethylene further, the shedding of petals from individual flowers treated with AVG was examined. The AVG caused the petals to be shed in ones and twos over a 3-day period when compared to the normal case of all petals being shed in synchrony over a 3 to 4 h period (Table 1). This resulted in most of the AVG treated flowers losing some petals but very few having lost them all when compared to the controls up to 68 h after collection.

The specificity of AVG action was examined by adding ethylene (40 μl l^{-1}) to AVG treated buds (Table 2). This resulted in abscission comparable to that in ethylene alone. A comparison of the control flowers with or without ethylene after 28 h showed that natural ethylene production was not saturating since the addition of ethylene accelerated both abscission and senescence. This accelerating effect of ethylene on petal senescence has been well documented (Halevy and Mayak, 1981; Nichols, 1984) although there are far fewer data on petal abscission.

Table 1 Effect of transpiring AVG (0.4 mM) on the pattern of petal abscission 44 h and 68 h after collection at stage 1. Sample size 50 flowers per treatment.

Treatment	% Flowers		% total petals lost
	Lost all petals	Lost some petals	
44 h dH_2O	53	80	78
AVG	13	27	18
68 h dH_2O	93	100	97
AVG	13	80	42

Table 2 Effect of transpiring AVG (0.4 mM) on petal abscission and senescence in the presence and absence of 40 µl l^{-1} ethylene. Abscission was measured 28 h after collection at stage 1 and senescence (as indicated by a brown area at the petal base) after 40 h. Sample size 50 flowers per treatment.

Treatment	% petals lost 28 h	% petals senescent 40 h
On water, in air	67	70
On AVG, in air	15	1
On water, in ethylene	88	96
On AVG, in ethylene	87	91

Ethylene production is regulated in other systems by the activities of ACC synthase and the ethylene forming enzyme (Yang and Hoffman, 1984). Supplying the ethylene precursor ACC (1-aminocyclopropane-1-carboxylic acid, 1 mM) through the transpiration stream to flowers at different stages of development resulted in very high rates of ethylene production in all cases (>100 nl/flower/h); it also accelerated petal fall. Hence ethylene production may be limited by the supply of ACC through the activity of ACC synthase or some earlier step in the pathway. The endogenous levels of ACC (assayed according to Lizada and Yang, 1979) mirrored the ethylene production rates increasing from 0.015 nmol/flower during sepal opening to 0.6 nmol/flower during the ethylene burst and declining to 0.2 nmol/flower after petal abscission.

It seems from these results that the ethylene burst serves to synchronise the shedding of petals from a flower by accelerating the abscission of the individual petals. There is evidence that ethylene may play a similar role in other abscission systems (Sexton *et al.*, 1985). If ethylene plays a role in the induction of abscission it must be the low background level that is responsible since the ethylene burst is clearly not essential for the process but simply regulates its rate.

REFERENCES

Armitage AM, Heins R, Dean S, Carlson W (1980) Factors influencing flower petal abscission in the seed propagated geranium. J Am Soc Hort Sci 105:562-564.

Burdon JN (1987) The role of ethylene in fruit and petal abscission in red raspberry (*Rubus idaeus* L. cv. Glen Clova). PhD Thesis, Stirling University, UK.

Burg SP, Dijkman MJ (1967) Ethylene and auxin participation in pollen induced fading of Vanda orchid blossoms. Plant Physiol 42:1648-1650.

Fitting H (1911) Untersuchungen über die vorzeitige Entblätterung von Blüten. Jahrb Wiss Bot 49:187-263.

Halevy AH, Mayak S (1981) Senescence and post-harvest physiology of cut flowers. Part 2. Hort Rev 3:59-143.

Halevy AH, Whitehead CS, Kofranek AM (1984) Does pollination induce corolla abscission of cyclamen flowers by promoting ethylene production? Plant Physiol 75:1090-1093.

Hill SE, Stead AD, Nichols R (1987) Pollination induced ethylene and production of 1-aminocyclopropane-1-carboxylic acid by pollen of *Nicotiana tabacum* cv. White Burley. J Plant Growth Reg. 6:1-14.

Lizada MCC, Yang SF (1979) A simple and sensitive assay for 1-aminocyclopropane-1-carboxylic acid. Analytical Biochem 100:140-145.

Nichols R (1984) Ethylene and flower senescence. In: Fuchs Y, Chalutz E (eds). Ethylene, Biochemical, Physiological and Applied Aspects. Martinus Nijhoff/Dr W Junk, The Hague, pp 101-110.

Sexton R, Lewis LN, Trewavas AJ, Kelly P (1985) Ethylene and abscission. In: Roberts JA, Tucker GA (eds). Ethylene and Plant Development. Butterworths, London, pp 173-196.

Sexton R, Struthers WA, Lewis LN (1983) Some observations on the very rapid abscission of the petals of *Geranium robertianum* L. Protoplasma 116:179-186.

Stead AD (1985) The relationship between pollination, ethylene production and flower senescence. In: Roberts JA, Tucker GA (eds). Ethylene and Plant Development. Butterworths, London, pp. 71-82.

Stead AD, Moore KG (1983) Studies on flower longevity in *Digitalis*. Planta 157:15-21.

Wallner S, Kassalen R, Burgood J, Craig R (1979) Pollination, ethylene production and shattering in geraniums. HortSci 14:446.

Wang CY, Baker JE, Hardenburg RE, Lieberman M (1977) Effects of two analogs of rhizobitoxine and sodium benzoate on senescence of snapdragons. J Am Soc Hort Sci 102:517-520.

Yang SF, Hoffman NE (1984) Ethylene biosynthesis and its regulation in higher plants. Ann Rev Plant Physiol 35:155-189.

PEROXIDASES AND CELL WALL GROWTH BY EPICOTYLS OF CICER ARIETINUM L. POSSIBLE REGULATION BY CALCIUM AND CALMODULIN AND RELATIONSHIP WITH CELL WALL LOOSENING

G. Nicolás, O.J. Sánchez, J. Hernandez-Nistal, E. Labrador
Departmento de Biología Vegetal
Universidad de Salamanca
Spain

INTRODUCTION

Peroxidases have been implicated in the regulation of growth, and several different biochemical pathways have been proposed to account for their mechanism of action. At first, it was thought that these enzymes controlled endogenous auxin levels (Srivastava *et al.*, 1973). Later, these substances were considered as growth regulators, exerting their action on ethylene levels (Gaspar *et al.*, 1985). Recently, Fry has suggested that cell wall peroxidases are able to catalyze the formation of diphenyl bridges in the cell wall favouring rigidity and the halting of growth (Fry, 1983). Suggestions have also been made recently that calmodulin antagonists are involved in growth processes. Chlorpromazine (CP) and trifluoroperazine (TFP) are able to inhibit auxin-induced cell elongation (Raghotama *et al.*, 1985). On the other hand, such calmodulin inhibitors can enhance or prevent morphological responses to auxin, gibberellin and cytokinin (Elliott *et al.*, 1983). The foregoing observations suggest a role for calmodulin in plant growth and development. In recent years, increasing evidence has been found to support the role of calcium ions as intracellular messengers in plants. Since the discovery of calmodulin, it has become clear that calcium messages are often transmitted by this protein (Poovaiah, 1985). In relation to the specific title of this volume, a signal mediated through the calcium-calmodulin complex could affect the abscission zone and induce cell separation by promoting cell wall loosening following an increase in cell expansion (Wright and Osborne, 1974).

In this present work our aim has been to study the relationship between peroxidases and growth and the effect of calmodulin antagonists on growth. the results allow us to examine the possibility that calmodulin might regulate elongation and cell wall loosening by influencing peroxidase activity.

NATO ASI Series, Vol. H 35
Cell Separation in Plants
Edited by D.J. Osborne and M.B. Jackson

RESULTS AND DISCUSSION

Peroxidase activity in the cell walls of *Cicer arietinum* epicotyls must be bound to the cell wall by ionic forces since no covalently bound peroxidase was detected. All the peroxidases located in the cell wall can be extracted by 1 M NaCl. The extent of this peroxidase activity is negatively correlated with growth capacity, since the maximum capacity for growth in the epicotyls during the early stages of growth is correlated with the lowest level of peroxidase activity (3 days, Fig. 1). Growth subsequently slows (data not shown) as peroxidase levels rise (Fig. 1). The amount of

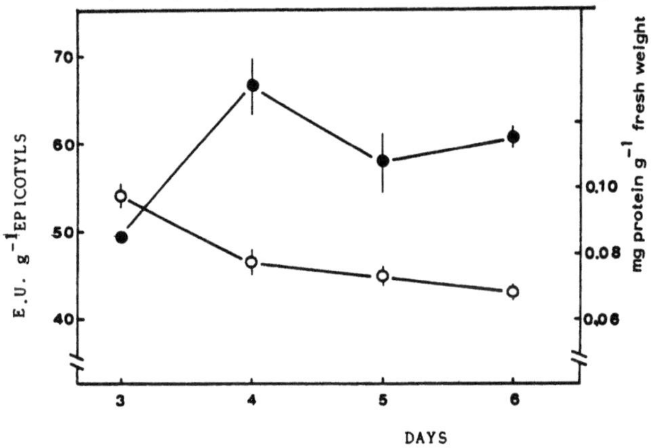

FIGURE 1. Cell wall peroxidase activity during growth of C. arietinum epicotyls (closed circles). Cell wall protein extracted by 1 M NaCl (open circles). One enzymatic unit increases 0.01 absorbance units per min at pH 5.5 at 25°C. Peroxidase was determined according to Putter (1974).

extractable cell wall protein decreased as growth progressed. For this reason, total peroxidase activity was estimated per unit weight of epicotyl tissue (Fig. 1). It is postulated that the action of peroxidase would be to confer rigidity to the cell wall and prevent later expansion involved in growth. Work carried out with other plant material is in agreement with such findings (Palmieri *et al.*, 1978; Lamport, 1986).

The role of cell wall peroxidases, except their participation in the lignification process, remains to be fully elucidated. Recent research by Fry (1986) has pointed to phenol as the natural substrate for cell wall peroxidases. Study of phenolic compounds in cell walls during growth (Fig. 2) revealed an increase in these compounds as the epicotyls age. At 6 days, the amount observed was twice that seen at 3 days. This increase is all the more remarkable considering that the amount of cell wall per unit fresh weight decreases over this time. The peroxidases would be expected to calatyse the oxidative coupling of phenols that in turn become bound to polysaccharides and glycoproteins (Lamport, 1986; Fry, 1986), thus inducing a more rigid structure to the cell wall. If this was true, the increased occurrence of phenols in cell walls may enhance the action of peroxidases, and it seems reasonable to assume that the increase in phenols detected during growth, promotes peroxidase activity.

The calmodulin inhibitors chlorpromazine (CP) and trifluoroperazine (TFP) decreased the growth of epicotyls of *C. arietinum*. CP led to a 38% decrease in fresh weight (Table 1). The decrease with TFP was 21% (data not shown). These calmodulin inhibitors also increased peroxidase activity compared to controls. The percentage increases were 30% with CP (Table 1) and 14% with TFP. A study of the effects of these drugs on phenolic

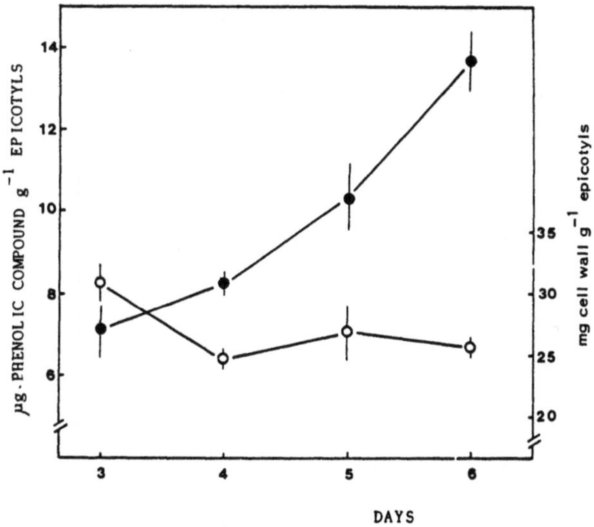

FIGURE 2. Phenolic compounds in the cell wall of C. arietinum epicotyls. (closed circles) μg of phenolic compounds. (open circles) μg of cell wall. Phenols were determined according to Swain and Hillis (1959)

TABLE 1. Effect of chlorpromazine on growth and peroxidase activity in epicotyls of Cicer arietinum

Concentration (μm)	Length (mm)	Fresh weight (mg/epicotyl	Peroxidase Activity U.E./g	Peroxidase Activity U.E./mg protein
Control	26.9±5.4	145.74	62.0±4	342±14
50	20.6±5.3	120.60	77.6±5.8	409±56
100	19.3±4.5	104.54	78.6±6.1	426±24
250	20.4±5.1	121.28	82.0±8	425±20
500	16.6±2.9	89.63	79.6±2.1	423±40

compounds in the cell wall gave similar results. Both compounds led to a considerable increase that was more pronounced with CP (data not shown). To check if the effect of calmodulin inhibitors was merely pharmacological, we studied the effects of several different concentrations on growth and peroxidase activity (Table 1). The effect of CP was appreciable at concentrations as low as 50 μM. We entertained the possibility that the inhibitory action of these agents on growth might be mediated by peroxidase activity. The increase in phenolics caused by calmodulin inhibitors may potentiate the increase in peroxidase activity. The above results suggest that calmodulin inhibitors favour peroxidase activity and inhibit growth, and also suggest that calmodulin is involved in epicotyl elongation. Pre-incubation of the enzymatic extract from cell walls with calmodulin (15 minutes) inhibited peroxidase activity (Fig. 3). This inhibition was proportional to the concentration of calmodulin used in the assay. Calmodulin might inhibit peroxidase activity *in vivo*, and thus regulate cell elongation through modulating peroxidase activity.

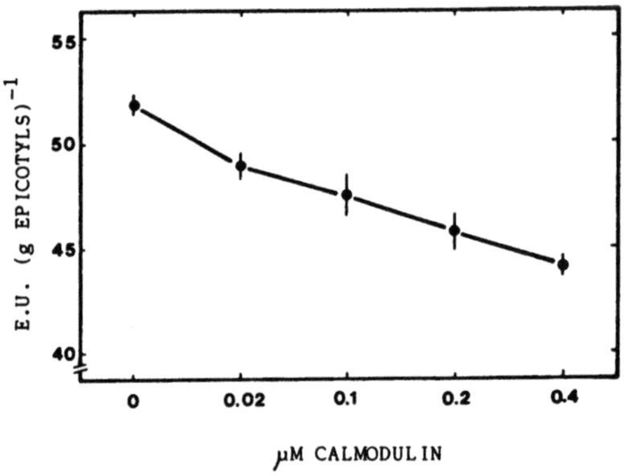

FIGURE 3. Effect of calmodulin on peroxidase activity

These results raise the question of whether calmodulin is present in cell walls. Although this location has received some attention in an earlier report (Biro *et al.*, 1984) no further experimental evidence has been forthcoming. To investigate if calmodulin was present in our material, the distribution of calmodulin within the cell wall was analysed (Table 2). Six cellular fractions were obtained, and calmodulin appears in protein extracted from cell walls with 3 M LiCl. What function calmodulin has there remains unclear. Calcium has long been known to confer rigidity to cell walls. One possible mechanism of action might be through the control of peroxidase activity (Penel, 1986). On the other hand, the effect of calcium on some parameters of senescence, such as dissolution of the middle lamella, is well established. Since (i) the biochemical action of calcium has been

TABLE 2. Calmodulin levels in subcellular fractions of embryonic axis (36 h) of C. arietinum

Fractions	ng calmodulin/embryonic axis
Cytosol	2660
Microsomes	197
Mitochondria	19
Chromatin	75
Nuclear sap	4
Cell wall	13

Calmodulin determined using a radioimmunoassay kit (Amersham, U.K.)

shown to be mediated by calmodulin and (ii) according to our results calmodulin is present in the cell wall, it appears possible that calcium could exert biochemical effects after binding with calmodulin (Paliyath and Poovaiah, 1984). We therefore propose that the calcium- calmodulin complex could regulate cell wall loosening, rigidity and cell-to-cell adhesion.

Acknowledgements. This work was supported by a grant from "Comisión Asesora de Investigación Científica y Técnica". Spain. No. 0395-84.

REFERENCES

Biro RL, Daye S, Serlin BS, Terry ME, Datta N, Sapory SK, Roux SJ (1984) Characterization of oat calmodulin, and radio-immunoassay of its cellular distribution. Plant Physiol 75: 382-386

Elliott DC, Batchelor SM, Cassar RA, Marinos NG (1983) Calmodulin-binding drugs affect responses to cytokinin, auxin and gibberellic acid. Plant Physiol 72: 219-224

Fry SC (1983) Feryloyllated pectins from the primary cell wall. Their structure and possible function. Planta 153: 111-123

Fry SC (1986) Polymer-bound phenol as natural substrate of peroxidases. In: Greppin H, Penel C, Gaspar T (eds) Molecular and physiological aspects of plant peroxidases. University of Geneva, Switzerland,

Gaspar Th, Penel C, Castillo FJ (1985) A two-step control of basic and acidic peroxidases and its significance for growth and development. Physiol Plant 64: 418-423

Lamport DTA (1986) Roles of peroxidase in cell wall genesis. In: Greppin H, Penel C, Gaspar T (eds) Molecular and physiological aspects of plant peroxidases. University of Geneva, Switzerland,
Paliyath G, Poovaiah BW (1984) Calmodulin inhibitor in senescing apples and its physiological and pharmacological significance. Proc Natl Acad Sci USA 81: 2065-2069
Palmieri S, Odoardi M, Soressi, Salamini F (1978) Indoleacetic acid oxidase activity in two high-peroxidase tomato mutants. Physiol Plant 42: 85-90
Penel C (1986) The role of calcium in the control of peroxidase activity. In: Greppin H, Penel C, Gaspar T (eds) Molecular and physiological aspects of plant peroxidases. University of Geneva, Switzerland,
Poovaiah BW (1985) Role of calcium and calmodulin in plant growth and development. HortSci 20: 347-351
Putter J (1974) Peroxidases. In: Bergmeyer HU (ed.) Methods of enzymatic analysis. Verlag Chemie, Academic Press,
Raghothama KG, Mizrahi Y, Poovaiah BW (1985) The effect of calmodulin antagonists on auxin-induced elongation. Plant Physiol 79: 28-33
Srivastava OP, Van Huystee RB (1973) Evidence for close association of peroxidase, polyphenol oxidase and IAA oxidase isoenzymes of peanuts suspension culture medium. Can J Bot 51: 2207-2215
Swain T, Hillis WE (1959) The phenolic constituents of *Prunus domestica* 1. The quantitative analysis of phenolic constituents. J Sci Food Agric 10: 63-68
Wright M, Osborne DJ (1974) Abscission in *Phaseolus vulgaris*. The positional differentiation and ethylene-induced expansion growth of specialised cells. Planta 120: 163-170

A THREONINE AND HYDROXYPROLINE-RICH GLYCOPROTEIN FROM MAIZE

Marcia J . Kieliszewski and Derek T.A. Lamport
MSU-DOE Plant Research Laboratory
East Lansing
MI 48824-1312
U.S.A.

INTRODUCTION

The extensins are hydroxyproline-rich glycoproteins (HRGPs) insolubilized in the cell walls of higher plants (Lamport, 1965). They are highly basic and contain about one third protein which is a flexible rod about 80 nm long having repeated glycosylated blocks of the pentapeptide SER-Hyp-Hyp-Hyp (Smith *et al.*, 1986; Van-Holst and Varner, 1984). The carbohydrate components are arabinoside oligosaccharides O-linked to the hydroxyproline residues (hydroxyproline arabinosides) (Lamport, 1967) and galactosyl-serine (Lamport *et al.*, 1973).

The information known about extensin has been gleaned from dicot model systems whose cell walls are rich in hydroxyproline. In contrast, the role of extensin in the monocots has been virtually ignored. This is probably for a very good reason: the walls of monocots, or at least of the graminaceous monocots, are notoriously hydroxyproline-poor (Lamport, 1965; Boundy *et al.*, 1967; Van Etten *et al.*, 1963). This led us to ask: does extensin occur in the graminaceous monocots?

RESULTS AND DISCUSSION

Our system is a *Zea mays* cell suspension culture (cv. Black Mexican, a gift from Dr Tom Hodges, Purdue University, Indiana, U.S.A.). Like a typical monocot, the walls of these cultures are hydroxyproline-poor although they contain the same amount of wall protein (approximately 10%) as the walls of many dicots. Table 1 compares the amino acid compositions of tomato and maize cells. A major difference is in the wall hydroxyproline content which is low in maize. Another difference is the overall hydrophobicity: the maize cell wall protein is much more hydrophobic than the tomato wall protein.

We isolated crude protein by eluting the maize cell surface with 100 mM $AlCl_3$ and purified maize HRGP via ion exchange chromatography and gel

NATO ASI Series, Vol. H 35
Cell Separation in Plants
Edited by D.J. Osborne and M.B. Jackson

TABLE 1. The amino acid compositions of the maize cell wall and tomato cell wall. The walls differ dramatically in their hydroxyproline content and overall hydrophobicity

Amino Acid*	Maize Cell Wall	Tomato Cell Wall
Hyp	1.1	28.5
Asx	10.4	4.0
Thr	5.1	4.5
Ser	6.9	14.3
Glx	9.3	2.8
Pro	3.7	3.9
Gly	10.7	3.3
Ala	10.6	3.2
Cys	0.0	0.0
Val	6.4	7.0
Met	1.7	0.3
Ilu	4.2	1.8
Leu	10.3	2.5
Tyr	1.9	6.3
Phe	4.0	1.3
His	2.1	2.7
Lys	6.2	10.5
Arg	4.7	1.2

* represented as mole %

filtration as described earlier (Kieliszewski and Lamport, 1987). Having purified an HRGP from maize, we wanted to answer the question: does this putative maize extensin look chemically like a typical dicot extensin? This involved amino acid analysis, neutral sugar compositions determination of the hydroxyproline-arabinoside profile, visualization by transmission electron microscopy, peptide mapping and sequence analysis and reaction with Yariv's artificial antigen (Jermyn and Yeow, 1975), a reagent that helps distinguish between extensin and gumlike arabinogalactan proteins (another class of HRGP).

The amino acid composition of the maize HRGP was extensin-like. It was highly basic (12 mol % lysine) and rich in hydroxyproline and proline (25 and 14 mol % respectively), however it contained less tyrosine and valine and was extremely rich in threonine (25 mol %).

Typically, extensins are about two thirds carbohydrate: 90 mol % arabinose and 10 mol % galactose (Showalter and Varner, 1988). In contrast, the maize threonine-rich HRGP (THRGP) is unusual because it is more lightly glycosylated (30% carbohydrate) than typical extensins, and all of the carbohydrate is arabinose. Table 2 shows the hydroxyproline arabinoside profile of the THRGP compared to tomato extensin. While most of the hydroxyprolines are arabinosylated with 3 and 4 arabinose residues, free hydroxyproline and hydroxyproline arabinoside 3 predominate in the THRGP.

Table 2. Hydroxyproline arabinoside profiles of an extensin from tomato compared to maize THRGP

Hydroxyproline Arabinoside	Tomato (% Hyp)	THRGP (% Hyp)
1	9	15
2	8	6
3	33	25
4	38	6
Free Hyp	12	48
Total Hyp	100	100

If the THRGP is an extensin, it must be a structural protein. Structural proteins often produce simple peptide maps due to short term amino acid sequence periodicities (North, 1968). The chymotryptic peptide map of the THRGP is very simple and dominated by only one major peptide: Thr-Hyp-Ser-Hyp-Lys-Pro-Hyp-Thr-Pro-Lys-Pro-Thr-Hyp-Hyp-Thr-Try. Although the Ser-Hyp-Hyp-Hyp-Hyp pentapeptide that is viewed as diagnostic (Showalter and Varner, 1988) of extensin is not a major repeating unit in this protein, Ser-Hyp-Hyp-Hyp-Hyp occurs in only one peptide of the THRGP: Thr-Hyp-Thr-Hyp-Hyp-Val-Ser-His-Thr-Hyp-Ser-Hyp--Hyp-Hyp-Hyp-Tyr. A structural role for the THRGP is corroborated by transmission electron microscopy, which visualizes it as a flexible rod about 70 nm long (Kieliszewski and Lamport, 1987). A final point that argues for monocot extensin is that the THRGP did not precipitate Yariv's artificial antigen, indicating that it is not an arabinogalactan protein, a class of HRGP that precipitates Yariv antigen (Jermyn and Yeow, 1975).

We have presented evidence for the occurrence of extensin in the graminaceous monocots; however, they occur as lesser components of a cell wall whose major protein component is not extensin.

ACKNOWLEDGEMENTS

It is a pleasure to acknowledge Melanie Corlew and Joseph Leykam for their contributions to the paptide sequence data. This work was supported by DOE Grant DE-AC02-76ERO-1338 and USDA Competitive Reseaerch Grant 88-37261-3682.

REFERENCES

Boundy JA, Wall JS, Turner JE, Woychik JH, Dimler RJ (1967) A mucopolysaccharide containing hydroxyproline from corn pericarp, isolation and composition. J Biol Chem 242: 2410-2415

Jermyn MA, Yeow YM (1975) A class of lectins in the tissues of plant seeds. Aust J Plant Physiol 2: 501-531

Kieliszewski M, Lamport DTA (1987) Purification and partial characterization of an HRGP from a graminaceous monocot, *Zea mays*. Plant Physiol 85: 823-827

Lamport DTA (1965) The protein component of primary cell walls. In: Preston RD (ed) Advan Bot Research 2, pp 151-218

Lamport DTA (1967) Hydroxyproline-O-glycosidic linkage of the plant cell wall glycoprotein extensin. Nature 216: 1322-1324

Lamport DTA, Katona L, Roerig S (1973) Galactosyl serine in extensin. Biochem J 133: 125-131

North ACT (1968) The structure and activity of lysozyme. In: Crewthers WG (ed) Symposium on fibrous proteins. Plenum Press, N.Y., pp 13-21

Showalter AM, Varner JE (1988) Biology and molecular biology of plant hydroxyproline-rich glycoproteins. In Marcus A (ed) The Biochemistry of plants: A Comprehensive Treatise Vol. 11: Molecular Biology. (In press)

Smith JJ, Muldoon EP, Willard JJ, Lamport DTA (1986) Tomato extensin precursors P1 and P2 are highly periodic structures. Phytochemistry 25: 1021-1030

Van Etten CH, Miller RW, Wolff IA, Jones Q (1963) Amino acid compositions of seeds from 200 angiospermous plant species. Agric Food Chem 11: 399-410

Van-Holst GJ, Varner JE (1984) Reinforced polyproline-II conformation in a hydroxyproline-rich cell wall glycoprotein from carrot root. Plant Physiol 74: 247-251

THE ROLE OF ENDOSPERM DEGRADATION IN THE GERMINATION OF TOMATO SEEDS

A. M. Haigh, S. P. C. Groot, I. Zingen-Sell, and C. M. Karssen
Department of Plant Physiology
Agricultural University
Arboretumlaan 4
6703 BD Wageningen
The Netherlands

INTRODUCTION

Tomato seeds are flattened and ovoid in shape, up to 4 mm in length, with a curved linear embryo embedded in a non-starchy endosperm. In such seeds, the tissues enclosing the embryo may influence germination by mechanically restraining the expansion of the embryo. Two mechanisms for overcoming this restraint have been proposed: the mechanical force of the embryo pushing against the endosperm, and an enzymatically induced weakening of the endosperm (Ikuma and Thimann, 1963).

While the stimulative effect of gibberellin (GA) application on germination has been known for some time, unequivocal study of the action of endogenous GA during germination has awaited the isolation of GA-deficient mutants. The isolation of such mutants in *Arabidopsis thaliana* has shown the essentiality of GA action for germination (Karssen and Lacka, 1986). The GA-deficient mutant of tomato has permitted the study of the mode of GA action during germination.

The germination of tomato seeds required the presence of endogenous GA (Groot and Karssen, 1987). Stimulatory treatments, such as red light and ethylene were dependent on GA action, but fusicoccin action was independent of GA. As with lettuce (Tao and Khan, 1979) and capsicum seeds (Watkins and Cantliffe, 1983), the germination of tomato seeds was associated with a weakening of the resistance offered by the endosperm opposing the radicle tip. This weakening did not occur in GA-deficient mutant seeds unless GA was supplied in the imbibition medium (Fig. 1).

The embryo of the wild-type tomato seed produced a factor (probably GA, but not definitively identified) which led to the induction of a degradation process in the endosperm. The degradation could be induced in de-embryonated seed halves of both wild-type and mutant by incubation in a medium containing

NATO ASI Series, Vol. H 35
Cell Separation in Plants
Edited by D. J. Osborne and M. B. Jackson

GA or the wild-type embryo. This degradative weakening of the endosperm had to occur before germination could result (Groot and Karssen, 1987).

RESULTS AND DISCUSSION

A study of the changes in the water relations of germinating tomato seeds has provided confirmatory evidence for the role of the endosperm (Haigh and Barlow, 1987). By regarding germination as a specialised growth phenomenon it was possible to determine the roles of the various tissues in the control of germination. During imbibition, water uptake continued until water potential equilibrium was established between the seed and the imbibitional solution. Changes in the properties of the embryo were found prior to radicle emergence. The embryo was capable of expansive growth prior to radicle protrusion, but the endosperm tissue enclosing the embryo was found to restrict the hydration level of the embryo prior to its emergence (Fig. 2). Further water uptake can only result from the removal of the restraining pressure imposed by the enclosing tissues. This pressure maintained the water potential equilibrium between the restrained tissue and the imbibing water and thus prevented further water uptake. For radicle emergence to occur weakening of the endosperm must be necessary.

The restraint imposed by the endosperm was lessened by processes involved in the degradation of the endosperm cells (Groot *et al.*, 1988). The weakening of the endosperm of de-embryonated GA-deficient mutant seeds was accompanied by the induction of endo-beta-mannanase activity and the release of mannan from the endosperm cell walls. GA treatment also increased the activity of mannohydrolase, an enzyme involved in the degradation of mannan-rich cell walls (Fig. 3).

The morphological changes occurring during germination of tomato seeds have also been studied recently (Haigh, 1988). The endosperm of tomato seeds consists of two identifiable cell types found in separate locations within the seed. At the micropylar end of the seed (surrounding the radicle) the endosperm cells have thin walls, whereas those in the rest of the seed have much thicker walls. During germination, protein body breakdown occurs in the endosperm cells of the micropylar region, while no change is observed in the thick- walled endosperm cells of the rest of the seed (Fig. 4).

CONCLUSIONS

Most work on the endosperm of germinating seeds has concentrated on the mobilization of reserves after radicle emergence. The similarity between the changes occurring in the cereal aleurone layer and the dicot endosperm cells has been recognised (Jacobsen, 1984), but little attention has been paid to the control of endosperm weakening during the germination of dicot seeds or to the role of the aleurone layer enclosing the cereal embryo prior to radicle emergence. It is probable that the weakening processes occurring in dicot endosperm also occur in the aleurone layer enclosing the embryo of cereals. This layer has been shown to control the dormancy of *Avena fatua* seeds by controlling the uptake of water by the embryo (Raju *et al.*, 1986).

REFERENCES

Groot SPC, Karssen CM (1987) Gibberellins regulate seed germination in tomato by endosperm weakening: a study with GA-deficient mutants. Planta 171:525-531

Groot SPC, Kieliszewska-Rokicka B, Vermeer E, Karssen CM (1988) Gibberellin-induced hydrolysis of endosperm cell walls in gibberellin-deficient tomato seeds prior to radicle protrusion. Planta 174:500-504

Haigh AM, (1988) Why do tomato seeds prime? Physiological investigations into the control of tomato seed germination and priming. PhD Thesis, Macquarie University, Sydney

Haigh AM, Barlow EWR (1987) Water relations of tomato seed germination. Aust J Plant Physiol 14:485-492

Ikuma H, Thimann KV (1963) The role of the seed-coats in germination of photosensitive lettuce seeds. Plant Cell Physiol 4:169-185

Jacobsen JV (1984) The seed: germination. In: Johri BM (ed) Embryology of angiosperms. Springer-Verlag, Berlin Heidelberg New York Tokyo, p 611

Karssen CM, Lacka E (1986) A revision of the hormone balance theory of seed dormancy: studies on gibberellin and/or abscisic acid-deficient mutants of *Arabidopsis thaliana*. In: Bopp M (ed) Plant growth substances 1985. Springer-Verlag, Berlin Heidelberg New York Tokyo, p 315

Raju MVS, Hsiao AI, McIntyre GI (1986) Seed dormancy in *Avena fatua*. III. The effect of mechanical injury on the growth and development of the root and scutellum. Bot Gaz 147:443- 452

Tao K-L, Khan AA (1979) Changes in the strength of lettuce endosperm during germination. Plant Physiol 63:126-128

Watkins JT, Cantliffe DJ (1983) Mechanical resistance of the seed coat and endosperm during germination of *Capsicum annuum* at low temperature. Plant Physiol 72:146-150

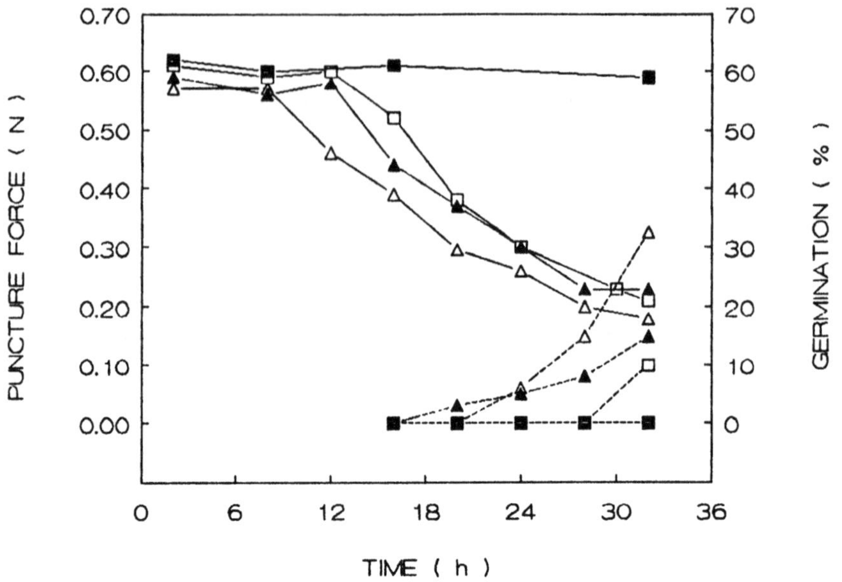

FIGURE 1. Change in the force required to puncture the layers opposing the radicle tip (solid lines) and of germination (broken lines) of wild-type (△▲) and GA-deficient (□■) seeds incubated in water (closed symbols) or 10µM GA_{4+7} (open symbols). See Groot and Karssen, 1987.

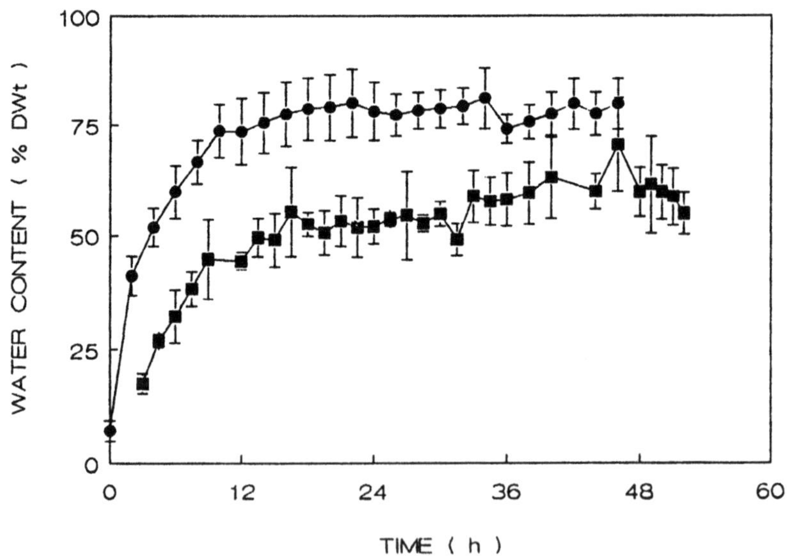

FIGURE 2. Water uptake patterns of whole tomato seeds (●) and embryos within seeds (■) during germination in water. See Haigh and Barlow, 1987.

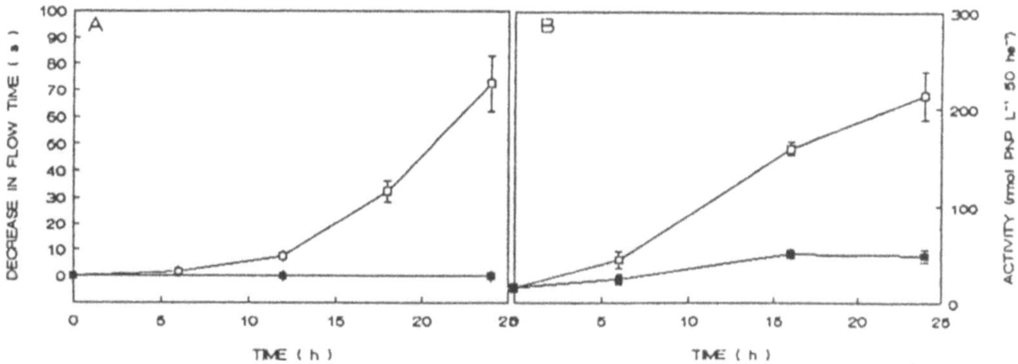

FIGURE 3. Activity of endo-beta-mannanase (A) and mannohydrolase (B) extracted from the de-embryonated seed halves of GA- deficient tomato seeds after incubation in water (■) or 10 µM GA_{4+7} (□). See Groot et al., *1988.*

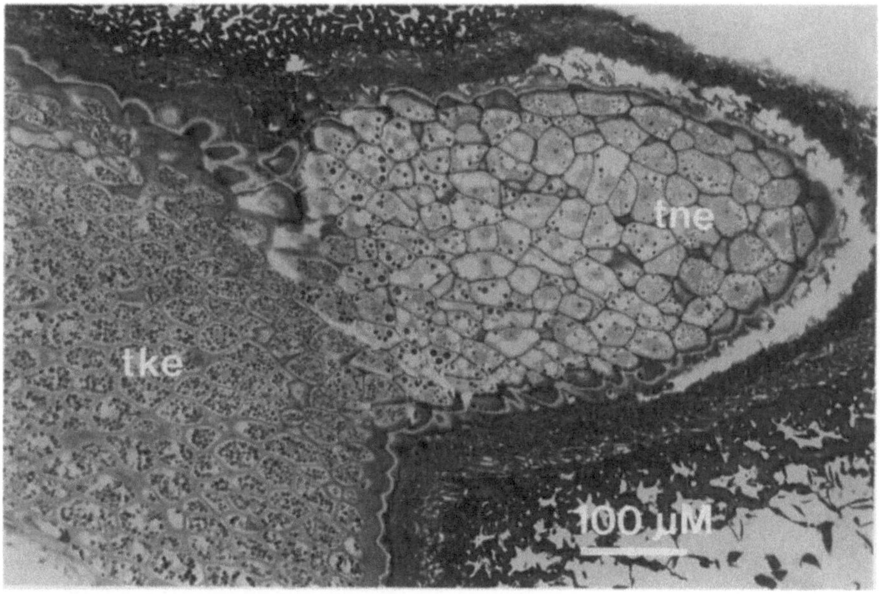

FIGURE 4. Endosperm cells of a tomato seed imbibed in water at 25°C for 48 hours. This section is from an ungerminated seed from a population of seeds of which 60% had germinated at this time. tne-thin-walled endosperm, tke-thick-walled endosperm.

AUXIN-DEPRIVED PEAR CELLS IN CULTURE AS A MODEL SYSTEM FOR STUDYING SENESCENCE

C. Balagué, J.M. Lelièvre and J.C. Pêch
Ecole Nationale Supérieure Agronomique
145, Avenue de Muret
31076 Toulouse Cédex - France

INTRODUCTION

The senescence of plant cells has been associated with the *de novo* synthesis of a great number of proteins (Laurière, 1983; Sabater,1986). It becomes more and more clear however, that the cessation of the synthesis of some proteins may also play an important role in the loss of some cellular functions. It has been shown, for instance, that several mRNAs present in mature tissues disappear during the senescence process (Schuster and Davies, 1983). A shift in the hormonal balance is probably responsible for these changes: increased synthesis or action of senescence-promoting hormones (ethylene and ABA) and decreased synthesis or action of senescence-retarding hormones (auxins and cytokinins).

Previous studies had shown that auxin-deprivation of plant cells in culture results in the development of physiological and ultrastructural changes similar to those occuring in senescing tissues (Balagué *et al.*, 1982). In this model, re-addition of auxins and ABA respectively delay and accelerate the senescence process (Balagué *et al.*, 1986) by reducing or increasing over a long period of time the synthesis of polypeptides associated with the maintenance or loss of cell integrity (Lelièvre *et al.*, 1987).

In this paper, we now examine the short term effects of 2,4-D and ABA in order to assess their direct action on the early stimulation or repression of the synthesis of specific polypeptides involved in the control of the senescence process.

MATERIAL AND METHODS

Quiescent cell suspensions of Passe-Crassane pear fruits (*Pyrus communis*L.) were cultured in a closed-continuous bioreactor as described by Pech and Romani (1979) and Balagué *et al.* (1982). Cell viability and ethylene production were measured as reported by Balagué *et al.*(1982). *In vivo* labelling of cells was performed by incubating 2.5 ml of cell suspension in 25 ml flasks 20 h at 25°C in the presence of 740 kBq ^{35}S methionine (41.1 $TBq.mmol^{-1}$ CEA, France). Rapid changes induced by hormones were estimated by removing samples at day 7 after 4 h of hormone treatment (2.3 µM 2,4-D or 7.6 µM ABA) followed by incubation for 4 h with ^{35}S methionine. Extraction of proteins and

NATO ASI Series, Vol. H 35
Cell Separation in Plants
Edited by D. J. Osborne and M. B. Jackson

two dimensional electrophoresis were previously described (Lelièvre *et al.*, 1987).

RESULTS

During culture in the continuous bioreactor in the absence of 2,4-D, no division occured and cells remained alive for at least 15 days and then rapidly died (Fig. 1A). Ethylene production decreased sharply during the first days and then remained constant at a very low rate of about 7 $pl.h^{-1}$ per 10^5 cells (Fig.1B). When the medium was supplemented with 2,4-D (2.3 µM) at day 7, cell division did not resume and the onset of cell death was postponed by at least 15 days (Fig.1A) even though ethylene production was greatly stimulated to reach a constant rate of 25 $pl.h^{-1}$ per 10^5 cells 2 days after the treatment (Fig.1A). In contrast, cell death was significantly hastened by 7.6µM ABA : only 20% of ABA-treated cells remained alive at day 15 as compared to 50% in the control (Fig.1A). In this case, ethylene production was always higher than in control cells (Fig. 1B).

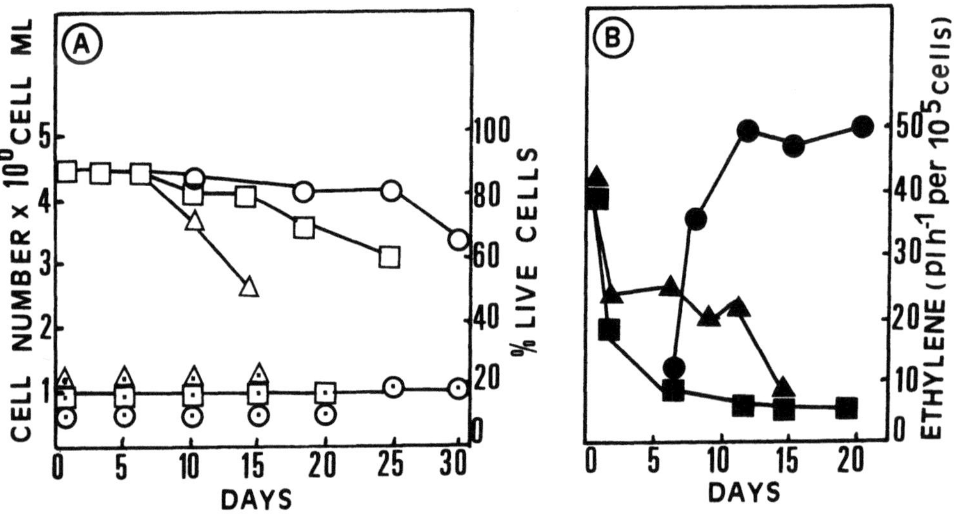

FIGURE 1 : Changes in total cell number (⊡,◬,⊙) live cells (□,△,○,A) and ethylene production (closed symbols, B) during the culture in continuous bioreactor of auxin-deprived cells in the absence (⊡,□,■) or in the presence of ABA (◬,△,▲); or after re-addition of 2,4-D at day 7 (⊙,○,●).

Two dimensional analysis of *in vivo* labelled polypeptides indicated that auxin starvation resulted in the stimulation or inhibition of the synthesis of a limited number of polypeptides (Fig. 2). Two groups could be characterized. The first group, whose synthesis increased during senescence (referred to as "Senescence-Related-Proteins, SRP, indexed as letters in Fig. 2B), included mainly acidic proteins with a molecular weight ranging from 14 to 80 kD. The second group whose synthesis was reduced (referred to as "Homeostasis-Related-Proteins, HRP, indexed as numbers) included mostly neutral polypeptides of 24 to 45 kD (Fig 2A).

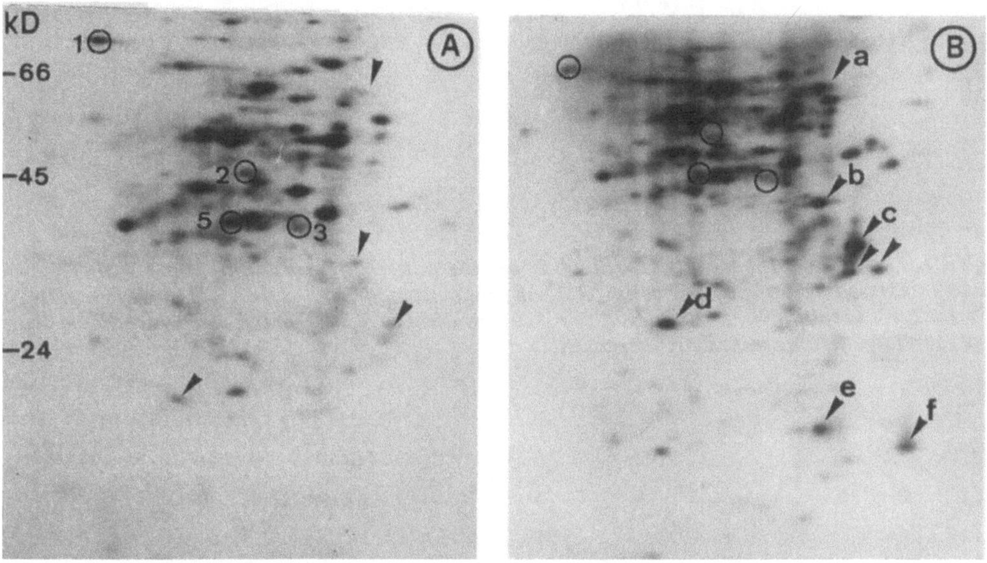

FIGURE 2. Pattern of labelled polypeptides during a culture of bioreactor in auxin-deprived cells . Labelling was performed for 20 h using ^{35}S methionine. A= day 7, B= day 18. Arrows and circles indicate, respectively, the polypeptides whose synthesis is increased or decreased.

A brief treatment of auxin-deprived cells (8 hours) with 2.3 uM 2,4-D resulted in the resumption of the synthesis of HRPs 2,3,4,5,6 (Fig. 3) but had no significant effect on the synthesis of SRPs. In contrast, ABA treatment caused a stimulation of the synthesis of SRPs b and c (Fig. 4).

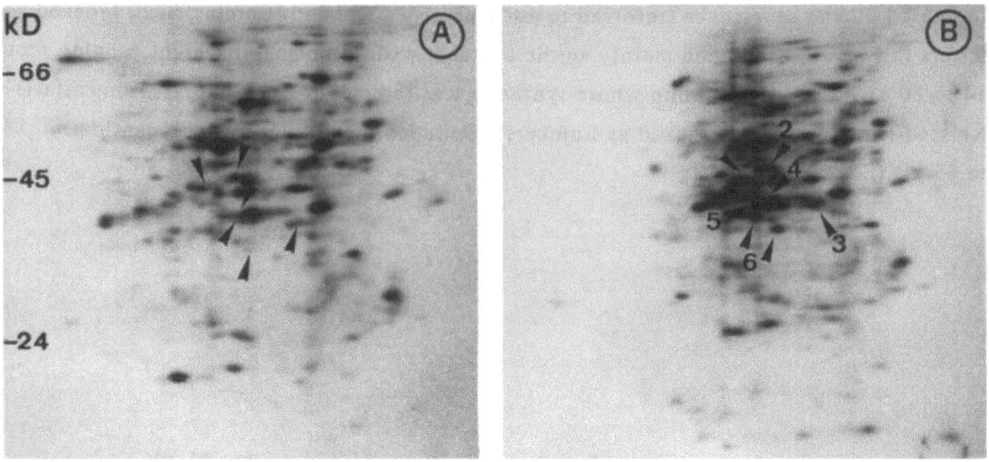

FIGURE 3. Short-time effects of 2,4-D on the pattern of polypeptides of 7 day-old auxin-deprived cells: Control cells (A) and cells treated with 2,4-D for 8 hours (B). Radioactive labelling was performed for 4 hours using ^{35}S methionine. Arrows indicate the polypeptides whose synthesis is increased.

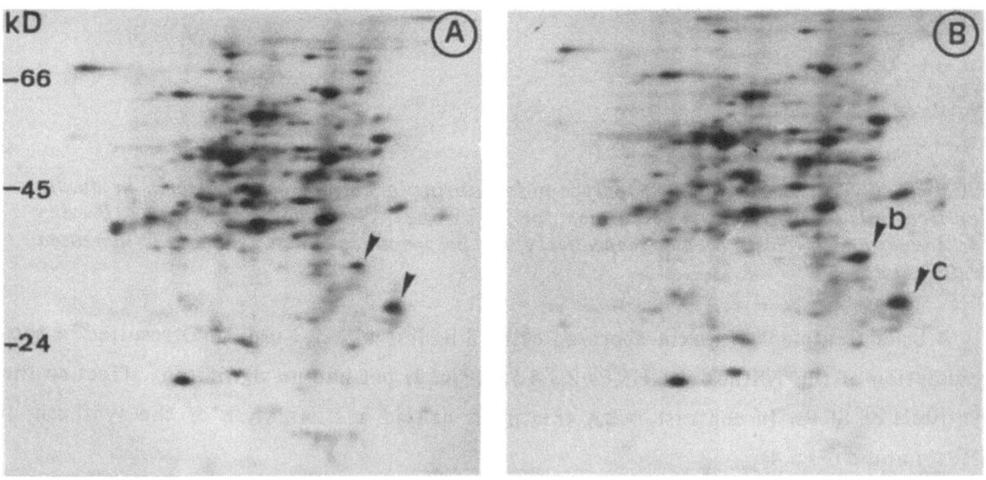

FIGURE 4 . Short-time effects of ABA on the pattern of polypeptides of 7 day-old auxin-deprived cells : control cells (A) and cells treated with ABA for 8 hours (B). Labelling was performed as in Figure 3.

CONCLUSIONS

In the present work, a model system of pear cells grown in continuous culture has been used. It readily allowed manipulation of the level of hormones to delay or accelerate cell senescence.

Our previous results indicated that auxin deprived cells underwent changes similar to those occuring in other senescing tissues (Balagué *et al.*, 1982). Nevertheless, in these system ethylene treatment had no influence in hastening the senescence process; but on the otherhand, ethylene had a marked effect on the reduction of the size of aggregates corresponding to a dissociation of the cell wall (Balagué *et al.*, 1986). We had also demonstrated (Lelièvre *et al.*, 1987) that auxin starvation was associated not only with an increase in the synthesis of some polypeptides (referred to as SRPs) but also with a decrease of some others (referred to as HRPs). The increased synthesis of some polypeptides has been observed in auxin-starved soybean cells (Leguay and Jouanneau, 1987) or during aging of different tissues (Skadsen and Cherry, 1983). A decrease in synthesis of several proteins or enzymes has also been reported during aging of others plant systems (Laurière, 1983; Skadsen and Cherry, 1983). We now present evidence for a direct effect of 2,4-D and ABA on the short term regulation (4 hours) of the synthesis of some of these polypeptides. The stimulation of the synthesis of several HRPs or SRPs, respectively by 2,4-D and ABA, could represent the primary events accounting for the delay or acceleration of cell death. Long term effects of these hormones (20 hours) had shown (Lelièvre *et al.*, 1987) that the synthesis of SRPs was decreased in the presence of auxin and increased in the presence of ABA. Inversely, the synthesis of HRPs was stimulated in the presence of 2,4-D and reduced in the presence of ABA. One could therefore speculate that senescence could be not only dependant upon the expression of new proteins (SRPs) but also upon the cessation of the expression of some preexisting proteins (HRPs). This concept fits with the observation that senescent cells, tissues or organs loose some cellular functions and become unable to maintain their homeostasis (Romani, 1984).

Preliminary results on protein synthesis in cell-free extracts of auxin-deprived cells indicated a significant increase in the synthesis of SRPs polypeptides "b, c and d" (data not shown). The similarities of protein syntesis *in vivo* and *in vitro* suggest that changes in synthesis arise from the modulation of specific mRNAs or from limited post translational processing.

The model system used in our studies can be a useful tool for studying hormone mediated control of gene expression. The elaboration of recombinant cDNA probes is in progress to ascertain whether the observed changes in protein and mRNA synthesis reflect increased rates of genes expression in response to hormones.

REFERENCES

Balagué C, Latché A, Fallot J, Pech JC (1982) Some physiological changes occuring during the senescence of auxin-deprived pear cells in culture. Plant Physiol 69: 1339-1343

Balagué C, Lelièvre JM, Pech JC (1986) Interrelations between ethylene and abscisic acid in the loss of membrane integrity of 2,4-D pear cells in culture. Physiol Vég 24:581-589

Laurière C (1983) Enzyme and leaf senescence. Physiol Vég 21:1159-1177

Leguay JJ, Jouanneau JP (1987) Patterns of protein synthesis in dividing and auxin starved soybean cell suspensions. Differential expression of a major group of Mr-17000 peptides. Plant Cell Rep 6:235-238

Lelièvre JM, Balagué C, Pech JC (1987) Protein synthesis associated with quiescence and senescence in auxin-starved pear cells. Plant Physiol 85:400-406

Pech JC, Romani R (1979) Senescence of pear fruit cells cultured in a continuously renewed auxin deprived medium. Plant Physiol 64:814-817

Romani R (1984) Respiration, ethylene, and homeostasis in an integrated view of postharvest life. Can J Bot 62:2950-2955

Sabater B, (1986) Hormonal regulation of senescence. In : Hormonal regulation in plants. Purdhit SS (ed) Martinus Nijhoff/Dr. W Junk, Dordrecht. pp. 169-217

Schuster A, Davies E (1983) Ribonucleic acid and protein metabolism in pea epicotyls. I. The aging process. Plant Physiol 73:809-816

Skadsen RW, Cherry H (1983) Quantitative changes in *in vitro* and *in vivo* protein synthesis in aging and rejuvenated soybean cotyledons. Plant Physiol 71:861-868

STUDIES ON THE RELEASE OF MESOPHYLL PROTOPLASTS FROM CINCHONA LEDGERIANA L.

Adele F. Pow and C.S. Hunter
Plant Sciences Group
Bristol Polytechnic
Frenchay
Bristol BS16 1QY, UK

INTRODUCTION

The tropical tree *Cinchona* is the source of more than 30 alkaloids including quinine and quinidine. The trees forming the main commercial plantations are outbreeders and show wide variation in alkaloid yield. High-yielding trees have resulted from extensive conventional breeding programmes, and clonal micropropagation of such trees is now undertaken (Hunter, 1988). Our interest focuses on the rapid generation of novel genotypes to introduce characters that are of agronomic value or will help to understand aspects of gene expression connected with the production of quinoline alkaloids *in vitro*. Our short-term pre-requisite is to isolate protoplasts and to regenerate plants therefrom. The conventional approach to isolating protoplasts from any new plant system is empirical (Bengochea and Dodds, 1986; Power and Chapman, 1985). The donor tissue is subjected to commercial preparations of macerating enzymes (principally pectinases) and cell wall hydrolysing enzymes (principally cellulases and hemicellulases) in an appropriate osmoticum, either sequentially or simultaneously and any resulting protoplasts collected. 'Successful' isolation is usually the result of painstaking alterations to enzyme concentrations and incubation conditions. The isolation of protoplasts impinges on areas of plant metabolism and biochemistry too complex and wide to be elucidated by most groups using tissue culture predominantly as a tool in other studies. The criticism has been levelled that plant tissue culturists publish or use little basic science in the development of their techniques. It is our thesis that the scientific understanding of the physiological processes of viable protoplast separation would be greatly enhanced by the better use of specialised inputs from workers in complimentary areas such as stress metabolism, cell wall structure, production and release of toxins from damaged cells and enzymology.

METHODS

The uppermost two pairs of leaves from light or dark-grown plants were

NATO ASI Series, Vol. H 35
Cell Separation in Plants
Edited by D. J. Osborne and M. B. Jackson

removed from *in vitro* cultures and subjected to various conventional protoplast-releasing procedures: pre-plasmolysis, preparation into thin strips, incubation in cell macerating and wall hydrolysing enzymes, the addition of various organic compounds to the preparations, separating and concentration of the fractions of the resultant digest.

TABLE 1. Range of enzymes and concentrations tested in various combinations for protoplast release

Cellulase R10	0.5 - 2.5% w/v
Cellulase RS	1.0 - 2.0% w/v
Cellulase YC	1.0 - 2.0% w/v
Cellulysin	0.5 - 2.5% w/v
Driselase	0.5 - 1.5% w/v
Macerase	0.25 - 1.5% w/v
Macerozyme R200	0.25 - 1.5% w/v
Meicelase	0.5 - 1.0% w/v
Pectolyase Y23	0.01 - 0.1% w/v
Rhozyme HP150	0.5% w/v

All enzyme mixtures were clarified by centrifugation and filter sterilised before use (Sihachakr and Ducreux, 1987)

These various processes have been examined for their effect on the release and 'quality' of protoplasts. 'Quality' was assessed by microscopic examination and by viability assessment with fluorescein diacetate (Widholm, 1972). Analysis of enzyme digests has been by paper and thin layer chromatography (Fry, 1988).

RESULTS

Generally, release of protoplasts was poor and varied from one experiment to the next, but two protocols (A and B) were sufficiently promising to form a basis for further development. There appears to be two age bands of source cultures from which reproducible leaf mesophyll protoplast release was most successful under the various conditions investigated. These bands are between 37-45 days post-subculture for system A and between 65-75 days for system B. Reproducible yields were variable - the minimum for A was 3×10^6 but the maximum for B was 1.4×10^5 protoplasts per gram fresh weight of leaf.

TABLE 2. Protocols A and B for protoplast release

System A

Pretreatment: CPW salts (Power and Chapman, 1985) for 15 mins

Enzymes (% w/v): Cellulysin (2)+Macerase (0.5)+Driselase (0.5) dissolved in B5 medium (Gamborg, 1970) containing glucose (68.4 g l^{-1}) + sucrose (0.25 g l^{-1} pH 5.0

Weight of leaves: vol of enzyme mix, 0.36:1

Incubation: stationary, darkness, 25°C for 10-14 hours

System B

Pretreatment: CPW salts + 13% w/v mannitol for 1 hour

Enzymes (% w/v): Cellulysin (1)+ Macerase (0.5)+ Driselase (1) dissolved in CPW salts + 13% w/v mannitol + 1% w/v histidine + 20 mM MES + 0.25% w/v PVP (MWt 44000), pH 5.0

Incubation: stationary, darkness, 25°C for 7-11 hours

Experiments aimed at developing a sequential enzyme system by treating first with macerating enzymes, washing the free cells and secondly with a preparation of wall hydrolysing enzymes were generally unsuccessful: digestion of both the middle lamella and primary cell wall to release protoplasts occurred with greater rapidity following a mixed enzyme treatment.

Although leaf maceration and protoplast release increased with increasing incubation up to about 14 hours, the viability of the protoplasts and hence the overall yield, decreased with incubation periods greater than 7-10 hours. Furthermore, the viable protoplasts recorded in enzyme digests at the end of the incubations were generally lost subsequently during harvesting by floatation on appropriate concentrations of Percoll, Ficoll or sucrose in Babcock bottles. The large amounts of cell debris present often displayed similar buoyant density and hence co-harvested with the protoplasts.

Preliminary comparisons of sorbitol to mannitol as the main osmoticum in system B, showed greater and more rapid release with sorbitol, especially in the absence of histidine, though the effects of these on long-term viability have not been studied.

DISCUSSION

The general lack of high yields of viable protoplasts is typical of similar work with many other woody species (Davey and Power, 1988) and reflects the lack of understanding of most of the underlying processes and physiology of protoplast release. The poor rate of cell separation by pectinase enzymes alone almost certainly reflects complexing of the primary cell wall with the middle lamella which, without the synergistic action of cellulases, xylanases *etc.*, remains substantially intact.

Studies on plant defence mechanisms suggest that oligosaccharides released from cell walls may act as endogenous elicitors of phytoalexin synthesis (Davis *et al.*, 1984). It is proposed (Hahne and Lorz, 1988) that such compounds may adversely affect the release of protoplasts. Indeed they may contribute to the overdigestion syndrome often present in our preparations. The addition of histidine, other amino acids and organic compounds has been found by many workers to be helpful in protoplast release (Ochatt, personal communication, 1988). The addition of PVP was to stabilise pH and to reduce the effect of any phenolic compounds released into the system during digestion treatments (Scragg *et al.*, 1988). Chromatographic analysis of digest fractions show that differences in oligosaccharides, sugar monomers and alditols do not yet explain the discrepancies in tissue hydrolysis, but that changes in phenolic compounds released may do so.

Associated with the enzymic hydrolysis processes of cell walls is the release of ethylene (see Mattoo, this volume): this is, in part, regarded as a response to stress. Generally there is a peak of ethylene release 2 to 3 hours after the start of enzyme treatment. Although ethylene is closely associated with natural separation of cells in abscission and senescence, it would still seem reasonable to try to reduce all stresses to minimise the physiological trauma imposed during protoplast release. The change from mannitol to sorbitol seems also to be associated with reducing stress (Mattoo, this volume). However, many research groups have been successful in protoplast work when using mannitol. Opinion is divided as to whether or not it is better replaced by sorbitol: further study in our system should clarify the position. It is increasingly apparent that in order to improve yields of viable protoplasts a greater understanding of *Cinchona* wall structure and its enzymic hydrolysis is needed. The traditional approach to protoplast isolation is limiting and no longer satisfactory. It is necessary to determine the signals and processes involved to predict better the most likely effects of changes in protocol. This is particularly important since numerous studies with dicot cell walls suggest that gross differences in composition alone are insufficient to explain the range in digestibility observed at different plant ages, between clones of the same species and at different stages in the cell cycle. The range of cell wall hydrolysing enzymes present in commercial fungal extracts is poorly defined as are the 'impurities' added as stabilisers by the manufacturers, which could have major inhibitory effects on protoplast release and viability. It is clear that much fundamental study on these processes remains to be done.

REFERENCES

Bengochea T, Dodds J (1986) Plant protoplasts. Chapman and Hall, London

Davey MR, Power, JB (1988) Aspects of protoplast culture and plant regeneration. In: Puite KJ, Dons JJM, Huizing HJ, Kool AJ, Koornneef M, Krens FA (eds) Progress in plant protoplast research. Kluwer, Dordrecht, p 15

Davis KR, Lyon GD, Darvill AG, Albersheim P (1984) Host-pathogen interactions XXV. Plant Physiol 74: 52-60

Fry SC (1988) The growing plant cell wall: chemical and metabolic analysis. Longman Scientific and Technical, Harlow

Gamborg OL (1970) The effects of amino acids and ammonium on the growth of plant cells in suspension culture. Plant Physiol 45: 371-375

Hahne G, Lorz H (1988) Release of phytotoxic factors from plant cell walls during protoplast isolation. In: Puite KJ, Dons JJM, Huizing HJ, Kool AJ, Koornneef M, Krens FA (eds) Progress in plant protoplast research. Kluwer, Dordrecht

Hunter CS (1988) *Cinchona*: micropropagation and the *in vitro* production of quinine. In: Bajaj YPS (ed) Biotechnology in agriculture and forestry, Vol. 4 "Medicinal and aromatic plants I" Springer Verlag, Heidelberg, p 367

Power JB, Chapman JV (1985) Isolation, culture and genetic manipulation of plant protoplasts. In: Dixon RA (ed) Plant cell culture. ITL Press, Oxford, p 37

Scragg A, Allan E, Morris P (1988) Investigation into the problems of initiation and maintenance of *Cinchona ledgeriana* suspension cultures. J. Plant Physiol 132: 184-189

Sihachakr D, Ducreux G (1987) Variation or morphogenetic behaviour and plant regeneration in cultured protoplasts of *Solanum nigrum*. Plant Sci 52: 117-126

Widholm JM (1972) The use of fluorescein diacetate and phenosafranine for determining viability of cultured plant cells. Stain Technol 47: 189-194 [illegible] is difficult and no longer satisfactory [illegible] necessary to [illegible] signals and [illegible] involved [illegible] possible [illegible] study [illegible] of changes in protocols. [illegible] with [illegible] positive [illegible] sufficient to [illegible] recent stages in the [illegible] [illegible]

CELL SEPARATION EVENTS IN POPLAR IN RESPONSE TO SULPHUR DIOXIDE AND OZONE: INVOLVEMENT OF ETHYLENE

Hariklia Kargiolaki
Department of Forestry and Natural Resources
Aristotelian University of Thessaloniki
THESSALONIKI 54006, GREECE.[1]

INTRODUCTION

The harmful effects of sulphur dioxide (SO_2) and ozone (O_3) on plants, are generally accepted, even if they are not always visible. Growth and yield have been found to decrease even in the absence of any obvious injury (Malhotra and Khan, 1984). Sulphur dioxide symptoms on leaves, in cases of visible injury, include a chlorotic appearance due to the destruction of chlorophyll (Malhotra and Khan, 1984). More advanced injury causes necrotic brownish areas close to the midrib (Barrett and Benedict, 1970). Ozone causes the appearance of sharply defined dot-like lesions or bleached areas. A shiny, oily or waxy appearance of either the upper or the lower surfaces also develops in some plant species. In the severest cases, bifacial necrosis occurs (Hill *et al.*, 1970). Premature leaf abscission has also been reported as a response to both SO_2 and O_3 (Taylor *et al.*, 1986), but no visual effects on the stem region have so far been reported, for any of the gases. Sometimes, the appearance of lesions or cankers on the main stem or trunk, has been attributed to secondary infections from fungus or bacteria, after plants were stressed by the effects of the gaseous air pollutants (Treshow, 1984).

This paper discusses the effects of SO_2 and O_3, when applied alone and in combination, upon cell organization and cell to cell association in abscission zones and stem intumescences in four genetically linked poplar clones. Differences in the response of the four clones to the two gases is investigated, when applied separately and together. Ethylene evolution from leaves of the four clones, is measured after fumigation with SO_2 and O_3.

---

[1] Present address: Department of Plant Sciences, Oxford University, South Parks Rd., Oxford, OX1 3RB, ENGLAND

NATO ASI Series, Vol. H 35
Cell Separation in Plants
Edited by D. J. Osborne and M. B. Jackson

MATERIALS AND METHODS

Statistical design and treatments. The statistical lay-out of the experiments was a split-plot design with four gas treatments. The gas treatments included charcoal scrubbed air, SO_2, O_3 and SO_2+O_3. The gas concentrations used were 262-291 $\mu g\ m^{-3}$ SO_2 (90-100 ppb), 153-175 $\mu g\ m^{-3}$ O_3 (70-80 ppb) and a mixture of the two. They are all concentrations likely to occur in urban areas (Bell, 1984). The clones used were two parental clones, *P. deltoides* (female) and *P. italica* (male), with their hybrid X3 (female) and K7 (male), which is a hybrid of the *P. italica* clone with *P. nigra* cultivar 'Serres' (female).

Experimental equipment. Experiments were carried out in closed fumigation chambers (floor area 2 m^2, 1.5 m high), in a greenhouse. The chambers were constructed of chemically inert Tedlar (PVF) plastic walls of low gas permeability and supported on enamelled steel frames. The air flow (1.25 m^3 sec^{-1}) gave an air change approximately every 2 minutes. The aerodynamic leaf boundary layer resistance of the chambers, with the plants inside was 1.9 s cm^{-1}. Thirty two plants were placed in each of the chambers, in four random groups. Each group included two replicates of each clone, giving a total of eight plants per gas treatment.

The SO_2 concentration in the chambers was achieved by mixing liquified sulphur dioxide (BDH Chemicals Ltd.) with charcoal filtered air in a Signal 852 Series VI-SL gas blender (flow rate= 1 l min^{-1}). A Meloy, FPD, Sulfur Dioxide Analyzer (SA 285) was used for monitoring the sulphur dioxide concentration. It was calibrated against a SO_2 calibrator, Model [SI] CS 10-3. Ozone was produced by a Wallace and Tierman electrical discharge ozonator tube and it was detected by using an AID (Analytical Instruments Development, Inc.) portable Ozone Analyzer (Model 560).

Description of the plants. The plants were propagated from cuttings, rooted in John Innes No 3 potting compost. They were pruned after one year at 15-20 cm height, so the plants used in the experiments had a two-year-old rooting system and small part of the stem, whereas the rest of the above ground part was one-year-old. Experiments were started after full leaf expansion either in August or September.

The plants of *P. deltoides* have a small number of large leaves (approximately 35 leaves per plant, mean leaf size 60.3 cm^2). X3 also is a large-leaved plant (70 leaves per plant, mean leaf size 47.7 cm^2). *P. italica* and K7 are both small-leaved clones. These plants had mean leaf size of 19.7 cm^2 and 12.5 cm^2 respectively and about 180 leaves each.

Light microscopy sample preparation. Cell organization and cell to cell association were investigated by light microscopy, in the parts of the stem showing the injury. Pieces of stem were embedded in water wax made up by mixing polyethylene glycol 4000 to polyethylene glycol 1500 (BDH Chemicals Ltd.) in a 3:1 ratio, and adding 1% glycerol. The stem samples were first dehydrated by keeping them in 10% acrolein (BDH Chemicals Ltd.,) for 24 hours. They were then immersed in increasing concentrations of water wax, for 24 hours in each. The wax concentrations used were 30%, 60%, 95%, 100%. The first two changes were performed at room temperature, whereas the last two at 45°C. The stem section was frozen at -2°C. After 24 hours, the frozen embedded stem was warmed to room temperature until sectioned with a microtome. The stem sections were 20-30 µm thick and were stained with 0.1% toluidine blue in 0.1 M phosphate buffer (pH 6.5).

Scanning microscopy sample preparation. Injured stem parts were fixed in 3% gluteraldehyde, for scanning electron microscopy (SEM) observations. They were kept in this fixative at room temperature for 1 year. They were then dehydrated by successively rinsing each for two hours, in different concentrations of ethanol (10%, 25%, 50%, 75%, 90% and twice in 100%), before critical point drying and mounting with epoxy resin and gold coating for 10 minutes in an SEM Coating unit E5000 (POLARON Equipment Ltd.) flushed with argon or nitrogen. A Cambridge Stereoscan 150 was used for the observations.

Ethylene evaluation. The effects of the gas treatments, on ethylene evolution from leaves of the four poplar clones were investigated with a 2 layer split-plot experiment, having a 4 x 4 x 3 factorial structure. Leaves were collected 6, 18, 30, 43, 55, 67, 92, 113 and 135 hours from the beginning of fumigation. Old and young leaves were picked in a random sequence, so that differences could be investigated. They were put in plastic bags and handled as gently as possible. After weighing, they were placed in air-tight glass jars (720 cm^3), fitted with a 'Suba-seal' sampling port. Vaseline was applied to all joints. The leaves were kept in the dark, at room temperature.

Two 1 ml air samples from each glass jar were analyzed for ethylene, using a Pye Unicam Series 104 gas chromatograph with an alumina F_1 column. Nitrogen was used as the carrier gas at a flow rate of 30 ml min^{-1} and an oven temperature of 110°C (Ward *et al.*, 1978).

RESULTS

Leaf abscission. The number of leaves retained by the plants of each clone during the experiments is given in Fig. 1.

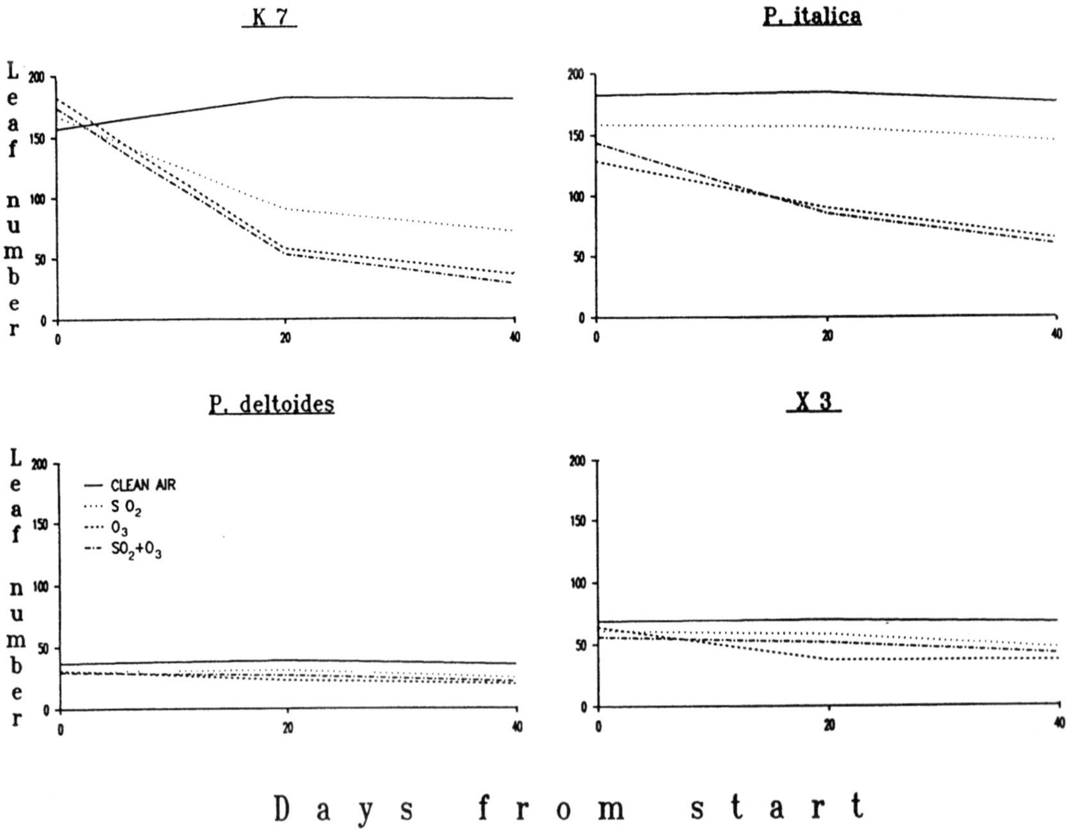

FIGURE 1. Leaf number after different fumigation treatments.

The shed of the leaves of *P. deltoides* and X3 was not statistically different from the control and among the three gas treatments, throughout the experiment. In contrast, *P. italica* and K7 showed a significant leaf loss in the fumigated plants. *P. italica* lost 49% and 58% of its initial leaf number, by the 40th day, during the O_3 and the SO_2+O_3 treatments respectively, whereas K7 lost 58% during the SO_2 and more than 80% in the two O_3 including treatments. Natural leaf abscission occurred only after the 50th day in the clean air treatments; a consequence of conducting the experiment in the autumn. By the 60th day, most of the leaves had fallen from all plants.

Stem intumescences. The abscission of the leaves exposed some stem swellings, first noted on the 72nd day. The number of plants showing intumescences given as a percentage of the number of treated plants (8), is given in Tables 1 and 2.

TABLE 1. Percentage of plants showing stem intumescences on the two-year old shoots.

Clones	Days after the beginning of the fumigation											
	Purified air			SO_2			O_3			SO_2+O_3		
	72	82	94	72	82	94	72	82	94	72	82	94
a. Swollen bark												
P. deltoides.	0	0	0	0	0	0	0	0	0	0	0	0
P. italica.	0	0	0	0	0	75	0	12	87	0	12	100
K 7.	0	0	0	12	62	75	0	25	25	25	25	25
X 3.	0	0	0	12	12	25	25	50	50	37	50	50
b. Bark cracking												
P. deltoides.	0	0	0	0	12	12	12	12	12	0	0	50
P. italica.	0	0	0	0	0	0	0	25	25	25	25	75
K 7.	0	0	0	50	75	87	25	25	25	0	12	25
X 3.	0	0	0	25	25	25	62	62	62	37	37	37
c. Severely swollen and cracked bark												
P. deltoides.	0	0	0	0	0	0	0	0	0	0	0	0
P. italica.	0	0	0	0	0	75	0	0	0	0	12	25
K 7.	0	0	0	12	12	12	0	0	0	0	0	0
X 3.	0	0	0	12	12	12	25	25	25	25	25	25

The swellings were initially found on the two-year-old part of the shoots, near the base of the current year growth, where they also appeared. The lesions gradually increased in size, resulting in bark cracking and the exposure of soft tissue, although the bark sometimes cracked without any previous swelling. Later, similar lesions developed in the area surrounding dormant buds and leaf scars. They were initially localized below or above the bud and the leaf scar and finally surrounded them. Buds midway up these stems showed first the lesions, which gradually developed around lower or higher buds, but without ever reaching the top ones. The lesions were called 'intumescences' after Wallace (1926, 1927), who also observed similar symptoms in woody shoots exposed to dilute ethylene gas. At a later stage, shoot abscission also occurred in *P. deltoides* and K7 at the current year shoot base.

TABLE 2. Percentage of plants showing stem intumescences on the current year shoots.

Clones	Days after the beginning of the fumigation											
	Purified air			SO_2			O_3			SO_2+O_3		
	72	82	94	72	82	94	72	82	94	72	82	94
a. Swollen bark												
P. deltoides.	0	0	0	0	0	0	0	0	0	0	0	0
P. italica.	0	0	0	0	0	25	0	0	0	0	0	12
K 7.	0	0	0	0	12	12	0	25	25	0	12	12
X 3.	0	0	0	0	12	12	0	25	25	0	50	87
b. Bark cracking												
P. deltoides.	0	0	0	0	0	0	0	0	0	0	0	25
P. italica.	0	0	0	0	0	25	0	0	0	0	0	12
K 7.	0	0	0	0	37	37	0	12	12	0	0	0
X 3.	0	0	0	25	25	25	50	62	62	25	50	62
c. Swollen bark around the buds.												
P. deltoides.	0	0	0	0	0	0	0	0	0	0	0	0
P. italica.	0	0	0	0	0	0	25	25	25	12	25	25
K 7.	0	0	0	37	50	50	0	0	0	0	12	12
X 3.	0	0	0	12	37	37	37	37	37	50	50	50

P. deltoides showed a rather small proportion of bark cracking in the single gas treatments and this was restricted to the two-year-old shoots (Table 1). The plants exposed to the SO_2+O_3 treatment however, developed the symptoms very late; They were apparent on the 94th day, in both two-year-old and current year's shoots (Tables 1, 2).

P. italica showed the highest proportion of injured two-year-old shoots, in all gas treatments (Table 1). In contrast, the current year's shoots were less affected by the treatments, with only a few plants showing swollen bark around the buds in the O_3 and the SO_2+O_3 treatments. A small percentage of plants also showed bark cracking (Table 2).

K7 plants showed greatest damage with SO_2 on two-year-old shoots (Table 1). Current year shoots, showed the highest degree of bark cracking, and swellings around the buds in response to SO_2 (Table 2).

X3 was one of the first clones to show the fumigation damage. Many large intumescences were present in all three gas treatments on the first sampling date and the majority of plants were injured by O_3 and SO_2+O_3; each type of lesions was observed in this clone (Tables 1, 2).

Examination of intumescences by light microscopy. Transverse, longitudinal and tangential sections were taken from intumescence tissue of the current year stem of X3 plants and stained in 0.1 % toluidine blue. The dye differentially stains cellulose blue, lignified cell walls and walls with bound phenolics green-turquoise, cutinized and suberized cell walls a fainter green-turquoise and fungal hyphae reddish purple. It was not possible to observe cell structure in the areas of the cracks, because in spite of the embedding procedure, these cells were collapsed and had been sloughed off.

Cell hypertrophy, followed by cell separation, was indicated as the cause of all the different intumescence forms (Fig. 2). The intumescence, under the light microscope, showed hypertrophied cells in the phloem side of the cambium and sometimes in the cortical cells, below the periderm. In the first stage of the injury, the cells separated and big intercellular spaces were formed. A lysigenous aerenchyma formation was shown in the stem sections taken from the area just below, or above the injury. Sections of injured tissue, showed the parenchyma cells in the phloem side of the cambium to be hypertrophied. Sections taken in more advanced stages of injury showed that after reaching gigantic sizes the cells finally collapsed, leaving very big intercellular spaces (Fig. 2). The bark cracking and the swollen bark around the buds were formed during these last two stages and were caused by the hypertrophy and the 'lysis' of the cambial and the cortical cells. Tangential and longitudinal sections showed that the hypertrophy occurred only in localized areas, especially in the regions surrounding axillary buds and the base of the current year shoots. The rest of the tissue appeared normal.

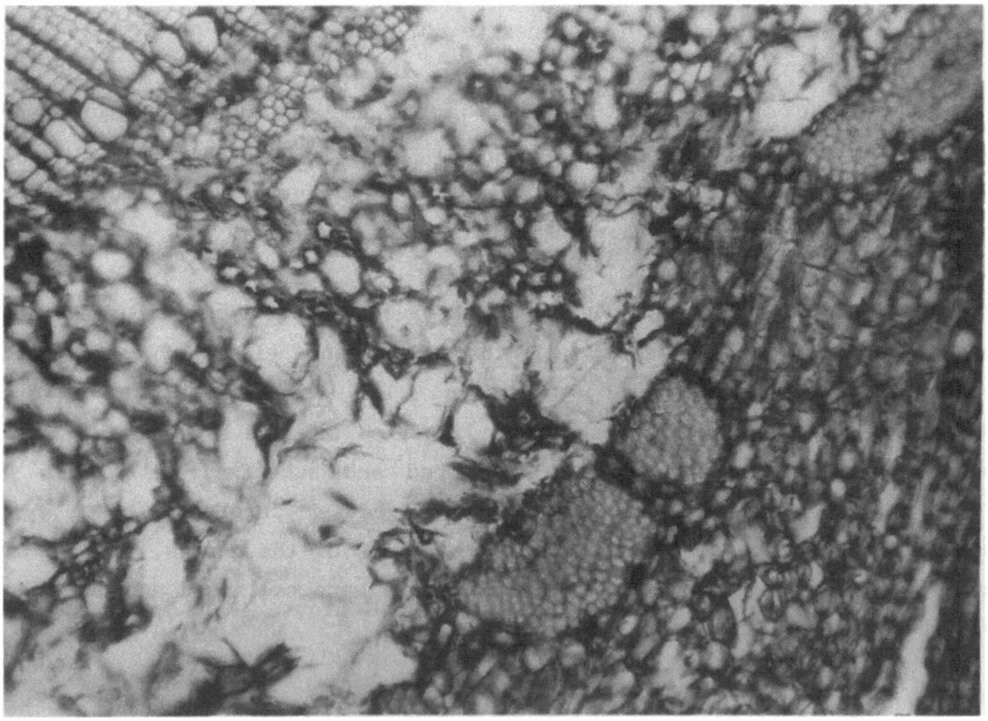

FIGURE 2. Stem intumescence as seen via light microscopy. Clone X3. SO_2+O_3 treatment. Scale bar 50 µm.

Examination of intumescences by scanning electron microscopy. Stem samples of the same clone (X3, SO_2+O_3) showing intumescences were sectioned and processed for scanning electron microscopy.

As in the light microscopy longitudinal sections showed cell proliferation and separation in the cortex of the region of the intumescence. The cell hypertrophy lifted the nearby bark, causing the externally visible swelling. The enlarged, hypertrophied cells were sac-like and separate. Development of empty cavities, between the bark and the cambium, was caused when these cells collapsed (Fig. 3, see arrow).

FIGURE 3. Longitudinal section of a stem intumescence, viewed by scanning electron microscopy. Clone X3. SO_2+O_3 treatment. Scale bar 100 µm.

Ethylene evolution. The mean rate of ethylene evolution in each of the gas treatments, from leaves of the different clones is given in Fig. 4. The leaf samples collected at 6, 18, 43, 67, 92 hours represent the values for old mature leaves, whereas the ones at 30, 55, 113 and 135 hours represent the ethylene evolution of young, still developing leaves.

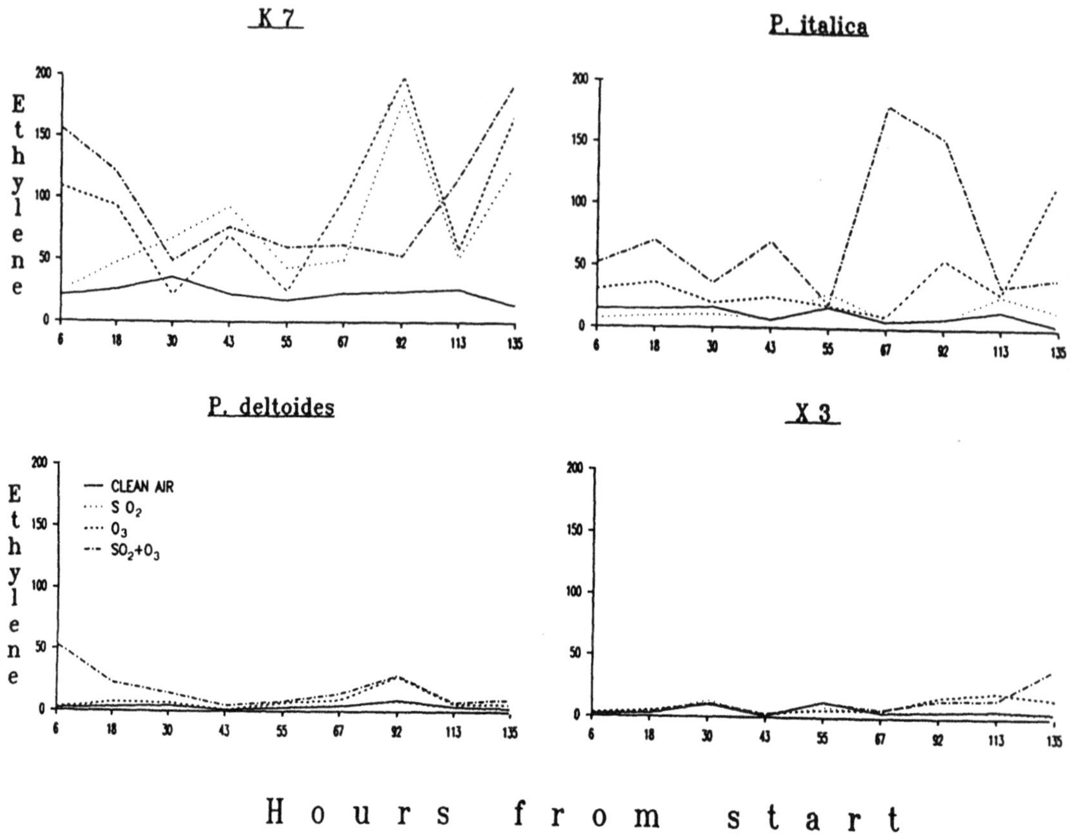

FIGURE 4. Ethylene evolution during different fumigation treatments (nl h^{-1} g^{-1} fresh weight).

The four clones showed large differences in ethylene production. K7, the most sensitive clone, produced the most ethylene in all three gas treatments. *P. italica* showed increased ethylene production in O_3 and SO_2+O_3 treatments. *P. deltoides* and its hybrid X3 produced only small amounts of ethylene regardless of treatments. Young leaves of K7 and *P. italica* evolved less ethylene than older ones, except at the later stages of fumigation (135 hours) when the leaves were about to abscind.

DISCUSSION

One can speculate about the factors causing the observed symptoms only by considering possible reactions taking place inside the cells after the entry of SO_2 and O_3. These gases can enter the plant via stomata or epidermis when the plants are in leaf (Garsed, 1984), and via lenticels when they are leafless (Fahn, 1982). The nature of the reactions in cells has long been under investigation. A theory involving free radicals produced by the reactions of the dissociation products of the gases has received considerable attention. Sulphur dioxide enters the cells of the leaves mainly as sulphite (SO_3^{2-}) (Alscher, 1984) or possibly also as the bisulphite (HSO_3^-) ion (Garsed, 1984). When sulphite is oxidized or reduced, free radicals such as the superoxide ($\cdot O_2^-$), bisulphite (HSO_3), hydrogen peroxide (H_2O_2), hydroxyl ($OH\cdot$) and singlet excited oxygen ($^1O_2$) are generated (Garsed, 1984; Tanaka and Sugahara, 1980; Asada, 1980). Ozone decomposition in water also gives hydroxyl and hydroperoxyl ($O_2H\cdot$) free radicals (Bennett *et al.*, 1984) besides the superoxide anion (O_2^-) and a singlet excited oxygen (Tingey and Taylor, 1982).

The presence of free radicals has been investigated by applying free radical scavengers to plants before SO_2 or O_3 fumigation. This type of investigation showed that treated plants suffered less injury after SO_2 (Shu-Wen *et al.*, 1982) or O_3 fumigation (Bennett *et al.*, 1984). Ethylene production can also be an indicator of the presence of free radicals, since such radicals can trigger ethylene formation (Yang and Hoffman, 1984). Sulphur dioxide has been shown to stimulate ethylene production in *Pinus resinosa* and *Betula papyrifera* seedlings (Kimmerer and Kozlowski, 1982); and larch plants showed a periodical oscillation in ethylene production after SO_2 fumigation (Bucher, 1981; 1984). *Eucalyptus globulus* has also shown increased ethylene production when treated with O_3 (Rodecap and Tingey, 1986).

The involvement of ethylene in abscission is a well documented phenomenon (Sexton *et al.*, 1985). During abscission, besides interacting with other growth substances, ethylene induces cell expansion of the specific cells of the abscission layer (Osborne, 1982). Moreover, cell hypertrophies in trees have been attributed to ethylene. They were initially studied by Wallace (1926, 1927), who performed experiments with externally applied ethylene, on apple and poplar twigs. He reported bark cracking and emergence of loose cells similar to the ones presented in this paper. The 'lysis' of cells and aerenchyma formation in roots, has also been attributed to endogenously produced ethylene during waterlogging (Cambell and Drew, 1983; Jackson, 1985; Kawase and Whitmoyer, 1980). Pieces of intumescence tissue from these experiments, were tested for evolution of ethylene, but the results showed insignificant levels of ethylene production. However, ethylene is not produced in material where membrane systems have been disrupted (Yang *et al.*, 1985), so this may apply to the hypertrophied tissue of the intumescences, where many cells were dying.

However, the results from the ethylene evolution experiment demonstrated clearly that extra ethylene is produced from poplar leaves in response to SO_2 and O_3 fumigation. A high and a low ethylene producing group represented by K7 and *P. italica* and by X3 and *P. deltoides* respectively, was found. The results show a high correlation between high ethylene evolution from leaves and enhanced leaf abscission. The possibility of a direct involvement of ethylene in intumescence formation is not yet resolved, as ethylene evolution from stems has not been measured in detail. However, it is likely that ethylene *is* the promoting factor in this case as well. Although X3 showed stem intumescences, its ethylene production from the leaves was small. This suggests that some mechanism exists of suppressing the rise in ethylene production in the leaves of this clone, or there is a different sensitivity to ethylene in the cortical and cambial cells of the stem compared with cells of the abscission zone.

CONCLUSION

Ethylene production, triggered by free radicals produced during the dissociation of SO_2 and O_3 in the cytoplasm, seems a likely cause of enhanced leaf abscission and of the cell hypertrophy and cell separation which result in stem intumescences on shoots of poplar. The ethylene signal seems to cause a response only in certain target cells. They include cells at the abscission layers and those at the cortex in the areas surrounding the bases of petioles and shoots. Mature parts of the tree seem more responsive possibly due to lower amounts of free radical scavenging substances compared with the younger parts. It could be that, the cortex provides target cells responding to ethylene by increased production or release of lytic enzymes that lead to the development of schizogenous aerenchyma in plant shoots. Poplar clones that were subjected to water stress that are known to produce ethylene have shown aerenchyma formation (Kargiolaki, unpublished). In future work, we hope to examine these hypotheses further and study specific enzymes involved either in suppressing ethylene production e.g. superoxide dismutase or induced by ethylene e.g. glucanhydrolases.

ACKNOWLEDGEMENTS

I wish to thank Mr F. B. Thompson and Dr D. J. Osborne for providing me with the equipment to carry out this research and for their encouraging advice. I also, wish to thank the Greek State Scholarships Foundation, the Greek Ministry of National Economy, and the Commission of the European Communities for their financial support.

REFERENCES

Alscher R (1984) Effects of SO_2 on light-modulated enzyme reactions. In : Koziol M J, Whatley F R (eds) Gaseous air pollutants and plant metabolism. Butterworths, London, pp 181-200

Asada K (1980) Formation and scavenging of superoxide in chloroplasts, with relation to injury by sulfur dioxide. Studies on the effects of air pollutants on plants and mechanisms of phytotoxicity. National Institute Environment Studies Japan, Research Report No 11, pp 165-179

Barrett T W, Benedict M H (1970) Sulfur dioxide. In :Jacobson J S, Hill A C (eds) Recognition of air pollution injury to vegetation: A pictorial atlas. Air Pollution Control Association, Informative Report No 1 TR-7, Agricultural Committee, Pittsburgh, Pennsylvania, pp C1-C17

Bell J N B (1984) Air Pollution Problems in Western Europe. In :Koziol M J, Whatley F R (eds) Gaseous air pollutants and plant metabolism. Butterworths, London, pp 3-24

Bennett J H, Lee E H, Heggestad H E (1984) Biochemical aspects of plant tolerance to ozone and oxyradicals: Superoxide dismutase. In :Koziol M J, Whatley F R (eds) Gaseous air pollutants and plant metabolism. Butterworths, London, pp 413-424

Bucher J B (1981) SO_2 induced ethylene evolution of forest tree foliage and its potential use as stress-indicator. Europ J Forest Path 11, 369-373

Bucher J B (1984) Emissions of volatiles from plants under air pollution stress. In :Koziol M J, Whatley F R (eds) Gaseous air pollutants and plant metabolism. Butterworths, London, pp 399-412

Cambell R, Drew M C (1983) Electron microscopy of gas space (aerenchyma) formation in adventitious roots of *Zea mays* L. subjected to oxygen shortage. Planta 157, 350-357

Fahn A (1982) Plant Anatomy. Third Edition. Pergamon Press, Oxford

Garsed S G (1984) Uptake and distribution of pollutants in the plant and residence time of active species. In :Koziol M J, Whatley F R (eds) Gaseous air pollutants and plant metabolism. Butterworths, London, pp 83-103

Hill A C, Heggestad H E, Linzon S N (1970) Ozone. In :Jacobson J S, Hill A C (eds) Recognition of air pollution injury to vegetation: A pictorial atlas. Air Pollution Control Association, Informative Report No 1 TR-7, Agricultural Committee, Pittsburgh, Pennsylvania, pp B1-B22

Jackson M B (1985) Ethylene and the responses of plants to soil waterlogging and submergence. Ann Rev Plant Physiol 36, 145-174

Kawase M, Whitmoyer R E (1980) Aerenchyma development in waterlogged plants. Am J Bot 67, 18-22

Kimmerrer T W, Kozlowski T T (1982) Ethylene, ethane, acetaldehyde and ethanol production by plants under stress. Plant Physiol 69, 840-847

Malhotra S S, Khan A A (1984) Biochemical and physiological impact of major pollutants. In :Treshow M (ed) Air pollution and plant life. John Wiley and Sons, Chichester, pp 113-157

Osborne D J (1982) The ethylene regulation of cell growth in specific target tissues of plants. In :Wareing R F (ed) Plant growth substances 1982. Academic Press, London, pp 279-290

Rodecap K D, Tingey D T (1986) Ozone-induced ethylene release from leaf surfaces. Plant Sci 44, 73-76

Sexton R, Lewis L N, Trewavas A J, Kelly P (1985) Ethylene and abscission. In :Roberts J A, Tucker G A (eds) Ethylene and plant development. Butterworths, London, pp 173-196

Shu-Wen Y, Yu L, Zhen-Guo L, Chang T, Zi-Wen Y (1982) Studies on the mechanism of SO_2 injury in plants. In :Unsworth M H, Ormrod D P (eds) Effects of gaseous air pollution in agriculture and horticulture. Butterworths, London, pp 507-508

Tanaka K, Sugahara K (1980) Role of superoxide dismutase in the defense against SO_2 toxicity and induction of superoxide dismutase with SO_2 fumigation. Studies on the effects of air pollutants on plants and mechanisms of phytotoxicity. National Institute Environment Studies Japan, Research Report No 11, pp 155-164

Taylor H J, Ashmore M R, Bell J N B (1986) Air pollution injury to vegetation (A guidance manual commisioned by HM Industrial air pollution inspectorate of the Health and Safety Executive). Imperial College Centre for Environmental Technology (ICCET) Imperial College of Science and Technology, IEHO, Chadwick House, Rushworth Street, London

Tingey D T, Taylor G E Jr (1982) Variation in plant response to ozone: A conceptual model of physiological events. In :Unsworth M H, Ormrod D P (eds) Effects of gaseous air pollution in agriculture and horticulture. Butterworths, London, pp 113-138

Treshow M (1984) Diagnosis of air pollution effects and mimicking symptoms. In :Treshow M (ed) Air pollution and plant life. John Wiley & Sons, Chichester, pp 97-112

Wallace R H (1926) The production of intumescences upon apple twigs by ethylene gas. Bulletin of the Torrey Club 53, 385-405

Wallace R H (1927) The production of intumescences in transparent apple by ethylene gas as affected by external and internal conditions. Bulletin of the Torrey Club 54, 499-543

Ward T M, Wright M, Roberts J A, Self R, Osborne D J (1978) Analytical procedures for the assay and identification of ethylene. In :Hillman J R (ed) Isolation of plant growth substances. Society for Experimental Biology Seminar Series 4, Cambridge University Press, pp 135-151

Yang S F, Hoffman N E (1984) Ethylene biosynthesis and its regulation in higher plants. Ann Rev Plant Physiol 35, 155-189
Yang S F, Liu Y, Su L, Peiser G D, Hoffman N E, McKeon T (1985) Metabolism of 1-aminocyclopropane-1-carboxylic acid. In :Roberts J A, Tucker G A (eds) Ethylene and plant development. Butterworths, London, p 9-21

CELL WALL PECTIC CONTENT AS AN EARLY SIGNAL FOR CELL SEPARATION

Roberto Jona
Istituto di Coltivazioni Arboree dell'Universita
Via P. Giuria, 15
1-10126 Torino
Italy

INTRODUCTION

Preliminary remarks

Softening is common to various processes that include ripening, several physiological disorders, changes due to pathological attack, wilting of flowers and, in a sense, also abscission. Chemically, these processes have in common, changes to pectic substances which comprise one of the main groups of compounds that make up cell walls.

Pectic substances

Pectic substances are polygalacturonides with non-uronide carbohydrates covalently bound to an unbranched chain of α-1,4-galacturonic acid units. The carboxyl groups of the galacturonic acid are partly esterified with methanol and free groups are partly neutralized as salts. Some of the hydroxyl groups on C_2 and C_3 may be acetylated (Pilnik and Voragen, 1970). An alternative classification is more extensive and groups pectic polysaccharides according to the polymers found in covalent association with galacturonosyl-containing polysaccharides, namely the rhamno- galacturonans, the arabans, the glacturonans and possibly the arabino- galacturonans (McNeil *et al.*, 1979). However, the definitions of the American Chemical Society are still useful (Kertesz, 1952 pp. 6-8) and are summarised as follows. 'Pectic substances' is a group-designation for those complex, colloidal carbohydrate derivatives which occur in, or are prepared from, plants and contain a large proportion of anhydrogalacturonic acid units which are thought to exist in a chain-like combination. The carboxyl groups of polygalacturonic acids may be partly esterified by methyl groups and partly or completely neutralized by one or more bases. *Protopectin.* The term protopectin is applied to the water-soluble parent pectic substance which occurs in plants and which, upon restricted hydrolysis, yield pectinic acids. *Pectinic acids.* The term pectinic acids is used for colloidal polygalacturonic acids containing more than a negligible proportion of methyl ester groups. Pectinic acids, under suitable conditions, are capable

NATO ASI Series, Vol. H 35
Cell Separation in Plants
Edited by D. J. Osborne and M. B. Jackson

of forming gels (jellies) with sugar and acid or, if suitably low in methoxyl content, with certain metallic ions. The salts of pectinic acids are either neutral or acid pectinates. *Pectin.* The general term pectin (or pectins) designates those water-soluble pectinic acids of varying methyl ester content and degree of neutralization which are capable of forming gels with sugar and acid under suitable conditions. *Pectic acid.* The term pectic acid is applied to pectic substances mostly composed of colloidal polygalacturonic acids and essentially free from methylester groups. The salts of pectic acid are either neutral or acid pectates.

CASE HISTORIES FOR CELL SEPARATION

(a) *Fruit maturation*

The ripening of fruits may be defined as the sequence of changes in colour, flavour and texture which lead to the state at which fruit is acceptable to eat. Moreover, the maturity of fruit is the stage beyond which decay begins. In both instances, the texture of fruits becomes generally softer. Ripening may occur before harvest, as in berries, stone fruits, figs and grapes, or largely after harvest as in pear, quince, late apples, avocado and persimmons. It may be more or less synonymous with maturity in non climacteric fruits, such as citrus which do not ripen after harvest. From this definition it appears that, though the patterns of ripening are different in the various fruits, a common feature of fleshy fruits is the softening of the pulp.

A scheme of the primary wall outlined by Knee and Bartley (1981) may also be relevant to the cell wall structure of fruit: it indicates that cellulose fibrils, together with hemicellulose, constitute the molecular skeleton of the cell wall. An interdispersed pectin material composed of rhamnogalacturonans is bridged to the cellulose-hemicellulose complex by neutral homogalacturonans, arabinans and galactans. However, it was demonstrated that this scheme, may not apply to certain fruit cell walls as cellulose and hemicellulose are hardly present, while pectins are the main polysaccharide component (Jona and Foa, 1979).

The fruit softening associated with ripening represents a complex process resulting both from cell expansion, which may lead to a mechanical separation of cells at the middle lamella region (Mohr and Stein, 1969;

Vickery and Bruinsma, 1973) and from the action of pectolytic enzymes resulting in the dissolution of the middle lamella material (Pesis *et al.*, 1978; Ben-Arie *et al.*, 1979). Knee and Bartley (1981) proposed also that the loss of cohesion between cells may result from S-adenosyl methionine (SAM) methylation of free carboxyl groups in the pectic material and thereby the disruption of calcium crosslinkages of adjacent polyuronides. Various pathways have been envisaged for the solubilization of previously insoluble protopectins, but none is clear.

Interference with the cross linking function of Ca^{2+} contributes to the wall degradation. Demethylation by pectin-methyl-esterase (PME) is known to prepare pectic substances for degradation by polygalacturonase (PG).

The correlation of PG (exo- and endo-) activity with the softening mechanism has been shown by various researchers (Duescher *et al.*, 1976; Pressey, 1977; Pressey *et al.*, 1978). Soft, stone-free peach cultivars exhibit both endo and exo PG activity, whereas firm cultivars contain mainly exo-PG. This appears logical considering that by stepwise hydrolysis of chain end-units, the latter type leads to a limited substrate dissolution, while endo-PG cleaves intramolecular bonds and markedly enhances the dissolution of pectic substrates. Some recent examples are discussed below.

Apples and pears. Tavakoli and Wiley (1968) reported that fruit of various cultivars harvested on consecutive dates decreased in firmness. Total pectin also decreased. On storage, there was a further loss of firmness and of total pectin. The protopectin fraction of total pectin increased before harvest and decreased after harvest. Raunhardt and Neukom (1964) determined total pectin in washed dried pomace from apples and pears picked at various dates and stored after harvest. During ripening on the tree and on storage, pectin content decreased. The extent of esterification remained constant at around 70% before harvest, but dropped after harvest to 60-65% (apples) and 50% (pears).

Peaches and apricots. Shewfelt (1965) described a storage experiment with stone-free peach varieties showing decreasing firmness. Total pectin (300 mg%) changed little but the water-soluble fraction increased while the versene-soluble fraction and protopectin decreased. In a clingstone peach cultivar their was almost no change of hardness, total pectin concentration was 300 mg per 100 g fresh wt.; of this total, 66% was protopectin.

Deshpande and Salunkhe (1964) investigated effects of maturity and storage on biochemical changes in apricots and peaches. For example, 'Redhaven' peaches contained 1270 (hard), 800 (firm mature) and 740 (soft) mg of total pectin per gram fresh wt., which decreased with maturity and upon storage.

Avocado. Dolendo *et al.* (1966) observed a decrease of protopectin to 25 mg per gram fresh wt. upon storage together with loss of firmness, while a water soluble fraction appeared. Percentage esterification of total pectin decreased from 85 to 45.

Tomatoes. Deshpande *et al.* (1965) found that vine ripe fruit graded for firmness had 60-100 mg total pectin per gram fresh wt. in firm, and 13-37 mg total pectin per gram fresh wt. in soft fruits. The latter had a higher protopectin fraction. Similar results were obtained with fruit ripened after harvest. There was a significant correlation between molecular weight of the pectin (as ratio carboxyl/pectin) and firmness as well as between total mineral content and firmness.

Citrus. Rouse *et al.* (1965) reported on changes in water soluble, oxalate soluble and NaOH soluble (protopectin) pectin fractions of component parts of 'Silver Cluster' grapefruit during a 10-month maturation cycle for 2 seasons. Total pectin varied during development (season 1961/62), increasing slightly with maturity, calculated on dry weight. In peel, the oxalate soluble fraction accounted for 70% and water soluble and protopectin for 15% each of total pectin. In membranes and juice sacs, the fractions were much nearer to each other. There were few changes with maturity. Rouse *et al.* (1964a) also studied the same component parts of pineapple orange during a similar maturation period, but used an acid extraction method (H-exchange). Average yields of galacturonic acid, calculated as an acid insoluble substance, were found to be 20% (peel), 29% (membrane) and 16% (juice sacs). Using jelly grade averages to measure the molecular weight, they found that there was very little change with maturity or in esterification averages. The same authors (1964b) obtained similar results with 'Silver Cluster' grapefruit, although the extracted pectins were of higher molecular weight and level of esterification.

Strawberries. Neal (1965) considered 3 stages of maturity for 2 cultivars. He found that on ripening, the middle lamella of the cortical parenchyma cells was separated into 2 layers, each of which remained attached to a cell wall. This effect could be imitated by the use of EDTA which had a macerating effect with a similar loss of firmness. By addition of Ca^{2+}, the EDTA effect could be reversed. De-esterification by immersion in alkaline alcohol and addition of Ca^{2+} also had a firming effect. Methylation on the other hand decreased firmness with no influence of Ca^{2+}. Both cultivars showed water soluble pectin to increase with maturity, with practically no change in the EDTA soluble fraction. Decreasing firmness is therefore attributed to methylation of pectin which inhibits the firming effect of Ca^{2+} in spite of the presence of PE in strawberries.

Kiwi fruit. More detailed analyses during kiwi fruit softening using several methods of analysis, confirmed the above reported results and consistently point to a decline of uronic acid associated with polysaccharides of the water-insoluble fraction (Arpaia *et al.*, 1987). Similarly, Yamaki *et al.* (1979) found that in Japanese pear fruit, the loss of arabinose and galactose (usually most closely associated with pectic polysaccharides) was prominent, similar to the findings of Knee (1973). Ahmed and Labavitch (1980) also found substantial losses of uronic acid and arabinose during the 'Bartlett' pear fruit ripening. By a different analytical method, Knegt *et al.* (1975) found that maturation and softening of tomato is accompanied by a decrease of soluble pectins. Also in nectarines (Jona 1985), fruit maturation is characterized by a loss of cell wall pectic substances. This seems to occur in all ripening fruit (Pilnik and Voragen, 1970), but Lidster *et al.* (1978) reported that this process may be partially reversed or at least arrested in cherry berries by a week-long dip in $CaCl_2$. Similar conclusions were reached by Burns and Pressey (1987) who analyzed Ca^{2+} content in cell wall middle lamella of ripening tomato and peach. Sterling and Kalb (1959) analyzing the methoxyl content of uronide material in peach also found continual decrease in both water soluble and acid soluble fractions, and concluded that there can be no question of a methylation process and that another mechanism must be involved in solubilizing pectic material. By contrast, the results obtained by Reeve (1959), by histochemical analysis provide a very accurate picture of the softening mechanism at the cell wall and middle lamella level. Reeve (1959) found a consistent increase of the degree of methylation of the cell wall in maturing peach fruit, followed by a sudden fall in the short time between hard ripeness and soft ripeness. This behaviour suggests that the maximum degree of esterification is reached just as the onset of ripening is preparing the substrate necessary for extensive PME activity which, in turn, becomes the softening agent: by degrading the pectins it is dismantling the cementing agent of cells which are ceasing to be a protective involucre for the seed and should fall apart rapidly. Ahmed and Labavitch (1980) did not find an increase in PME in the ripening pear fruit, but found, on the opposite, a sharp increase in the PG activity of the onset of maturation.

This contrast may be due either to the inevitable mixing of various parts of the cells when they are analysed in homogenates, or to a slightly different chemical pathway (which however leads to the same result) between stone fruit and pomaceae.

(b) *Senescence of flower corolla*

There are few analytical data on pectic content and variations in petals of harvested carnations and other flowers. From research performed mostly on fruit we learn that the ratio of maximum ethylene production to the basal rate is positively correlated with PG activity at full development (Ramina *et al.*, 1984). Furthermore, several enzymes and enzyme mixtures characterized by pectolytic activity are capable of inducing ethylene biosynthesis in tobacco leaf discs (Anderson *et al.*, 1984). Low ethylene apples have been found consistently firmer than fruits stored in a cabinet without an ethylene scrubbing system (Dover, 1985). Paulin (1988) stresses the advantage of increasing the level of Ca^{2+} in the tissue in order to decrease ethylene biosynthesis.

From the above reported experiments we may safely assume that the increased ethylene formation and pectin decrease in tissues, even if not causally connected, are at least concurrent. In many instances, it has been demonstrated that flower senescence is accompanied by a rapid upsurge of ethylene biosynthesis (Nichols, 1973a,b; Nichols and Ho, 1975). Thus it seems possible that pectin decay occurs during flower senescence.

By histochemical analysis, it was found that wilting of cut carnation flowers occurs much earlier in a cultivar whose flower cell-walls contained less pectic substances. Moreover, in this cultivar, the decay of pectic substances is much more rapid and extensive than in the long-keeping cultivar (Jona *et al.*, 1981).

(c) *Fruit abscission*

Fruits of most flowering plants have relatively short life spans and abscind at regular and predictable intervals. Fruit abscission involves two distinct types of structural change. The first type relates to detachment of the fruit from the plant. In this connection more than one abscission zone may form (Fehér, 1925; Bradbury 1929; Stösser *et al.*, 1969a & b; Bukovac, 1971). The second type, which is less interesting in the framework of our review, is associated with protection of the surface area exposed on the plant after fruit fall. Detachment of fruit is facilitated by developmental changes within the abscission zone. The extent of cellular differentiation within the zone varies from species to species, the degree of differen- tiation is particularly interesting because it bears on the relative ease with which fruits abscind naturally or are detached mechanically. Sour cherry (*Prunus cerasus)* develops a distinct abscission zone between fruit and pedicel which is delineated anatomically and histochemically and thus, sour cherry fruits separate readily from the pedicel. Fruits of sweet cherry (*Prunus avium),* which do not develop a distinct abscission zone at maturity are separated at this position only with difficulty and often with injury (Stösser *et al.*, 1969a; Bukovac *et al.*, 1971). Various elements are associated with abscission in some fruits, but the most distinctive and crucial are early symptoms of demethylation of pectins which could be revealed at a histochemical level several months before horticultural maturity of oranges and well before abscission occurred (Wilson and Hendershott, 1968). When abscission was allowed to progresss naturally or was induced artificially with iodoacetic acid, pectic changes in cells of abscission zone, both in mature and immature fruits, were the same. Wittenbach and Bukovac (1972) observed loss of pectic substances along the abscission layer of the sour cherry fruit during abscission. Stösser *et al.* (1969b) reported a low affinity of cells of the pedicel-fruit abscission zone for haematoxylin and the low level of birefringence of the cells of *Prunus cerasus.* Sequential extraction of various cell wall constituents followed by staining with the periodic acid Schiff's reagent (Jensen, 1962) indicated that the abscission zone contained less total polysaccharides than the surrounding tissues and that separation was accompanied by a partial breakdown of non-cellulosic polysaccharides and cellulose (Stösser *et al.*, 1969a & b, 1971). It is probable that the breakdown of cell wall constituents in sour cherry is caused by enzymatic action. This suggestion derives from the fact that PME, PG and cellulase activity have all been demonstrated in the abscission zone of other plants (Osborne, 1958; Rasmussen, 1965; Horton and Osborne, 1967; Morré, 1968). In addition, to facilitate separation of primary-walled cortical cells of the abscission zone, the enzymes (particularly cellulases) might also be involved in the rupture of non-lignified vascular elements (Osborne, 1968). Walls of the sour cherry, abscission zone cells also lose calcium and magnesium during separation (Stösser *et al.*, 1969a & b). Crystals of calcium oxalate are also present in parenchyma cells of the abscission zone of many fruits at the time of separation (Scott *et al.*, 1948). Based on evidence of this loss, plus data on pectin degradation in the abscission zone, cell separation might reasonably account for abscission since pectins act as cementing substances between cells and are linked together by polyvalent cations such as Ca^{2+} (Frey-Wyssling, 1959; Setterfield and Bailey, 1961). Furthermore, the molecular chains of pectic acids are linked to hemicellulose and cellulose through calcium and magnesium bridges; indeed the role of Ca^{2+} or Mg^{2+} in maintaining cell wall integrity is well established (Tagawa and Bonner. 1957; Preston, 1964).

This evidence notwithstanding, Bukovac (1971) retains an air of scepticism, pointing out that the loss of Ca^{2+} and the progressively diminishing capacity of cherry abscission zone cells to bind Ca^{2+} as separation progresses may be primarily a reflection of general wall degradation. In point of fact, speculation regarding Ca^{2+} changes in abscission zone cells has been prevalent in the literature since Sampson's (1918) and Facey's (1950) observations on changes in the abscission zones of *Coleus* and ash. Based on these studies and ensuing investigations of the role of hormones in abscission, one might extrapolate to suggest that auxin which is associated with changes in wall plasticity (Bonner, 1960), may (1) exert its action through enzymes which alter cell wall characteristics; or (2) serve to remove Ca^{2+} which crosslinks the carboxyls of cementing polymers; or (3) alter the arrangement of cellulose fibrils in the walls. The hormonal control of the abscission mechanism does not affect the situation that the basic mechanism of abscission is cell separation through alterations of pectic substances.

(d) *Radiation induced softening*

Softening of fruits and vegetables after ionizing radiation has been observed and reported (Clarke, 1959; Maxie, 1963). Since chemical changes in macromolecular polymers, such as the pectins, have been shown to occur after exposure to ionizing radiations (Kertesz *et al.*, 1956) the softening effect of irradiation has been postulated to be the result of the degradation of pectins *in situ.* In fact, pears and peaches exposed to gamma radiation exhibited textural changes which corresponded to a protopectin content decrease and an increase in pectin and pectate fractions. Somogyi and Romani (1964) found an increased PME activity just after irradiation, followed four days after irradiation by a considerable reduction of the enzyme activity. They postulated direct radiation effects combined with enzyme activity. On the other hand, the enzyme appeared to be insensitive to radiation (Pansolli, 1970). Moreover, Echandi and Massey (1970) found that gamma irradiated tissues leak electrolytes, especially Ca^{2+}. Romani *et al.* (1971) found, by enzyme assay, that pectin is modified by radiation and becomes undegradable by enzymes. However, Belli-Donini (1973) found that pectin degradation in irradiated strawberries is mainly enzymatic. As hinted by Somogyi and Romani (1964), an inhibitor of polygalacturonase is present in pears and other fruits and this may be the molecule most damaged by radiation.

In histochemical studies (Foa *et al.*, 1980; Jona and Vallania, 1976) the results varied according to the kind of fruit irradiated. In tomato, grape and raspberry, cell wall pectic substances were decreased while in peach and cherry, irradiation did not affect pectic substances. Evidently, the contradictory results of Romani *et al.* (1971) and of Belli-Donini (1973) resulted from the use of different species.

(e) *Physiological disorders*

Apple and orance will be considered here. Apple may suffer from various disorders such as dark brown pits, watery patches, senescent breakdown, bitter pit, lenticel breakdown, water core (Perring, 1968). Perring found in various experiments that a low level of calcium, or eventually an unbalanced ratio of cations (Ca^{2+} and others such as Mg^{2+} and K^{+}) with comparatively low levels of Ca^{2+}, accompanies these physiopathies. Even parasitic rotting by *Gleosporium*, is accompanied by a low level of Ca^{2+} in the pulp of apples.

Oranges can suffer from 'creasing' which causes the breakdown of the albedo, the inner white peel of the fruit. Cutuli (1968) found that a low level of Ca^{2+} in the fruit characterizes the affected fruit.

These analyses, though performed on mineral elements show the importance of pectins, because, clearly the lack of Ca^{2+} is likely to destabilize pectins through lack of saltification of polygalacturonic chains.

In agreement with these findings are the results of Monselise *et al.* (1976), who found a sharp increase in PME in the albedo (which almost totally comprises pectic substances) of creasing oranges. On the other hand Jona *et al.* (1989) found that fruit most susceptible to the disease are comparatively low in pectin content in the cell walls, even in the very early phase of their development (October). Moreover, a very early treatment (July) with GA_3 to the affected trees keeps the pectic substances in the cell wall of the albedo at levels significantly higher than the affected controls from early in the season until picking in the following April (Jona *et al.*, 1989, Moselise *et al.*, 1976). Parallel to this increase is the disappearance of the physiological disorder.

In both cases, the cell walls are very rich in pectic substances, espeecially in the albedo of the orange. If Ca^{2+}, which renders the pectic substances insoluble in water is lacking or is in chemical competition with other more abundant cations, the protopectin structure is affected and physiological disorders result. However, if the availability of Ca^{2+} and or/the synthesis of new pectic substances is assured by adequate growth regulator supply (GA_3), the disorder is overcome.

ANALYTICAL METHODS

Two conceptually different ways to investigate and analyze the composition of cell wall material have been employed, namely chemical and histochemical.

(a) *Chemical*

The most common is based on sequential extraction with chemical agents which selectively dissolve groups of components. Then the extracting solution can be analyzed and the composition of the extracted material studied with the most advanced chemical tools. Two substantially similar sequences of extraction (Yamaki *et al.*, 1979) and Arpaia *et al.*, 1987) have been devised to analyze respectively pear fruit and kiwi fruits. They are based on solubilization of various fractions of polysaccharides followed by acid hydrolysis for uronic acid and monosaccharides (Yamaki *et al.*, 1979). Arpaia *et al.* (1987) used a slightly different sequence to solubilize the various fractions, with more extended use of enzymes, followed by hydrolysis and analysis of uronic acid and sugars. Two major problems are faced using these methods: a) the exact delimitation of the region of the fruit from which the sample is taken and (b) the lack of evenness within the excised region. Though there are procedures which should discriminate between cell sap and cell wall, it is very difficult to avoid cross-contamination, mainly because there is a continuous exchange of substances between the two parts. Furthermore, vascular or other structures may "contaminate" the material to be analyzed thus compromising the results, even if they are obtained with great analytical accuracy.

(b) *Histochemical*

The technique is based on the same principles as the previous section, but the analyses are performed on the insoluble remnant, not on the extracted solutions. This limits the analysis, because it it not possible to ascertain the molecular details since colour has to be produced in presence of a specific stain. This tool, though not providing highly detailed analysis of the chemical structure is extremely powerful and refined in terms of cell localization.

The methodology is schematized in Fig. 1 and can be summarised as follows:

1) Fixed and paraffin infiltrated tissues are section and affixed to a microscope slide.

2) After removal of paraffin, a batch of slides is set aside and the bulk of the remainder submitted to pectin extraction with 1% pectinase for 16 hours and hot (90°C) ammonium oxalate for 2 hours.

3) A group of slides treated with pectinase is then set aside and the remainer treated for 12 hours at room temperature with 4% NaOH in order to extract hemicellulose.

4) Half of these slides are set aside and the remainder treated with 17.5% NaOH in order to remove non-cellulosic polysaccharides, thus leaving only cellulose in the cell wall.

Then, the extracted, partially extracted and totally extracted slides are stained by periodic acid Schiffs reagent (Hotchkiss, 1948) reaction, and the stain intensity is measured photometrically on uniform areas of cell wall. The whole sequence is described in detail by Jona and Foa (1977).

The numerical value thus obtained cannot be defined as optical density because the light path is less than 1 cm. Instead, it is called the

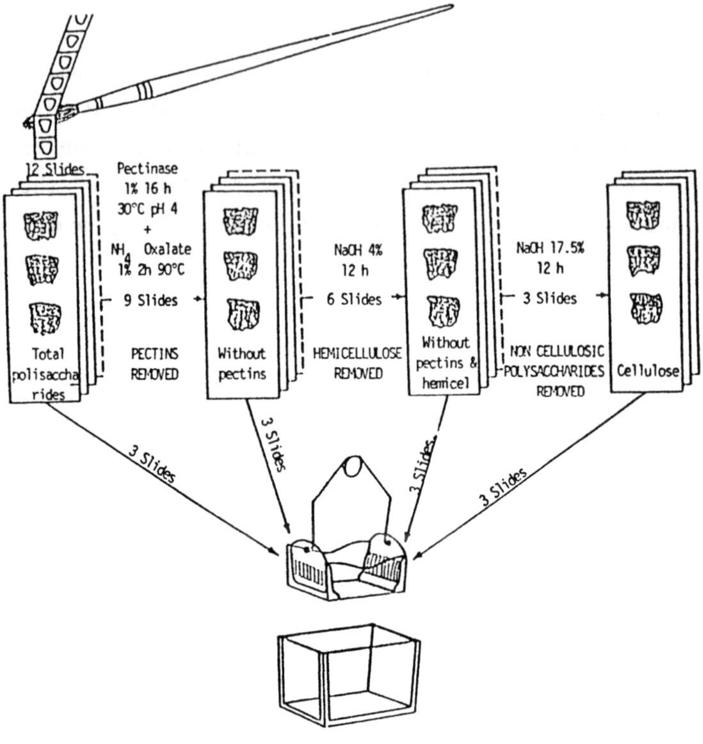

FIGURE 1. Diagrammatic representation of the histochemical procedure

extinction value (EV) and is quantitative, and proportional to the density of material present on a given area of the cell wall. This method is very sensitive and allows fluctuations of various groups of substances to be followed under different physiological or experimental conditions.

Additional information can be gained by use of a TV camera combined with a computer; this makes it possible to measure the area of the cell lumen, thus allowing the surface of the entire cell wall to be calculated. The eventual decrease of the EV per surface unit, may be accompanied by an increase in the size of the cell, resulting in an increase of polysaccharide content of each cell. This tool is invaluable for analyzing the results correctly and to improve information on the growth patterns of various cells and tissues.

A rather sophisticated histochemical method is also available for the evaluation of methylation of cell wall pectic substances using ferric hydroxylamine (Reeve, 1959). This method is especially successful with fresh slices but some problems are encountered with fixed material; however Albersheim and Killias (1963) applied it successfully to thin sections used for ultrastructural analysis with the electron microscope.

A further development, underway in our laboratory, is to subject the extracting solutions, used to selectively remove cell wall components from the microscopic slides, to accurate chemical analysis. In this way, the advantage of both methodologies are combined: on the one hand excellent localization of the undissolved materials is made possible, on the other a very detailed structural analysis of the extracted molecules can be performed.

DISCUSSION AND CONCLUSIONS

From the behaviour of the various structures mentioned above it emerges clearly that breakdown of pectic chains is instrumental in cell separation. However, breakdown of pectic chains can take place in more than one way since pectin is neither homogeneous nor uniform, with its composition varying from structure to structure. But, two mechanisms are of prime importance for their stability (i) interchain Ca^{2+} bonds and (ii) the presence of non-esterified acidic radicals of galacturonic molecules. Taking into account these elements, we find that pectic substances may be degraded principally by saponifying and depolymerizing enzymes. PME splits methylester groups of polygalacturonic acids. The depolymerizing enzymes (PG) are numerous but they all act by transeliminative cleavage of the α-1,4-glycosidic bonds of the polygalacturonic chains. They have been classified by Neukom (1963) and by Koller (1966). Moreover conditions that remove Ca^{2+} bonds also favour pectic solubilization. PME has been found in many fruits while PG is found frequently in yeasts, moulds and bacteria. Plant PG has been much less investigated than microbial PG, the only exception being tomato PG. Moreover in some cases there is the suspicion that plant PGs may be of microbial origin. All the above reported mechanisms collectively bring polygalacturonic chains into solution in the cell sap.

From the examples cited it is evident that various and separate processes are accompanied by changes in pectin content and pectin structure. These include the degree of esterification, molecular weight, neutral sugar components, and acetylation. Such changes may be enzymic or chemical in nature, but all entail pectic changes. It is important to stress that the process of pectic transformation is usually very precocious: consequently it can be used not only to explain the process, but also to predict an early symptom of the developing process before it becomes evident macroscopically.

REFERENCES

Ahmed AE, Labavitch JM (1980) Cell wall metabolism in ripening fruit: I. Cell wall changes in ripening 'Bartlett' pears. Plant Physiol 65: 1009-1013

Albersheim P, Killias V (1963) Histochemical localization at the electron microscope level. Am J Bot 50: 732-745

Anderson JD, Chalutz E, Mattoo AK (1984) Purification properties of the ethylene-inducing factor from the cell wall digesting mixture, cellulysin. In: Fuchs Y and Chalutz E (eds). Ethylene. Martinus Nijhoff/Dr W Junk, The Hague, pp 188-198

Arpaia ML, Labavitch JM, Greve C, Kader AA (1987) Changes in the cell wall components of kiwifruit during storage in air controlled atmosphere. J Am Soc Hort Sci 112: 474-481

Barnell, E (1939) Studies in tropical fruits. V. Some anatomical aspects of fruit-fall in two tropical arboreal regions. Ann Bot 3: 77-89

Belli-Donini ML (1973) Relazione tra contenuto in ioni calcio e solubilizzione della sostanze pectiche di fragole irradiate e conservate. Agrochimica 17: 370-376

Ben-Arie R, Kislev N, Frenkel C (1979) Ultrastructural changes in the cell walls od ripening apple and pear fruit. Plant Physiol 64: 197-202

Bonner J (1960) The mechanical analysis of auxin induced growth. Z Schweiz Forstv 30: 141-159

Bradbury D (1929) A comparative study of the developing and aborting fruits of *Prunus cerasus*. Am J Bot 16: 525-545

Buescher RW, Sistrunk WA, Tigchelaar EL, Ng TJ (1976) Softening, pectolytic activity and storage life of *rin* and *nor* tomato hybrids. HortSci 11: 603-604

Bukovac MJ (1971) The nature and chemical promotion of abscission in maturing cherry fruits. HortSci 6: 385-388

Bukovac MJ, Zucconi F, Wittenbach VA, Flore AJ, Inoue H (1971) Effects of (2-chloroethyl)phosphonic acid on development and abscission of maturing sweet cherry (*Prunus avium*) fruit. J Am Soc Hort Sci 96: 777-781

Burns JK, Pressey R (1987) Ca^{2+} in cell walls of ripening tomato and peach. J Am Soc Hort 112: 783-787

Clarke ID (1959) Possible application of ionizing radiation in the fruit, vegetable and related industries. Intl J Appl Rad and Isotopes 6: 175

Cutuli G (1968-69) I "creasing" delle arance. Ann Ist Sperim Agrumicoltura Vol 1-2

Deshpande PB, Salunkhe DK (1964) Effects of maturity and storage on certain biochemical changes in apricots and peaches. Food Technol 18: 1195-1198

Deshpande PB, Klinker WJ, Draudt HN, Desrosier NW (1965) Role of pectic constituents and polyvalent ions in firmness of canned tomatoes. J Food Sci 30: 594-600

Dolendo AL, Luh BS, Pratt HK (1966) Relation of pectic and fatty acid changes to respiration rate during ripening of avocado fruit. J Food Sci 31: 332-336
Dover CJ (1985) Commercial scale catalytic oxydation of ethylene as applied to fruit stores. In: Roberts JA, Tucker GA (eds) Ethylene and plant development. Butterworths London, pp 373-383
Echandi RJ, Massey LM Jr (1970) Loss of electrolytes and amino acids from gamma irradiated carrots. Rad Res 43: 372-378
Facey V (1950) Abscission of leaves in *Fraxinus americana* L. New Phytol 49: 103-116
Fehér D (1925) Untersuchungen über den Abfall der Früchte liniger Holzpflanzen. Deut Bot Gesell Ber 43: 52-61
Foa, E, Jona R, Vallania R (1980) Histochemical effects of gamma radiation on soft fruit cell walls. Environ and Expt Bot 20: 47-54
Frey-Wyssling A (1959) Die pflanzliche Zellwande. Springer Verlag Berlin
Horton RF, Osborne DJ (1967) Senescence, abscission and cellulase activity in *Phaseolus vulgaris.* Nature 214: 1086-1088
Hotchkiss RD (1948) A microchemical reaction resulting in the staining of polysaccharide structures in fixed tissue preparations. Arch Biochem 16: 131-141
Jensen WA (1962) Botanical histochemistry. Freeman, San Francisco.
Jona R, Vallania R (1976) Histochemical analysis of gamma radiation effects on rubus fruits. Acta Hort 60: 81-88
Jona R, Foa E (1977) Histochemical determination of polysaccharides in fruit pulp cell walls. Cytologia 42: 495-500
Jona R, Foa E (1979) Histochemical survey of cell wall polysaccharides of selected fruit. Sci Hort 10: 141-147
Jona R, Accati E, Mayak S (1981) Senescence processes as reflected in change in polysaccharidic cell wall components. Acta Hort 113: 153-158
Jona R (1985) Pectin fluctuation during nectarine fruit development. Acta Hort 173: 131-138
Jona R, Goren R, Marmora M (1989) Effect of gibberellin on cell-wall components of creasing peel in mature 'Valencia' orange. Sci Hort (in press)
Kertesz, ZI (1951) The pectic substances. Interscience, New York pp 6-8
Kertesz ZI, Morgan BH, Tuttle LW, Lavin M (1956) Effect of ionizing radiation on pectin. Rad Res 5: 372-381
Knee, M (1973) Polysaccharide changes in cell walls of ripening apples. Phytochem 12: 1543-1549
Knee M, Bartley LM (1981) Composition and metabolism of cell wall polysaccharides in ripening fruits. In: Friend J, Rhodes MJC (eds) Recent advances in biochemistry of fruits and vegetables. Academic Press, London, pp 133-148
Knegt E, Kramer SJ, Bruinsma J (1975) pectin changes and internal concentrations in ripening tomato fruit. Colloque Intern du CNRS, Paris 238: 355-358
Koller, A (1966) Dissert No. 3774 ETH Zürich. Cited by Pilnik W and Voragen AGJ (1970). In: Hulme AC (ed.) Biochemistry of fruits and their products. Vol 1. Academic Press, London, p 83
Lidster PD, Porrit SW, Tung MA (1978) Texture modification of 'Van' Sweet Cherries by Postharvest Calcium Treatments. J Am Soc Hort Sci 103: 527-530
Maxie EC (1963) Radiation technology in conjunction with postharvest procedures as a means of extending the shelf life of fruits and vegetables. Annual Rept Contract No AT (11-1)-34, Project Agreement No 80. Univ. of California, Davis

McNeil M, Darvill AG, Albersheim P (1979) The structural polymers of primary cell walls of dicots. In: Herr W, Grisebach M, Kirby GW (eds) Progress in the chemistry of organic natural products. Springer Verlag, Berlin
Mohr WP, Stein M (1969) Fine structure of fruit development in tomato. Can J Plant Sci 49: 549-553
Monselise SP, Weiser M, Shafir N, Goren R, Goldschmidt EE (1976) Creasing of orange peel: physiology and control. J Hort Sci 51: 341-351
Morré, DJ (1968) Cell wall dissolution and enzyme secretion during leaf abscission. Plant Physiol 43: 1543-1559
Morris DA (1964) Capsule dehiscence in *Gossypium.* Emp Cotton Growers Rev XLI: 167-171
Neal GE (1965) Changes occurring in the cell walls of strawberries during ripening. J Sci Food Agric. 16: 605-611
Nelmes BJ, Preston RD (1968) Wall development in apple fruits: a study of the life history of a parenchyma cell. J Exp Bot 19: 496-518
Neukom H (1963) Schweiz Landw Forsch 2: 112-121. Cited by Pilnik W and Voragen AGJ. In: Hulme AC (ed.) The biochemistry of fruits and their products, Vol 1. Academic Press, London, p 84
Nichols R (1973a) Senescence of the cut carnation flower: respiration and sugar status. J Hort Sci 48: 111-121
Nichols R (1973b) Investigations on the cold storage of cut flowers. Proc. 12th Int Cong Refrig 3: 329-337
Nichols R, Ho LC (1975) Effects of ethylene and sucrose on translocation of dry matter and C-sucrose in the cut flowers of the glasshouse carnation (*Dianthus caryophyllus*) during senescence. Ann Bot 39: 287-296
Osborne DJ (1958) Changes in the distribution of pectin methyl esterase across leaf abscission zones in *Phaseolus vulgaris.* J Exp Bot 9: 446-457
Osborne DJ (1968) Hormonal mechanism regulating senescence and abscission. In: Wightman F, Setterfield G (eds) Biochemistry and physiology of plant growth substances. Runge Press, Ottawa, p 815-840
Pansolli P (1970) Radiation effects on pectinase solutions. Experientia 26: 499-500
Paulin A (1988) La lutte contre le veiellissement. Application à la conservation des produits horticoles. (in press)
Perring MA (1968) Mineral composition of apple. VII and VIII. J Sci Fd Agric 19: 186-192 and 640-645
Pesis E, Fuchs Y, Zauberman G (1978) Cellulase activity and fruit softening in avocado. Plant Physiol 61: 416-419
Pilnik W, Voragen AGJ (1970) Pectic substances and other uronides. In: Hulme AC (ed.) Biochemistry of fruits and their products Vol 1. Academic Press, London, pp 53-87
Pressey R (1977) In: Ory RL, Angelo A St (eds) Enzyme in food and beverage processing. ACS Washington DC. ACS Symposium Series 47, pp 172-191
Pressey R, Avants JK, Dull GG (1978) Difference in polygalacturonase and freestone peaches. XXth Hort Congr Sydney Australia, Abstr 1096
Preston RD (1964) Structural plant polysaccharides. Endeavour 23: 153-159
Ramina A, Masia A, Vizzotto G (1984) Ethylene and auxin transport and metabolism in peach fruit abscission. In: Fuchs Y, Chalutz E (eds) Ethylene. Martinus Nijhoff/Dr W Junk, The Hague, pp 279-280
Rasmussen HP (1965) Chemical and physiological changes associated with abscission layer formation in the bean (*Phaseolus vulgaris* L cv Contender). PhD Thesis Mich State Univ. East Lansing USA
Raunhardt O, Neukom H (1964) Mitt. Geb. Labensmittelunters v. Hyg 55: 446-455. Cited by Pilnik W, Voragen AGJ In: Hulme AC (ed.) Biochemistry of fruits and their products. Vol. 1. Academic Press, London p 85
Reeve RM (1959) Histological and histochemical changes in developing and ripening peaches. II. The cell walls and pectins. Am J Bot 46: 241-247

Romani RS, Somogyi LP, Manalo J (1971) Irradiated pectin: a substrate for pectic enzymes. Rad Bot 11: 383-387
Rouse AH, Atkins CD, Moore EL (1964a) Evaluation of pectin in component parts of Pineapple oranges during maturation. Proc Florida St Hort Soc 77: 271-274
Rouse AH, Moore, EL, Atkins CD (1964b) Evaluation of pectin in component parts of Silver Cluster grapefruit during maturation. Proc. Florida St Hort Soc 77: 274-278
Rouse, AH, Atkins CD, Moore EL (1965) Food Technology 19: 673-676. Cited by Pilnik W and Voragen AGJ. In: Hulme AC (ed.) The biochemistry of fruits and their products. Vol. 1. Academic Press, p 85
Sampson HC (1918) Chemical changes accompanying abscission in *Coleus blumei.* Bot Gaz 66: 32-53
Scott FM, Schroeder MR, Turrell FM (1948) Development cell shape, suberization of internal surface and abscission in the leaf of Valencia orange *Citrus sinensis.* Bot Gaz 109: 381-411
Setterfield G, Bailey S (1961) Structure and physiology of cell walls. Ann Rev Plant Physiol 12: 35-62
Shewfelt AL (1965) Changes and variations in the pectic constitution of ripening peaches as related to product firmness. J Food Sci 30: 573-576
Somogyi LP, Romani RJ (1964) Irradiation-induced textural change in fruits and its relation to pectin metabolism. J Food Sci 29: 366-371
Sterling C, Kalb AJ (1959) Pectic changes in peach during ripening. Bot Gaz 121: 111-113
Stösser R, Rasmussen HP, Bukovac MJ (1969a) Histochemical changes in the developing abscission layer in fruits of *Prunus cerasus* L. Planta 86: 151-164
Stösser R, Rasmussen HP, Bukovac MJ (1969b) A histological study of abscission layer formation in cherry fruit during maturation. J Am Soc Hort Sci 94: 239-243
Stösser R, Rasmussen HP, Bukovac MJ (1971) Localization of RNA and protein synthesis in the developing abscission layer in fruit of *Prunus cerasus* L. Z Pflanzenphysiol 64: 328-334
Tagawa T, Bonner J (1957) Mechanical properties of the *Avena* coleoptile as related to auxin and ionic interaction. Plant Physiol 32: 207-212
Tavakoli M, Wiley CR (1968) Relation of trimethylsilyl derivatives of fruit tissue polysaccharides to apple texture. Proc Am Soc Hort Sci 92: 780-787
Vickery RS, Bruinsma J (1973) Compartments and permeability for K in developing fruits of tomato. J Exp Bot 24: 1261-1270
Wilson WC, Hendershott CH (1968) Anatomical and histochemical studies of abscission of oranges. Proc Am Soc Hort Sci 92: 203-210
Wittenbach VA, Bukovac MJ (1972) An anatomical and histochemical study of abscission in maturing sweet cherry fruit. J Am Soc Hort Sci 97: 214-219
Yamaki S, Machida Y, Katiuchi N (1979) Changes in cell wall polysaccharides and monosaccharides during development and ripening of japanese pear fruit. Plant Cell Physiol 20: 311-321

CONTRIBUTOR INDEX

SUBJECT INDEX

[Pages underlined refer to figures, plates or tables]

A

B

NATO ASI Series H

Vol. 1: **Biology and Molecular Biology of Plant-Pathogen Interactions.** Edited by J. A. Bailey. 415 pages. 1986.

Vol. 2: **Glial-Neuronal Communication in Development and Regeneration.** Edited by H. H. Althaus and W. Seifert. 865 pages. 1987.

Vol. 3: **Nicotinic Acetylcholine Receptor: Structure and Function.** Edited by A. Maelicke. 489 pages. 1986.

Vol. 4: **Recognition in Microbe-Plant Symbiotic and Pathogenic Interactions.** Edited by B. Lugtenberg. 449 pages. 1986.

Vol. 5: **Mesenchymal-Epithelial Interactions in Neural Development.** Edited by J. R. Wolff, J. Sievers, and M. Berry. 428 pages. 1987.

Vol. 6: **Molecular Mechanisms of Desensitization to Signal Molecules.** Edited by T. M. Konijn, P. J. M. Van Haastert, H. Van der Starre, H. Van der Wel, and M. D. Houslay. 336 pages. 1987.

Vol. 7: **Gangliosides and Modulation of Neuronal Functions.** Edited by H. Rahmann. 647 pages. 1987.

Vol. 8: **Molecular and Cellular Aspects of Erythropoietin and Erythropoiesis.** Edited by I. N. Rich. 460 pages. 1987.

Vol. 9: **Modification of Cell to Cell Signals During Normal and Pathological Aging.** Edited by S. Govoni and F. Battaini. 297 pages. 1987.

Vol. 10: **Plant Hormone Receptors.** Edited by D. Klämbt. 319 pages. 1987.

Vol. 11: **Host-Parasite Cellular and Molecular Interactions in Protozoal Infections.** Edited by K.-P. Chang and D. Snary. 425 pages. 1987.

Vol. 12: **The Cell Surface in Signal Transduction.** Edited by E. Wagner, H. Greppin, and B. Millet. 243 pages. 1987.

Vol. 13: **Toxicology of Pesticides: Experimental, Clinical and Regulatory Perspectives.** Edited by L. G. Costa, C. L. Galli, and S. D. Murphy. 320 pages. 1987.

Vol. 14: **Genetics of Translation. New Approaches.** Edited by M. F. Tuite, M. Picard, and M. Bolotin-Fukuhara. 524 pages. 1988.

Vol. 15: **Photosensitisation. Molecular, Cellular and Medical Aspects.** Edited by G. Moreno, R. H. Pottier, and T. G. Truscott. 521 pages. 1988.

Vol. 16: **Membrane Biogenesis.** Edited by J. A. F. Op den Kamp. 477 pages. 1988.

Vol. 17: **Cell to Cell Signals in Plant, Animal and Microbial Symbiosis.** Edited by S. Scannerini, D. Smith, P. Bonfante-Fasolo, and V. Gianinazzi-Pearson. 414 pages. 1988.

Vol. 18: **Plant Cell Biotechnology.** Edited by M. S. S. Pais, F. Mavituna, and J. M. Novais. 500 pages. 1988.

Vol. 19: **Modulation of Synaptic Transmission and Plasticity in Nervous Systems.** Edited by G. Hertting and H.-C. Spatz. 457 pages. 1988.

Vol. 20: **Amino Acid Availability and Brain Function in Health and Disease.** Edited by G. Huether. 487 pages. 1988.

NATO ASI Series H

MIX
Papier aus verantwortungsvollen Quellen
Paper from responsible sources
FSC® C105338

If you have any concerns about our products,
you can contact us on
ProductSafety@springernature.com

In case Publisher is established outside the EU,
the EU authorized representative is:
Springer Nature Customer Service Center GmbH
Europaplatz 3, 69115 Heidelberg, Germany

Printed by Libri Plureos GmbH
in Hamburg, Germany